BSc Zoology Series

Volume V

Evolution, Comparative Anatomy, Biometry, Economic Zoology and Animal Development

ABOUT THE AUTHOR

B N Pandey is Professor and Head at the Postgraduate Department of Zoology, Purnia College (affiliated to Bhupendra Narayan Mandal University, Madhepura), Purnia, Bihar. He obtained his BSc (Hons.) and MSc degrees in Zoology from Bhagalpur University, Bhagalpur and PhD from L N Mithila University, Darbhanga, Bihar. While working for his PhD, Dr Pandey collaborated with Prof. L I Korochkin Institute of Cytology and Genetics, Novisibrisk, Moscow (USSR). He has also worked in collaboration with Prof. Susan Dawson of Utah State University, USA, on Population Health.

Dr Pandey has extensive teaching and research experience. Twelve research scholars have successfully completed their PhD degrees and several students are doing research under his learned supervision.

He has been awarded a Certificate of Merit for Excellence in Teaching by the Vice Chancellor, B N Mandal University, Madhepura, Bihar; another certification as Teacher of Outstanding Merit by the Ministry of Youth Affairs, Govt. of India; and a Gold Medal by the Zoological Society of India, for his contribution to research in this field. Dr Pandey has organised national seminars and symposia as well as the reputed All India Zoological Congress. He has also delivered lectures in Vidyasagar virtual classrooms. Major research projects sanctioned by the Ministry of Environment and Forests, Indian Council of Medical Research, etc., have benefited from his contributions. He is also a fellow of the Zoological Society of India and the Society of Environmental Sciences, and a member of the editorial boards of many journals.

Dr Pandey has published about 100 research papers in national and international journals. His areas of research interest include: Aquatic Biology, Population Genetics, and Population Health.

BSc Zoology Series
Volume V
Evolution, Comparative Anatomy, Biometry, Economic Zoology and Animal Development

B N Pandey
Professor and Head
Postgraduate Department of Zoology, Purnia College
(Affiliated to Bhupendra Narayan Mandal University, Madhepura)
Purnia, Bihar

Tata McGraw Hill Education Private Limited
NEW DELHI

McGraw-Hill Offices
New Delhi New York St Louis San Francisco Auckland Bogotá Caracas
Kuala Lumpur Lisbon London Madrid Mexico City Milan Montreal
San Juan Santiago Singapore Sydney Tokyo Toronto

Tata McGraw-Hill

Published by the Tata McGraw Hill Education Private Limited,
7 West Patel Nagar, New Delhi 110 008

BSc Zoology Series: (Volume V)—Evolution, Comparative Anatomy, Biometry, Economic Zoology and Animal Development

ISBN (13) : 978-1-25-902088-9
ISBN (10) : 1-25-902088-1

Vice President and Managing Director: *Ajay Shukla*

Head—Higher Education Publishing and Marketing: *Vibha Mahajan*
Publishing Manager—SEM & Tech Ed.: *Shalini Jha*
Editorial Executive—Sponsoring: *Smruti Snigdha*
Development Editor: *Renu Upadhyay*
Sr Copy Editor: *Nimisha Kapoor*
Sr Production Manager: *Satinder S Baveja*
Proof Reader: *Yukti Sharma*

Marketing Manager—Higher Education: *Vijay Sarathi*
Product Specialist: *Sachin Tripathi*
Graphic Designer— Cover: *Meenu Raghav*

General Manager—Production: *Rajender P Ghansela*
Production Manager: *Reji Kumar*

Typeset at Print-O-World, 2579, Mandir Lane, Shadipur, New Delhi 110 008, and printed at Pushp Print Services, B-39/12 A, Gali No.-1, Arjun Mohalla, Moujpur, Delhi – 110 053

Cover Printer : SDR Printers

RZLACRXHRBZYQ

The McGraw Hill Companies

CONTENTS

PREFACE

This *BSc Zoology Series* of five volumes will be useful for all undergraduate students of life sciences. The series has been developed to follow a unique test-friendly approach to especially assist undergraduate-level students in exam preparation. Besides, the applicants of CSIR-NET, GATE, Civil Services and other competitive examinations will also find this series very helpful.

About The Series

The following five volumes collectively structure this series:

Volume 1: Animal Diversity

Volume 2: Cytology, Genetics and Molecular Genetics

Volume 3: Biochemistry, Physiology and Endocrinology

Volume 4: Ecology and Animal Behaviour

Volume 5: Evolution, Comparative Anatomy, Biometry, Economic Zoology and Animal Development

These volumes cover the latest syllabi, as per the UGC curricula, of BSc courses taught across different Indian universities. Each part of a volume in the series contains a synopsis which briefly introduces the theme and then details important features topic-wise. This is followed by a comprehensive section on objective-type questions which includes short-answer questions, long-answer questions, multiple-choice questions, fill in the blanks, true or false questions, and questions based on reasoning and diagrams.

This arrangement has been ideated to first get the students acquainted with a chapter by going through the synopsis and then attempt to answer different sets of questions based on that chapter. Such a flow seeks to encourage self-study and aids quick revision of the topics in a lesson. While the synopsis provides a clear framework and considerable depth to topic-wise study of the syllabi, the stupendous variety in exercises covers a broad spectrum of learning tools.

What Makes This Series Unique?

The changing pattern of syllabus of academic life-science courses has induced a change in the type of questions appearing in undergraduate-level examinations of major universities and noted competitive tests. A distinct alteration in the nature of objective questioning has been identified. Objective questions, now part of compulsory questions, include the variations mentioned above. It then becomes imperative that the students be made fully conversant with this new pattern.

However very few books, adequately containing the required pedagogical features, are available to facilitate such a pattern of study. Recognising the growing interest of students and a need for a comprehensive yet basic-level text, I have authored this *BSc Zoology Series* to aid test-ready academic study.

Besides students, this series will amply assist various faculty members in the design and preparation of periodical tests for internal evaluation, question papers for undergraduate-level university examinations as well as CSIR-NET, GATE and Civil Services examinations, etc.

Salient Features Of The Series

- Apposite theory to aid quick revision for examinations
- Wide range of chapter-end exercises designed as per undergraduate examinations
- Surplus artwork to help develop a holistic understanding of concepts

Volume V : Evolution, Comparative Anatomy, Biometry, Economic Zoology and Animal Development

Introduction

Evolution is the slow process of change from simple to complex. According to Dobzhansky (1973), 'Nothing in biology makes sense except in the light of evolution'. Evolution lends an understanding to how organisms change over time and helps us correlate this understanding with how they appear in their present form. Several organisms with complex organisations have been discovered today due to a deep study of evolutionary processes. *Comparative Anatomy* deals with the similarities and dissimilarities in different parts of different organisms, especially vertebrates. It is more closely related to Evolutionary Biology and Phylogeny wherein homologous and analogous structures are the two main themes of study. Comparative anatomy is known to provide useful evidences of evolution.

Biometry deals with the use of statistical methods in analysing various types of biological data. A basic knowledge of biometry is essential for any research in fields such as Life Sciences, Medical, Agriculture and Veterinary. The branch of Zoology that deals with the study of culturing of different animals for their large-scale production for human use as well as eradication or control of those animals that are harmful to humans is known as *Economic Zoology. Animal Development* elucidates the process of development of a single cell zygote into a fully functional complex organism through the study of cell division and cellular differentiation. Its findings give us an insight into the various kinds of developmental abnormalities.

Highlight

This volume elucidates all the important topics such as Evolution, Basic Patterns of Evolution, Digestive System, Respiratory System, Mean, Median, Mode, Pisciculture, Sericulture, Apiculture, Embryonic Development, etc.

Organisation of Volume V

This volume has been classified in to five parts, viz., *Evolution, Comparative Anatomy, Biometry, Economic Zoology and Animal Development.* The first part, *Evolution* discusses topics such as **Origin of Life**; **Lamarckism**—critical analysis, neo Lamarckism; **Darwinism**—criticism, neo Darwinism; **Organic Evolution**—evidences from comparative anatomy, connecting links, embryology, paleontology, geographic distribution of animals, biochemistry, physiology, atavism and taxonomy; **Basics of Organic Evolution**—patterns (divergent, convergent, co-evolution and parallel), and factors (natural selection, mutation, genetic drift, founder effect, bottleneck effect, gene flow and nonrandom mating); **Rules of Organic Evolution**; **Mutation Theory**; **Isolation**; **Genetics of Speciation**—Hardy–Weinberg law; **Evolution of Horse**—evolutionary trends (Eohippus, Orohippus, Epihippus, Mesohippus, Miohippus, Parahippus, Merychippus, Pliohippus, Plessippus and Equus); **Evolution of Man**—evolutionary trends, early evolution of primates—Australopithecus, origin of genus Homo (Homo Erectus), and origin of Homo Sapiens; **Geological Time Scale**—Azoic, Archaeozoic,

Proterozoic, Palaeozoic, Mesozoic and Cenozoic era; **Adaptation**—structural and behavioural; **Zoogeographical Regions**—Neotropical, Palaearctic, Ethiopian, Australian, Oriental and Nearctic regions, Wallace line and Weber's line; and **Mimicry**.

The second part, *Comparative Anatomy* elucidates topics such as **Integument and Its Derivatives**—structure of skin (epidermis and dermis), integument in different classes of vertebrates (cyclostomata, fishes, amphibians, reptiles, birds and mammals), and derivatives of integument (epidermal glands, epidermal scales and scutes, dermal scales and scutes, digital cornification, hair, feathers and teeth); **Digestive System**—mouth and oral cavity, pharynx, oesophagus, stomach, intestine, and glands associated with the digestive system; **Respiratory System**—adaptations for external respiration, respiratory organs (cutaneous respiration, gills and swim bladder), and lungs and associated structures (larynx, trachea, syrinx and lungs); **Heart**—heart in different groups of vertebrates (two-chambered heart, three-chambered heart and four-chambered heart), and aortic arches (trend of evolution and aortic arches in different groups of vertebrates); **Brain**—comparative account of brain in different classes of vertebrates (cyclostomes, fishes, amphibians, reptiles, aves and mammals); **Jaw Suspension**—autodiastylic, amphistylic, hylostylic and craniostylic; and **Urinogenital System**—excretory system (kidneys, urinary bladder, genital system, ovaries, testes, female genital ducts, male genital ducts and copulatory organs).

The third part, *Biometry* describes topics such as **Measures of Central Tendency**—mean, median, and mode; **Distribution**—types (binomial, Poisson and normal); **Correlation**—types (positive, negative, simple, partial, multiple, linear and nonlinear), degree (perfect, limited and absent), calculation and product-moment correlation; **Regression Analysis**; **Hypothesis Testing and Test of Significance**—null and alternative hypothesis, errors in testing, level and test of significance; **Standard Error of Mean**; **Standard Error of Standard Deviation**; **Student t-Test**—application; **Chi Square Test**—characteristics and uses; and **Analysis of Variance**—one way and two way.

The fourth part, *Economic Zoology* discusses topics such as **Apiculture**—species, life cycle, diseases, enemies and predators, and honey; **Sericulture**—life cycle, culture, diseases, and properties and uses of silk; **Lac Culture**—life cycle, cultivation, enemies, and properties and uses; **Pearl Culture**—formation and culture; **Pisiculture**; **Pests of Sugarcane**—Pyrilla perpusilla, Aleurolobus Barodensis and Chilo Infuscatellus (Sugarcane stem borer); **Pests of Paddy**—Paddy Gall Fly (*Pachydiplosis oryzae*), Rice Bug (*Leptocorisa Varicornis*), and *Tryporyza Incertulas* (The stem borer); **Pests of Wheat**— *Tanymecus Indicus* (Gujhia Weevil), *Mathymna Separata* (ArmyWorm), and *Schizaphis Graminum* (Wheat Aphids); and **Pests of Stored Grains**—Rice Weevil (*Sitophilus Oryzae*), Rice Moth (*Corcyra Cephalonica*) and Wheat Weevil (*Trogoderma Granarium*). The section on pests provides details on nature, damage and control for all categories.

The fifth part, *Animal Development* elucidates topics such as **History of Embryology**, **Parthenogenesis**—types (natural and artificial) and significance; **Fertilisation**—mechanism (approximation, fertilizine, anti-fertilizin incompatibility, sperm penetration, activation of ovum and amphimixis), abnormal fertilisation and significance; **Types of Eggs**—amount of yolk, distribution of yolk and basis of development); **Cleavage**—characteristics and patterns (based on symmetry, amount and distribution of yolk, and fate of development of blastomeres); **Blastulation**—characteristics of blastula; **Gastrulation**—characteristics, significance and fate of germ layers (ectoderm, mesoderm and endoderm); **Extra Embryonic Membranes**—yolk sac, amnion, chorion and allantois); **Placenta**—classification (nature of foetal membranes, nature of maternal and foetal tissues, fate of uterine endometrium at birth, contra-deciduous placenta, and nature and arrangement of Villi) and functions of placenta; and **Gestation Period**.

Online Learning Centre

For further interesting resources and supplements, please visit *http://mhhe.com/pandey/ecabad1/vol5*

Acknowledgements

Writing this series has been a tremendous yet fulfilling endeavour. All the volumes have taken a final shape after endless inputs of time and effort. Though many teachers and students assisted me in compiling this book, I must especially mention the effort made by my colleague, O P Ambasta who extended immense support in myriad ways for bringing out the series in its present form. I am also indebted to A K Jha for his many valuable contributions.

I am grateful to the following reviewers for their helpful suggestions for improving the contents of this series.

A K Ojha – *Rajendra College, Balangir, Odisha*

P N Pandey – *SSPG College, Ayodhya, Uttar Pradesh*

S P Sinha – *Bhagalpur University, Bhagalpur, Bihar*

Budhadeb Manna – *University of Calcutta, Kolkata, West Bengal*

I am thankful to the team at Tata McGraw Hill Education, most notably Smruti Snigdha for giving me the opportunity to author this series and Renu Upadhyay for helpful suggestions to improve the quality of the content and regular reminders for timely completion of the project. It has been a pleasure to work with Nimisha Kapoor and Yukti Sharma, who took great care during the copy-editing and production processes of all the volumes.

I welcome all feedback, criticisms and suggestions for improvements in all the volumes from teachers, students and all other readers of this series. You can write to me at *b.n.pandey@hotmail.com*.

B N Pandey

Publisher's Note

Do you have a feature request? A suggestion? We are always open to new ideas (the best ideas come from you!). You may send your comments to *tmh.sciencemathsfeedback@gmail.com* (Don't forget to mention the title and author's name in the subject line).

EVOLUTION

ORIGIN OF LIFE

- The earth is also called 'the blue planet' or 'the planet with life'. It originated about 4.8 billion years ago.
- The primitive atmosphere of the earth contained nitrogen, hydrogen and carbon, but no oxygen or ozone, all found in the atomic state due to extreme heat and ultraviolet insulation.
- Chemical evolution continued for about 2,000 million years, by which various inorganic and organic compounds appeared.
- The chief inorganic compounds that evolved during this period were N_2, H_2, CH_4, NH_3, CO, CO_2, H_2O, etc., but no O_2 or O_3 as the primitive environment was reducing.
- Miller in 1953 during his experiments on molecular evolution found that when a mixture of H_2, CH_4, NH_3 and H_2O is exposed to high electric discharge for weeks, a gaseous phase of advanced molecules including CO, CO_2, N_2, etc., and a liquid condensate phase, including a variety of amino acids like glycine, valine, etc., and other organic molecules were obtained.
- During this period, certain compounds that favoured molecular evolution survived, while the compounds disfavouring it were destroyed by predator molecules like hydrated electrons, poly anions, etc.

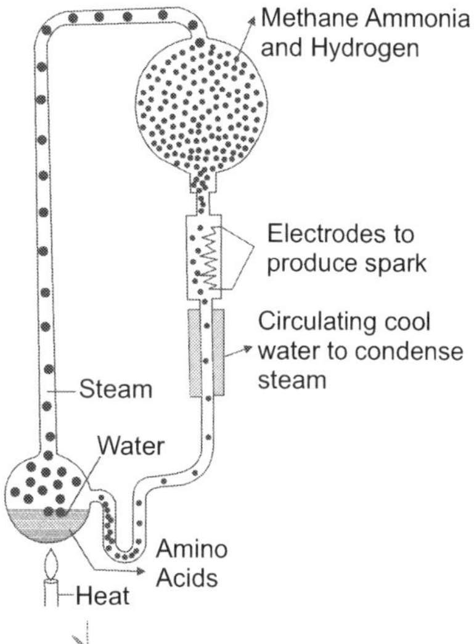

Fig. 1 Miller and Urey's experiment

- The culmination of chemical evolution was the polymerisation of monomers. As a result, certain biopolymers like proteins, polysaccharides, lipids and polynucleotides were formed.
- Certain polymers like histones and protamines are positively charged and called polycations and when such polycations enclosed polynucleotides, the latter were protected from the predator molecules.
- The earliest life originated in a primitive broth, which provided easy accessibility to the polymers essential for self-replication.
- The first life form had the unique property of self-replication and it became the first and foremost characteristic of life.

- Although the sun is the ultimate source of energy, the first transformed energy was obtained from the mobility of the molecules.
- The second primitive characteristic of tiny life was to use the ATP, the condensed energy source for all life activities.
- The two other prerequisites of life were a definite catalytic system and an information transfer system, also formed by the conversion of the two polymers.
- Certain theories and principles regarding origin of life are given in the following table:

Table 1

S. No.	*Theories*	*Founders*	*Principles*
1.	Theory of Autogenesis	Plato, Aristotle	Life originated spontaneously (from nonliving substances).
2.	Theory of Special Creation	Religious Mythology	Life originated at the mercy of a supernatural power.
3.	Theory of Catastrophism	Cuvier (1776)	Life originated due to sudden changes on the earth.
4.	Cosmozoic Theory	Richer (1865), Helmholtz (1884)	Life originated somewhere else in the cosmos and was blown to the earth in the form of spores, germs, etc.
5.	Theory of Conservation	Oparin (1938)	Life originated from inorganic substances in sequential stages.
6.	Theory of Organic Evolution	Lamarck, Darwin	Life originated slowly and steadily from simpler to complex forms.

- The first life form produced food from inorganic substances by the photoreduction reaction in which sulphur or other organic groups were produced as byproducts.

  $$CO_2 + 2H_2X \longrightarrow (CH_2O) + H_2O + 2X$$

- The above hypothesis has been confirmed by the tracer carbon technique and chromatographic identification of the products (Arnon, 1960).
- The evolution of chlorophyll molecule was a landmark in the origin of photosynthetic life because the magnesium porphyrin of chlorophyll molecule is capable of emitting electrons, when excited by the radiant energy of even visible range.
- The evolution of photosynthetic life could change the reducing environment to oxidising due to release of oxygen.

  $$6CO_2 + 6H_2O \longrightarrow C_6H_{12}O_6 + 6CO_2$$

- At the time of origin of animals, the atmosphere contained N_2, CO_2 and O_2 gases, while the land mass contained soil rich in soluble nitrates, sulphates, phosphates and organic materials.
- The evolution of the environment having definite biosphere zones facilitated the evolution of higher animals including humans.
- The overall sequence of evolution as suggested by Oparin (1938) may be summarised in following phases:

  (a) Atomic phase

(b) Molecular phase
(c) Organic phase
(d) Colloidal phase
(e) Autocatalytic phase
(f) Primordial cellular phase
(g) Autotrophic phase
(h) Prokaryotic phase
(i) Eukaryotic phase

LAMARCKISM

- Jean Baptiste de Lamarck (1744–1829) was a French scientist who for the first time put forth an evolutionary theory.
- He published his ideas in the book, *Philosophie Zoologique* in 1809.
- Lamarckian theory is also known as 'Theory of Inheritance of Acquired Characters' or the 'Theory of Use or Disuse of Organs'.
- The basic ideas of Lamarckism are as follows:
  (a) Complex plants and animals are produced from the simpler ones in due course of time under the influence of nature.
  (b) Living organisms have a tendency to continuously increase in size.
  (c) There is a change in the needs of an animal due to changes in the surrounding environment. As a result, new organ(s) is/are formed to cater the needs of the animal.
  (d) If an organ is used continuously, it will be more powerful. If it is not used for a long time, it gradually weakens and finally disappears.
  (e) The characters acquired during the lifetime of individuals will be transmitted to the next generation. This is called the 'inheritance of acquired characters'.

Critical Analysis

- Lamarck based his idea of use and disuse on the evidences in the form of vestigial organs and rudimentary organs. The idea is true, but it cannot be treated as a basis of evolution.
- As far as the theory of inheritance of acquired characters is concerned, it cannot be accepted.
- Weismann (1890) for the first time provided experimental proof against this assumption.
- Weismann cut off the tails of mice for many generations and kept them in breeding, but not a single mouse was born without a tail.
- Weismann proposed the 'Theory of Continuity of Germplasm'.

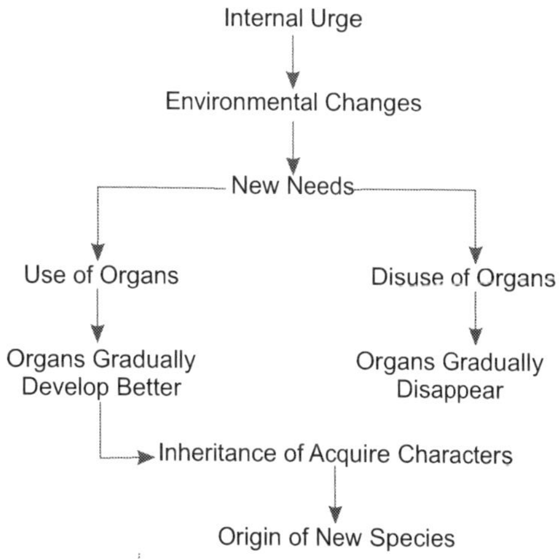

Fig. 2 Lamarek's theory of evolution

- It is now quite clear that only those changes are inherited which are in the germ cells.
- No somatic changes can be inherited. Whatever one acquires in one's lifetime passes away with the end of one's life.

Neo Lamarckism

- A few scientists notably, Herbert Spencer, Haeckel, Gadow, etc., have supported Lamarckism in a modified form and this is called Neo Lamarckism. Neo Lamarckism can be explained as listed below.
  (a) Only those acquired characters are inheritable, which are incorporated in the genetic makeup.
  (b) Environmental changes create needs in an organism. Such needs rarely cause changes in the genetic makeup. Somatic changes are not heritable.
  (c) It may be that environmental changes and thereby the needs emanating from them may not reach up to a mutational threshold and genetic effect will not be expressed. The inheritance of acquired characters depends upon its reaching the mutational threshold.

DARWINISM

- Charles Darwin (1809–1882) was the first individual who categorically explained organic evolution in detail.
- Charles Darwin joined a sea voyage that continued for five years in a ship named *HMS Beagle* and collected flora and fauna of many continents and islands.
- The view of A R Wallace regarding organic evolution was much similar to that of Darwin.
- The ideas of both scientists were greatly influenced by the writings of T R Malthus on the growth of population.
- Darwin published his detailed theory on organic evolution in 1859 in his book entitled *Origin of Species.*

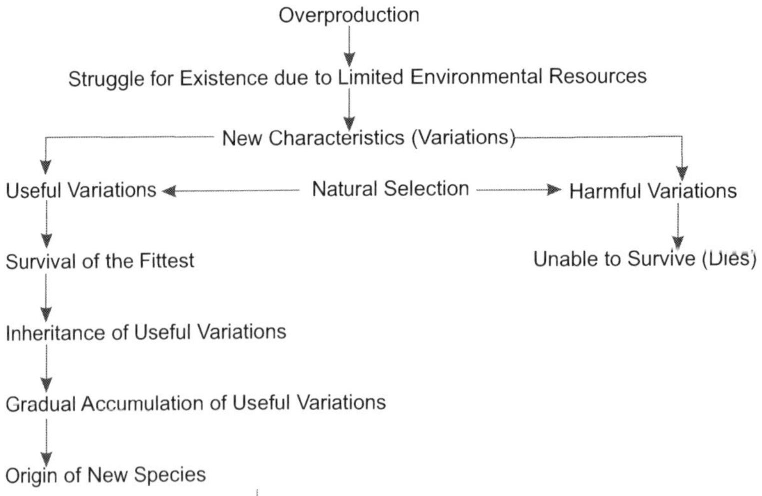

Fig. 3 Darwin's concept of evolution

- He argued that natural selection is the main cause of origin of new species.
- The presentation of Darwin's theory was based on the following facts:
  - (a) **Overproduction –** Organisms have the capacity of enormous reproduction, i.e., they produce much more than required.
  - (b) **Limited Environmental Resources –** The population of each species tends to increase in a geometric ratio but remains almost constant.
  - (c) **Struggle for Existence –** To check the increase in the number of each species, there is a struggle for existence, which may be intraspecific, interspecific and environmental. Of these, the intraspecific struggle is very severe.
  - (d) **Variations –** Darwin thought that each member of a sexually reproducing species is different from the others. The cause of difference among the members of a species is variation and the variations are heritable.
  - (e) **Survival of the Fittest (Natural Selection) –** Among these variations, some are useful and some are harmful. Useful variations make some individuals more suited to the environment than others in the struggle for existence. In this struggle, the most suitable individuals survive. Thus, the paradigm of 'survival of the fittest' is in operation. The survival of the fittest is made possible due to useful or beneficial variations, which is why the survival of the fittest is by natural selection.

Criticism of Darwinism

- Darwin could not explain the origin and cause of inheritance of variations. Further, he could not differentiate heritable and nonheritable variations.
- Although Darwin termed as 'sports' the sudden change in the genetic material, he was unable to explain it.
- There are examples of overgrowth/overspecialisation of useful organs that become instrumental in the extinction of that species.
- Darwin did not explain the occurrence of vestigial organs.

Neo Darwinism

- Neo Darwinism is the modification of Darwin's theory in the light of modern genetic findings.
- Neo Darwinism is based on three facts, viz., genotypic variations, natural selection and isolation.
- Neo Darwinism assumes that natural selection acts on the heritable variations in individuals in populations and mutations provide the main source of these genetic variations.
- Neo Darwinism separates the Darwinian idea of natural selection from his hypothesis of pangenesis.
- Neo Darwinism states that organic evolution takes place by natural selection of inherited characters.
- According to Neo Darwinism, evolution occurs because the environment is slowly changing.
- This exerts a selection pressure on individuals within a population.
- The characteristics that adapt to a new environment are able to survive and pass on to the next generation.
- These result in variations and accumulate in the gene pool of the organisms.
- With the passage of time, the genetic makeup of the population changes and ultimately a new species is formed.

ORGANIC EVOLUTION

- The slow process of change from simple to complex is known as evolution.
- Evolution assumes that all living things are interrelated.
- According to Darwin, evolution is descent with modification.
- The term 'evolution' was coined by the British philosopher, Herbert Spencer.
- According to Dobzhansky (1973) 'Nothing in biology makes sense except in the light of evolution'.
- Organic evolution is the change in the characteristics of a population of organisms over the course of generations, resulting in forms that are more complex from pre-existing simpler forms.
- There are several evidences in favour of organic evolution.

I. Evidences from Comparative Anatomy

- Comparative anatomy is the study of the different internal organs of different animals.
- The study of different systems of organisms like digestive, circulatory, respiratory, muscular, skeletal, etc., in various animals shows similarities in their structure and functions.
- These similarities in the structure and functions of different organs in different animals indicate that they have originated from a common ancestor.
- Evidences from comparative anatomy are as follows:

1. **Homologous Organs –** Organs having similar basic structure but different functions are known as homologous organs. Homologous organs of different animals provide evidence for evolution.
   (a) The forelimbs of frogs, birds and humans are built on the same basic design of bones, but perform different functions.
   (b) The presence of homologous forelimbs in frogs, birds and humans indicate that all these forelimbs have evolved from a common ancestor and during the course of evolution they became adapted for different functions.
   (c) Thus, it can be concluded that frogs, birds and humans have evolved from a common ancestor.
2. **Analogous Organs –** Organs having a different basic structure but similar appearance and functions are known as analogous organs.
   (a) The wings of insects and birds have a different structural plan, but perform the same function of flying.
   (b) The development of analogous organs in animals of unrelated groups to become adapted for performing the same function for their survival under hostile environmental conditions is known as convergent evolution.
3. **Vestigial Organs –** Vestigial organs are those organs of living beings, which are functionless and useless now (but used to function in their ancestors).
   (a) Vestigial organs are believed to be well developed and functional in ancestral forms but due to changed modes of life, are gradually disappearing in living forms.
   (b) The nictitating membrane is functionless in humans but is still functioning in birds and provides protection. Thus, it indicates that human beings have evolved from those ancestors that had a functioning nictitating membrane.

(c) The coccyx present at the end of vertebral column in humans is formed by the fusion of a few caudal vertebrae, indicating that the ancestors of humans were tailed.

(d) The vermiform appendix in the large intestine in human beings has no function but it is still functioning in herbivorous ruminant mammals and helps in the digestion of cellulose. The presence of the vestigial form of vermiform appendix in humans suggests that human beings evolved from mammals that were herbivorous. During the course of evolution, when humans changed their food, i.e., became omnivorous from herbivorous, it gradually reduced in size and became functionless.

(e) Third molars are regarded as wisdom teeth in humans. These do not erupt in all humans, but in other primates they are well developed and fully functional, and erupt early. This indicates that humans have descended from some primates.

(f) The presence of vestigial hindlimbs in the python and boa suggests that snakes have descended from ancestors that had well-developed limbs.

II. Evidences from Connecting Links

- Connecting links are those organisms that possess characteristics of two different groups. These organisms provide continuity in the series of organisms by proving that one group has evolved from the other.
- *Peripatus* is a connecting link between Annelida and Arthropoda.
- Its Annelidan characters are as follows:
  (a) Worm-like body with soft cuticle and continuous muscle layers in the body wall
  (b) Unjointed legs and simple eyes
  (c) Excretory organs are nephridia
- Its Arthropodan characters are as follows:
  (a) Presence of antennae
  (b) Presence of tubular heart
  (c) Presence of trachea as respiratory organ
  (d) Presence of haemocoel.
- *Neoplina* is a connecting link between Annelida and Mollusca.
- Its Annelidan characters are as follows:
  (a) Presence of metamerism
  (b) Presence of segmentally arranged gills
  (c) Presence of nephridia
  (d) Spiral cleavage
  (e) Larva trochophore
- Its Molluscan characters are as follows:
  (a) Presence of shell
  (b) Soft and flat body
  (c) Presence of mantle
  (d) Presence of foot
- Some other examples of connecting links are given below.

Table 2

	Organisms	*Connecting link between*
(a)	Viruses	Living and nonliving
(b)	*Euglena*	Plants and animals
(c)	*Proterospongia*	Protozoa and Porifera
(d)	*Balanoglossus*	Nonchordata and Chordata
(e)	*Chimera*	Cartilaginous and bony fish
(f)	Dipnoi	Fishes and amphibians
(g)	*Archaeopteryx*	Reptiles and birds
(h)	Prototheria	Reptiles and mammals

III. Evidences from Embryology

- The developing embryos of animals exhibit striking similarities in their structure.
- The embryos of vertebrates are so similar in their structure in their early stages of development that it is very difficult to differentiate one from another.
- The early embryos of vertebrates like fish, salamander, tortoise, chicks and humans look similar. It indicates that all these animals have evolved from a common ancestor.
- Based on these similarities, the developmental pattern in the embryos of vertebrates, Ernst Haeckel proposed the Biogenetic Law, which states that 'ontogeny recapitulates phylogeny, i.e., during the development of the embryo of any organism, its complete history is repeated'. For example, embryonic forms of human beings are similar to fishes, amphibians, reptiles and birds.
- Thus, on the basis of similarities in embryos of different vertebrates, it can be concluded that vertebrate animals have evolved from a common ancestor.

IV. Evidences from Palaeontology

- The remains of dead plants and animals that lived in the past are called fossils and the study of fossils is called palaeontology. Fossils help to trace evolution of a particular animal as they fill the gap between two animals.
- Fossils provide evidence in favour of evolution. For example, *Archaeopteryx* is a fossil bird and is a connecting link between reptiles and birds.
- It had a skeleton, beak with teeth, a tail like reptiles but wings with feathers and furcula like birds. Thus, it suggests birds too have evolved from reptiles.

V. Evidences from Geographical Distribution of Animals

- Animals are distributed throughout the entire earth, which has been divided into six different regions.
- Different regions of the world have different animals and plants. Different geographical regions, at various places, are separated or isolated from others due to geographical barriers.
- Due to different environmental and climatic conditions, the separated regions gave rise to different animals and plants.
- Prototheria and metatheria are found only on the Australian continent, which was separated from the Asian continent before the appearance of eutherians.

- Eutherian mammals, therefore, could not reach the Australian region.
- As a result of the evolution of eutherian mammals in the Asian continent, prototheria and metatheria disappeared from that region as eutherians were more efficient. On the other hand, in the absence of eutherian mammals in the Australian region, prototheria and metatheria had many opportunities to flourish.
- The camel and the llama are closely related and are found in Asia and South America, respectively. It suggests that Asia and South America were once a continuous landmass. Consequent to the separation of these two landmasses, the ancestors of the camel and the llama inhabiting Asia adapted themselves to desert life and evolved into the camel, while those that remained in South America underwent an evolutionary change to the llama, according to environmental conditions.
- Thus, every region of the world has its own specific plants and animals and such distribution provides proof in favour of organic evolution.

VI. Evidences from Biochemistry and Physiology

- Protoplasm is the physical basis of life and is made up of carbohydrate, protein and fat. Oxygen, hydrogen, nitrogen and carbon are its main components. The percentage of these components is found to be the same in the protoplasm of all living worlds. Thus, it suggests that all living beings have originated from a common ancestor.
- In animals, the same types of enzymes have similar reactions (e. g., trypsin acts on proteins in all animals). Thus, the occurrence of the same enzyme in different animals indicates their common origin.
- Blood protein test has revealed that humans are closer to apes (chimpanzees and gorillas) than monkeys.
- Hormones of all vertebrates show similar a chemical nature and function.
- In the entire living world, the chromosome has the same biochemical organisation. The octamer of nucleosome has the same protein in all animals, indicating their common origin.
- All organisms utilise the same DNA triplet base and the same 20 amino acids in their proteins. Many organisms share the same introns and types of repeats. These similarities suggest descent from a common ancestor.

VII. Evidences from Atavism

- The sudden reappearance of an ancestral characteristic is known as atavism or reversion. The reappearance of such features in newborns is suggestive of the fact that these features were present in their remote ancestors, but lost during the course of evolution to their present-day form.
- Some common examples of atavism are as follows:
  (a) Appearance of tail in the newborns of human beings.
  (b) Occurrence of cervical fistula in some human bodies, which are the remnants of gill slits found in ancestors.
  (c) Occurrence of long and pointed canine and thick body hair in human babies (Lion's boy of Russia).

VIII. Evidences from Taxonomy

- Depending on the basis of similarities and differences, animals are grouped as phyla, classes, orders, families, genus and species.

- The similarities revealed their common ancestry, while differences among them are due to different environmental conditions that operate on them.

BASICS OF ORGANIC EVOLUTION

Patterns of Organic Evolution

Following are the four basic patterns of evolution:

I. Divergent Evolution

(a) The pattern of evolution in which two closely related species gradually become increasingly dissimilar in different habitats of the same area is called divergent evolution.

(b) Divergent evolution is confirmed by DNA analysis in which the species that diverged is genetically similar to the form from which it diverged.

(c) Adaptive radiation is an example of divergent evolution.

(d) Any genus of plants or animals can show divergent evolution.

(e) A good example of divergent evolution is Darwin's finches. Presently there are 80 species of Darwin's finches, which have diverged from a single finch.

(f) A very common example of divergent evolution is the vertebrate limbs. Whale flippers, forelimbs of frogs and human arms perform different functions but share a common evolutionary origin.

(g) Divergent evolution results in speciation.

II. Convergent Evolution

(a) The pattern of evolution in which two unrelated species become similar to each other due to favourable changes in morphology, living in the same area due to natural selection is known as convergent evolution.

(b) Convergent evolution often results in analogous structures.

(c) The striking similarity in hummingbird moths and hummingbirds is an example of convergent evolution.

(d) The evolution of functionally similar but distinct antifreeze proteins in divergent species of fishes (one group living near Antarctica and the other found in the Arctic) is an example of convergent evolution

III. Co-evolution

(a) Combined changes in two or more species which are in close interaction, usually dependent upon each other, is known as co-evolution.

(b) A good example of co-evolution is shown by insects in which the shape and structures of the body as well as the size of proboscis is of different types, according to the structures of flowers.

(c) Plants and animals that pollinate them are a common example of co-evolution.

IV. Parallel Evolution

(a) The pattern of evolution in which independent development of similar characteristics in two groups of animals having common ancestry occurs due to similar environmental conditions is known as parallel evolution.
(b) Parallel evolution is more frequent between structures.
(c) In both parallel and convergent evolution, similar adaptations arise in different species. But in parallel evolution, the two species have a relatively recent common ancestor.
(d) The form of leaves in plants is a good example of parallel evolution. A similar pattern of leaves is found in separate genera and families.
(e) Several instances of parallel evolution are found in placental and marsupial mammals. Placental mammals like the anteater, house mouse, wolf, flying squirrel and groundhog are parallel in evolution to marsupials like the wombat, marsupial mouse, Tasmanian wolf, native cat, etc.

Factors of Organic Evolution

Following are the main factors of evolution:

I. Natural Selection

(a) Natural selection is the differential reproduction of genes or genotypes. It is one of the basic mechanisms of evolution.
(b) Natural selection is not responsible for evolution. Actually, it is a factor in change.
(c) Organisms are dependent on environmental factors for their survival. As there is competition among organisms, only those traits favouring their survival will do so.
(d) Natural selection is not responsible for reproduction.
(e) Natural selection selects only those traits in organisms that have survival value.

II. Mutation

(a) A sudden change in the characteristics of an organism is called mutation.
(b) Mutation involves change in genes or chromosomes or genetic material.
(c) Mutations cause variation in a population and are the source of all new alleles within the population.
(d) Mutations simply change the allele frequencies and do not determine the direction of evolution.
(e) Mutation is a very slow process that causes a change in the genetic composition of a population gradually.
(f) Mutation occurs in any direction.
(g) Mutations are mostly harmful.
(h) Mutations play a key role in the process of evolution as they are responsible for genetic variation, which acts as raw material for evolution.

III. Genetic Drift

(a) Genetic drift is the sudden change in the gene frequency in a population.
(b) Genetic drift has greater impacts on smaller populations.

(c) Genetic drift may also occur through random sampling error.

(d) Genetic drift becomes fixed when one allele is entirely replaced by another or the death of one allele occurs.

(e) The process of genetic drift may increase by bottleneck effect or founder effect as genetic drift works more properly in small populations.

(f) Genetic effect does not work to produce adaptation.

IV. Founder Effect

(a) Founder effect is an important example of genetic drift in humans.

(b) The concept of founder effect was discussed by Ernst Mayr (1952).

(c) When a small group of individuals becomes separated from the main population due to geographical barrier, the founder effect occurs. Thus, the formation of a new genotype in a new settlement area is known as founder effect.

(d) In rare instances, the founder effect may lead to speciation and subsequent evolution of a new species.

(e) The founder effect is common in island populations, which become cut-off from the main population.

(f) The founder effect is also known as allopatric speciation.

V. Bottleneck Effect

(a) A sudden reduction in the number of individuals in a population from natural disasters like earthquakes, habitat destruction and predation is known as the bottleneck effect.

(b) The bottleneck effect prevents the majority of genotypes from participating in the production of the next generation.

(c) Generally, the bottleneck effect is not associated with genetic advantage.

(d) Natural selection, mutation and genetic drift are the major factors of evolution. They cause changes in genotypes and phenotypes over a period of time.

VI. Gene Flow

(a) Movement of genes from one population to another is known as gene flow.

(b) It is also known as migration.

(c) Mutation developing in one population may spread to other populations through gene flow. It leads to introduction of new allele in the population.

(d) Gene flow may prevent evolution by preventing adaptation or accelerate evolution by spreading new genes or a combinations of genes.

VII. Nonrandom Mating

(a) A mating in which mates are selected on the basis of physical or behavioural traits is called nonrandom mating.

(b) It is also known as assortative mating.

(c) Assortative mating of similar or dissimilar genotypes leads to formation of excess of homozygotes or heterozygotes, respectively.

(d) Any departure from random mating upsets the equilibrium distribution of genotypes in a population.

Rules of Organic Evolution

- Some rules associated with evolution are as follows:
  1. **Allen's Rule** – Allen's Rule states that the extremities such as ears, tails and limbs of animals inhabiting colder regions are smaller than those of their counterparts living in warmer regions.
  2. **Bergman's Rule** – Warm-blooded animals living in colder regions have larger body parts than those living in a hot climate.
  3. **Cope's Rule** – Animals have a tendency to increase in body size during the course of evolution. However, this is not true for all taxonomic classes or clades.
  4. **Dollo's Rule** – Dollo's rule states that evolution does not occur in the opposite direction. Characteristics once developed do not develop again.
  5. **Gause's Principle** – It states that two or more species of organisms having similar environmental need cannot exist for a long time in an area due to similar needs; competition will develop between them in due course of time.
  6. **Gloger's Rule** – The colouration of some animals is effected by temperature, light and moisture. Many mammals, birds and insects living in a warm, humid climate have darker colour than their counterparts living in a dry climate.

Mutation Theory

- The Mutation Theory was proposed by Hugo de Vries (1901).
- de Vries was a Dutch botanist and one of the discoverers of Mendel's law.
- He pointed out that new species arise from the existing species by a sudden change called mutation.
- de Vries suggested that large-scale variations occurring in a population are the result of mutations that take place frequently, leading to formation of new species.
- Mutations are sudden, unpredictable and inheritable changes in an individual organism and are the main source by which new species arise during the course of evolution.
- de Vries worked on the plant evening prime rose (*Oenothera lamarckiana*).
- The mutation observed by de Vries in *Oenothera lamarckiana* was due to chromosomal variants.
- de Vries found several types of mutations that occurred in plants:
  - (a) **Progressive** – Introducing a wholly new character and generally resulting in the formation of new species.
  - (b) **Retrogressive** – Loss of a trait.
  - (c) **Digressive** – Activation of a trait which is long latent in a species.
  - (d) de Vries found retrogressive and digressive mutations as following Mendel's law, while progressive mutations did not follow Mendel's law.

Isolation

- Isolation is a phenomenon by which a species is divided into groups among which free interbreeding is not possible and the mechanism by which isolation is done is called isolating mechanism.
- The term 'isolating mechanism' was introduced by T Dobzhansky (1930).
- The isolation may be:

1. **Geographical Isolation –** The separation of two populations or part of a population by geographical barriers (such as forests, mountains, deserts, seas, etc.) is called geographical isolation.
2. **Spatial Isolation –** It is the isolation due to great distance. If a species is well distributed over a wide range of territory without any barrier, it will be difficult for a population situated on the extreme end to interbreed. Such an isolation is called spatial isolation.
3. **Reproductive Isolation**
   (a) Isolating mechanism which prevents fruitful reproduction is called reproductive isolation.
   (b) Reproductive isolation may be pre-zygotic or post-zygotic.

(i) Pre-Zygotic Isolating Mechanisms – Pre-zygotic isolating mechanisms are those which prevent interspecific crosses. They may be:
   (a) Temporal isolation – Individuals of different species do not mate because they are active at different times of the day or in different seasons.
   (b) Behavioural isolation – Potential mates meet but choose members of their own species.
   (c) Mechanical isolation – Copulation is attempted but transfer of sperm does not take place due to differences in genital parts.
   (d) Ecological isolation – Populations have different habitat preferences.

(ii) Post-Zygotic Isolating Mechanisms – These isolating mechanisms are effective after mating. Post-zygotic isolating mechanism may be:
   (a) Gametic incompatibility – Transfer of sperms occurs but eggs are not fertilised.
   (b) Zygotic mortality – Fertilisation of egg takes place but the zygote dies.
   (c) Hybrid unviability – Formation of hybrid embryo occurs but of reduced viability.
   (d) Hybrid sterility – The hybrid is viable but the resulting adult is sterile.
   (e) Hybrid breakdown – F_1 (first generation) hybrids are viable and fertile but further hybrid generations (F_2 and back crosses) may be unviable or sterile.
   - Whatever may be the cause of isolation, there is no doubt about the fact that it is one of the most important factors in the process of speciation.

GENETICS OF SPECIATION

- Species may be defined as a group of interbreeding population.
- Speciation is the formation of a new species.
- The whole story of speciation results in new species.
- Almost all the factors affecting the course of evolution are responsible for speciation.
- During the course of species formation, the race is formed first.
- Gradually inheritable adaptations accumulate, changing environmental act and the race is isolated.
- This isolated race forms the basis of a new race.
- It completely depends on natural selection, which, in turn, is dependent on genetic and ecological variability.
- When one species is transformed into another during the course of time, it is called phyletic speciation.
- When one species splits into two or more species, it is called speciation.
- The speciation may, in turn, be caused by sudden changes in the chromosome structure or may arise due to gradual application of evolutionary forces.

Hardy-Weinberg Law

- The Hardy–Weinberg Law is a basic concept in the population genetics developed independently in 1908 by the English mathematician, G H Hardy and German physician, W Weinberg.
- The Hardy–Weinberg law states that the gene frequencies and genotypic ratios in a natural breeding population remain constant from generation to generation if:
  (a) The population size is large.
  (b) There is no mutation.
  (c) There is no immigration or emigration.
  (d) Mating is random.
  (e) There is random reproductive success.
- The relationship between gene frequency and genotype frequency can be described by The equation $p^2 + 2pq + q^2 = 1$, where p^2 represents the frequency of dominant gene, 2pq is the frequency of heterozygote genotype and q^2 is the frequency of the homozygous recessive genotype.
- The Hardy–Weinberg law forms a theoretical base line for measuring evolutionary change as it predicts that there is no evolution.
- Whenever evolution occurs, the Hardy–Weinberg equilibrium is disturbed as a change in gene frequencies and indicates that evolution is in progress.
- The Hardy–Weinberg law is used to determine whether the number of harmful mutations in a population is increasing.
- In nature, the Hardy–Weinberg law is impossible.

EVOLUTION OF HORSE

- The evolutionary history of the horse gives a very clear picture of gradual progressive evolution.
- In the entire evolutionary history of animals, none is so beautifully arranged as the horse.
- The record of the evolutionary line of the horse is almost complete.
- The primary centre of the evolution of the horse was in North America, especially in the region of the Great Plains.
- The history of the horse family equidae began during the Eocene Epoch.
- The evolution of the horse provides a good example of Cope's law (the trend of evolution to lead to larger animals).
- The fossil lineage of the horse provides a remarkable demonstration of directional succession.
- Horses share a common ancestry with tapirs and rhinoceroses.
- Fossils of *Equus* are found in every continent except Australia and Antarctica.
- *Equus* is the only surviving genus in the once diverse family of horses.
- The horse belongs to:

(a)	Phylum	Chordata
(b)	Subphylum	Vertebrata
(c)	Class	Mammalia
(d)	Sub class	Eutheria

(e) Order Perissodactyla
(f) Sub order Hippomorpha
(g) Family Equidae
(h) Genus *Equus*

Evolutionary Trends

(a) An increase in body size.
(b) An increase in the size and complexity of teeth.
(c) An increase in the size of the head.
(d) Reduction in the number of toes on the front foot from four to one.
(e) Change of foot posture from semiplantigrade to unguligrade.
(f) Perfection of the hoof.

The recorded fossil history of the horse is as follows:

1. *Eohippus*

(a) It was of lower Eocene and four-toed form.
(b) It had low crowned teeth.
(c) It was a forest-dwelling animal that browsed on foliage.

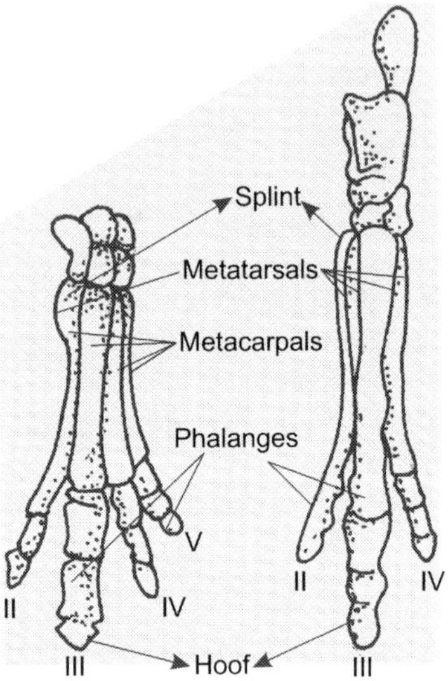

Fig. 4 Bones of forelimb and hindlimb of *Eohippus*

2. *Orohippus*

(a) It was of middle Eocene form.
(b) The forelimbs had four toes while the hindlimbs had three toes.
(c) The middle digits in both the limbs were dominant.
(d) It was a browser.

3. *Epihippus*

(a) It was of upper Eocene form.
(b) The forelimbs had four digits and the hindlimbs had three digits.
(c) It was a browser.
(d) They became extinct by the end of Eocene and were replaced by *Mesohippus*.

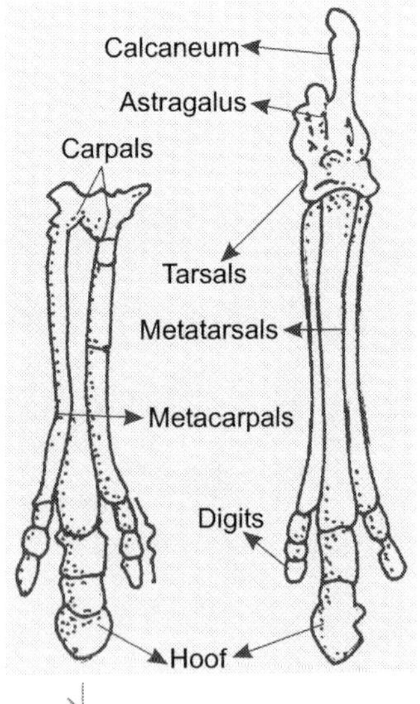

Fig. 5 Bones of forelimb and hindlimb of *Mesohippus*

4. *Mesohippus*

(a) It was of sheep-sized Oligocene form.
(b) It had only three toes on the forefoot.
(c) The teeth were low crowned.
(d) It became extinct by the end of middle Oligocene epoch.

5. *Miohippus*

(a) It was of upper Oligocene form.
(b) The forelimbs and hindlimbs were three toes.
(c) Toes were broad and spreading.
(d) The teeth were low crowned and it was a browser.
(e) It was a forest-dweller.
(f) *Miohippus* is the ancestor of many equine forms.

6. *Parahippus*

(a) It was of lower Miocene form.
(b) It was in the direct line of equine evolution.
(c) The third digit in the forelimbs and hindlimbs was more prominent.
(d) Only the median digit was effective in locomotion.
(e) The side toes were slender.
(f) The dentition was of a hyposodont type.

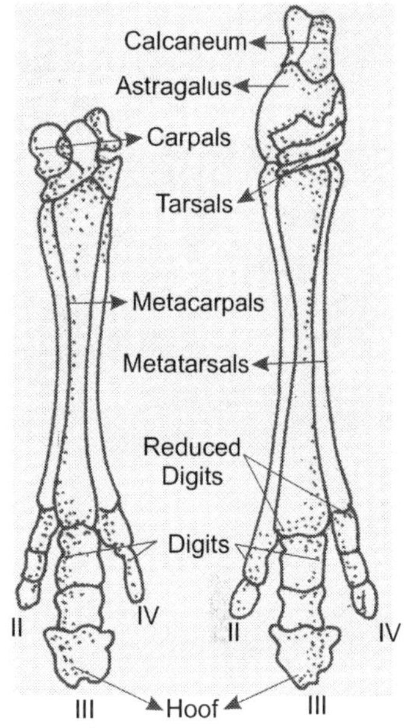

Fig. 6 Bones of forelimb and hindlimb of *Parahippus*

7. *Merychippus*

(a) *Merychippus* represents a milestone in the evolution of horses.
(b) It had a long face.
(c) Its long legs allowed it to escape from predators and migrate over long distances to feed.
(d) It was three-toed but the central one bore the most body weight.
(e) The teeth were highly crowned, coated in cement and had a more complex chewing surface.
(f) The ligaments of muscles were highly developed.
(g) The footpad was lacking.
(h) It was the first of three-toed grazers.

8. *Pliohippus*

(a) It was of late Miocene to Pliocene form.
(b) The side toes became vestigial, although some species are known to have had three toes.
(c) It was the first single-toed horse.
(d) It was well adapted for treeless plains.
(e) It is believed that *Pliohippus* have given rise to *Hippidion* and *Onohippidon* genera and *Dinohippus*.

9. *Plessippus*

(a) It was of upper Pliocene form.
(b) It was one-toed.
(c) The teeth were like those of the modern horse.

10. *Equus* (Modern horse)

(a) It originated in the upper Pliocene epoch.

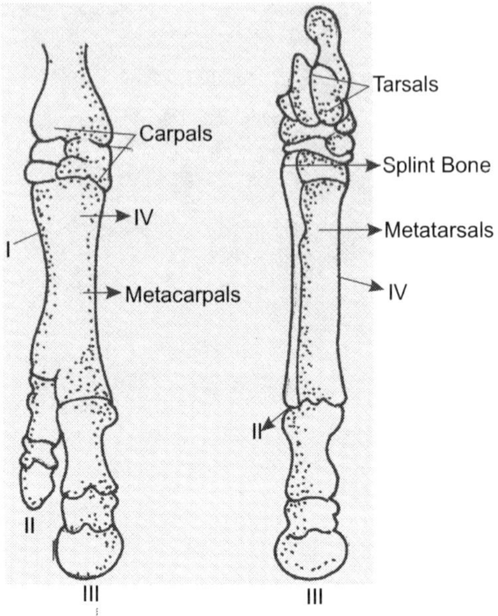

Fig. 7. Bones of forelimb and hindlimb of *Pliohippus*

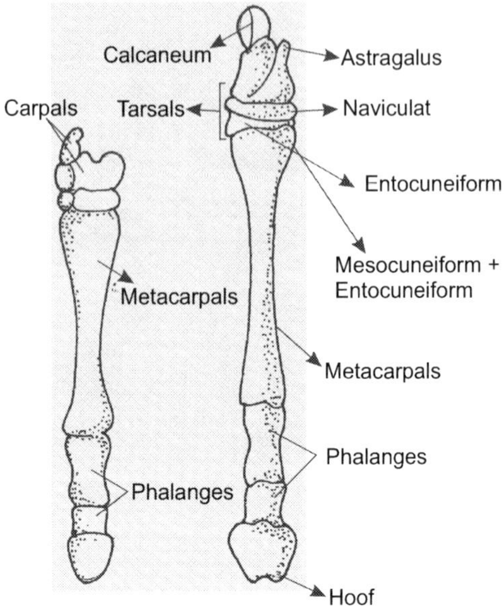

Fig. 8 Bones of forelimb and hindlimb *Equus*

(b) Some side branches are seen which became extinct in the Pleistocene epoch.
(c) The height is about 60".
(d) There is only one toe in each foot.
(e) A well-developed hoof is present.
(f) The molars have a very long crown.
(g) It has a large brain with grooved cerebral hemispheres.

- The evolution of the horse involves changes in height, toes and teeth as well as it becoming a grazer from a browser. Species of *Equus* lived from 5 million years ago until the present times. The horse was domesticated about 3,000 years ago and had profound impact on human history.

EVOLUTION OF MAN

- The fossil records of primates are fragmentary.
- All primates, including humans, must have had a common ancestry.
- According to Pough et al. (1996) 'Man descended from arboreal ancestors that lived in early tertiary forests about 65 million years ago'.
- Arboreal, shrew-like insectivores show the first indications of the evolutionary line of primates.
- Ancestral lemurs and lorises showed some remarkable features of primate evolution.
- Tarsius was more advanced and showed characteristics intermediate between lemurs and anthropoids.
- Humans belong to:

(a) Phylum Chordata
(b) Subphylum Vertebrata
(c) Class Mammalia
(d) Subclass Eutheria
(e) Order Primates
(f) Sub-order Anthropoidea
(g) Family Hominidae
(h) Genus *Homo*
(i) Specie *sapiens*

- The anthropoid ancestors of humans:
  1. The evolutionary line of the old world monkey led to the evolution of apes and humans.
  2. Four types of apes are found today, among which the gibbon and the orangutan are found in Asia, while the gorilla and the chimpanzee are found in Africa.
  3. The gibbon is the smallest ape while the gorilla is the largest.
  4. The chimpanzee is considered to be the closet relative of modern human beings.
  5. Chromosome numbers and banding pattern studies have revealed the common origin of humans and chimpanzees.
  6. Central Africa is regarded as the centre of evolution of humans, as maximum number of fossils of primitive humans has been excavated from that region.

Evolutionary Trends

- Shifting from arboreal life to ground life
- Towards erect postures and freeing of hand
- Bipedal locomotion
- Increase in cranial capacity and complexity of brain
- Reduction in the size of canines and incisors
- Development of the chin
- Loss of jaw power
- Development of intelligence
- Forward shifting of the foramen magnum
- Orthognathous skull with high forehead without projecting eyebrow ridges
- Lack of simian gap and simian shelf

Early Evolution of Primates

- Evolution of human apes began in the Oligocene epoch and a common stock has been recognised, called *Propliopithecus*.
- In the Miocene epoch, a group of ape humans called *Dryopithecus* had evolved.
- Humans, gorillas and other apes originated diverging from *Dryopithecus*.
- *Dryopithecus* had broadened jaws, semi-erect gait, large canines and five-cusped molars.

I. Australopithecus

- Australopithecus was the immediate forerunner of the genus *Homo*.
- Its fossil was discovered by Dart (1924), who named it the 'southern ape'.
- It had a mosaic form of humans and apes, and can therefore be considered a connecting link between humans and apes.

1. Human Characteristics *of Australopithecus*

- Erect posture and bipedal locomotion.
- Dental arch with smoothly rounded parabola.
- Vertebral column with a distinct lumber curve.
- The simian gap was absent and canine teeth were not projecting over the head.
- Foramen magnum forward under the base of the skull.
- They used weapons made of bones.

2. Ape Characteristics of *Australopithecus*

- The teeth were larger than modern humans.
- The face was prognathous.

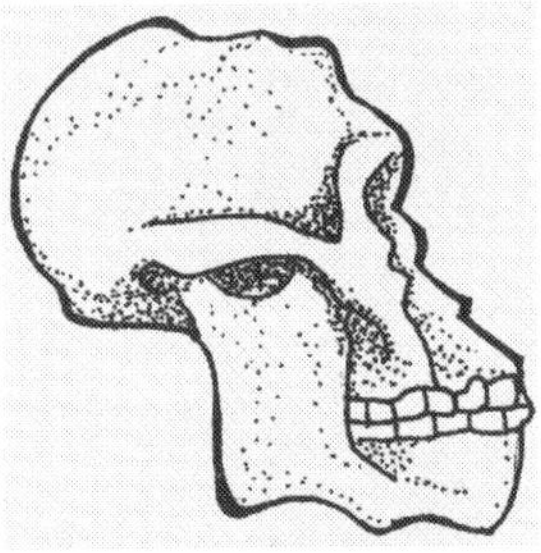

Fig. 9 Skull of *Australopithecus*

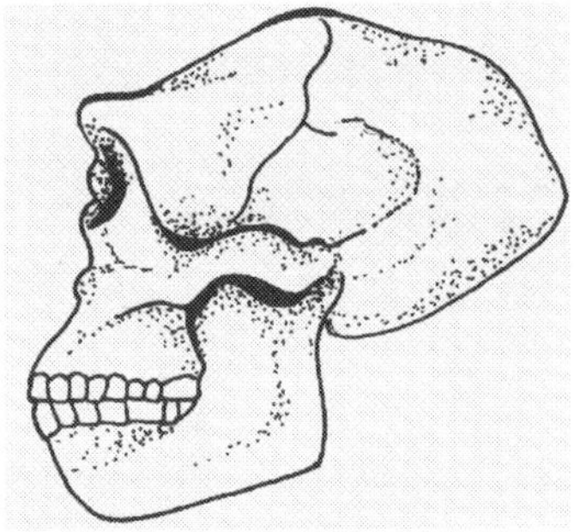

Fig. 10 Skull of *Pithecanthropus*

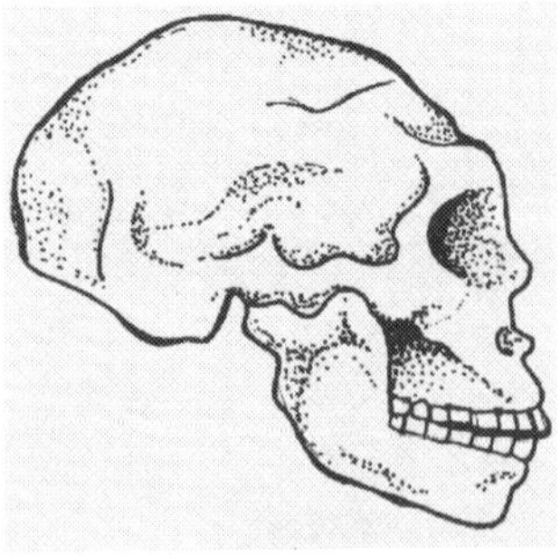

Fig. 11 Skull of *Neanderthal*

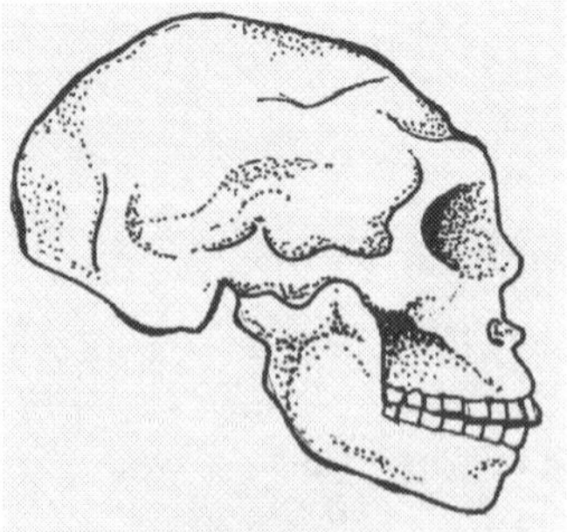

Fig.12 Skull of *Cro-Magnon*

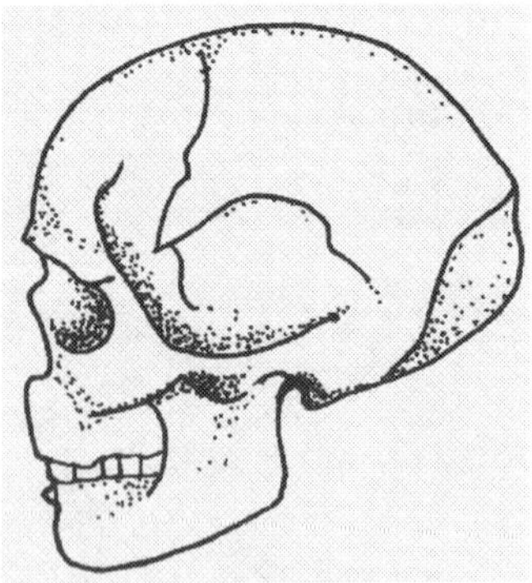

Fig. 13 Skull of *Homo sapiens*

- Eyebrow ridges projected over the eyes.
- Cranial capacity was 450 to 600 cc.
- There were several types of *Australopithecus*, like the gracile type (*A. africanus*), the robust type (*A. robust*) and the lucy (*A. afferensis*). The gracile type had a lightly body structure indicating an omnivorous form, but the robust type had strong cheek teeth, indicating an herbivorous diet.

II. Origin of Genus *Homo (Homo Erectus)*

- In the middle of Pleistocene epoch, the Java human and the Peking human originated. As both shared common basic features, they were included in the genus *Homo erectus*. They had the following characteristics:
  (a) Habitually erect and bipedal
  (b) Skull flat with little or no forehead
  (c) Large prognathous jaws
  (d) Broad nasal aperture
  (e) Larger teeth and eyebrow ridges
  (f) No simian shelf
  (g) No chin
  (h) Cranial capacity 725 cc to 90 cc (Java man) and 915 cc to 1225 cc (Peking man)
  (i) Used fires for protection and cooking
  (j) *Homo erectus* represents a level of organisation that permits its possessors to spread into new niches
  (k) *Homo erectus* was not only confined to Asia but also spread through Africa as *Homo habilis*

III. Origin of *Homo Sapiens*

- The *Homo erectus* gradually evolved into *Homo sapiens*. In this transitional event, two types of humans were found in later deposits of the Pleistocene epoch.
- Of these, one was identified as the primitive human called the *Neanderthal man*, who has been labelled as *Homo sapiens neanderthalensis* and the other, the modern human being called *Homo sapiens sapiens*.

1. *Homo Sapiens Neanderthalensis*

- Its fossil was discovered by C Fuhlrott (1856) from the Neandertal river valley (Germany).
- They were abundant in Europe.
- Their forehead was low and slanting.
- They had heavy eyebrow ridges.
- They had a long prognathous and narrow face.
- Cranial capacity was about 1200 cc to 1600 cc.
- They lived in caves.
- They buried dead bodies. The burial of the dead reveals the emergence of religious beliefs in human records.
- They were capable of big game hunting.
- They used tools.

- Neanderthal humans were succeeded by *Homo sapiens* of modern times; the first of these were Cro-Magnon humans.

2. *Homo Sapiens Fossilis* (Cro-Magnon man)

- They were taller than Neanderthal man.
- They had teeth and jaws like modern humans.
- Their cranial capacity was 1600 cc.
- They had an orthognathous skull.
- The appendicular skeleton is adapted for an upright posture and gait.
- The foramen magnum faces directly downwards.
- They used to domesticate dogs and hunt in groups.
- From the Cro-Magnon man, due to a great cultural evolution, the modern human (*Homo sapiens sapiens*) has resulted. Modern human beings are still acquiring more complex structures and their evolution is still a work in progress.

GEOLOGICAL TIME SCALE

- Geological time scale refers to the various stages in the history of life on the earth.
- On the basis of fossils found in rocks, the geological time scale has been divided by scientists into different intervals that are characterised by the most significant changes that occurred in the organisation of organisms.
- The largest defined unit is the Eon. An Eon is divided into Eras and Eras are, in turn, divided into Periods and Epochs.
- Nicholas Steno in the late 17th century laid down the principle underlying the geological time scale.
- The geological time scale has been divided into six major divisions called Eras, which are as follows:

I. Azoic Era

(a) Earliest time in the history of the earth.
(b) Complete absence of living organisms.
(c) The earth was formed during this period.

II. Archaeozoic Era

(a) Age of invisible life.
(b) Prokaryotes and eukaryotes originated.
(c) Oxygen begins to accumulate in the atmosphere.
(d) Age of primitive marine invertebrates.

III. Proterozoic Era

(a) Extended from 1,600 to 6 million years ago.
(b) Age of primitive marine invertebrates.

(c) The Archaeozoic and Proterozoic eras are together designated as the Cryptozoic era or 'Precambrian time'.

IV. Palaeozoic Era

(a) Popularly known as the 'cradle of ancient life'.
(b) Duration of about 370 million years, beginning about 600 million years ago and ending by about 230 million years ago.
(c) At the beginning of Palaeozoic era, life existed only in or near the oceans.
(d) It was also characterised by mass extinction of many life forms.
(e) Fossils of first vertebrates appeared.
(f) The Palaezoic era has been divided into the following six periods:

1. **The Cambrian Period**
   - Age: 600 to 500 million years ago.
   - Age of trilobites
2. **Ordovician Period**
   - Age: 438 to 505 million years ago.
   - The first vertebrate appeared.
3. **Silurian Period**
   - Age: 405 to 425 million years ago.
   - The first jaw vertebrate appeared.
4. **Devonian Period**
   - Age: 360- 408 million years ago.
   - Age of fishes.
5. **Carboniferous Period**
   - Age: 285 to 320 million years ago.
   - Formation of coral beds.
   - Origin of reptiles.
6. **Permian Period**
   - Age: 245 to 285 million years ago.
   - Trilobites disappear completely.
   - Crinoids became rare.
   - Appearance of first reptiles, cotylosaurs.

V. Mesozoic Era

- It is one of the three eras of the Phanerozoic era.
- The Mesozoic era lasted about 180 million years from about 245 million years to about 65 million years ago.
- It is called the age of dinosaurs.
- At the beginning of Mesozoic era, all of the world's continents were joined into the super-continent of Pangea.
- By the end of this era, most continents had separated into their present forms.

- The Mesozoic era is divided into three periods:

1. **Triassic Period**
   - Age: 208 to 225 million years.
   - Egg-laying mammals made their first appearance.
   - The first dinosaurs appeared.
   - Primitive amphibians had disappeared.
2. **Jurassic Period**
   - Age: 144 to 188 million years ago.
   - Dinosaurs were dominant.
   - Marsupial mammals originated.
   - At the end of this period, the *Archaeopteryx* (primitive bird) had originated.
3. **Cretaceous Period**
   - Age: 63 to133 million years ago.
   - Giant reptiles became extinct.
   - Eutherian mammals made their appearance.

VI. Cenozoic Era

- It is the last major division of the geologic time scale.
- It is the present age (65 million years ago to the present).
- It is the age of mammals.
- It is divided into the following two periods:

1. **Tertiary Period**
   - Duration: 1 to 63 million years ago.
   - More than 95 per cent of the Cenozoic era belongs to the Tertiary period.
   - Modern orders of class mammalians originated.
   - The Tertiary period is divided into following five epochs:
     - (a) **Paleocene Epoch**
       - Age: 58 to 63 million years ago.
       - Primitive primates appeared.
     - (b) **Eocene Epoch**
       - Age: 37 to 58 million years ago.
       - The horse originated.
     - (c) **Oligocene Epoch**
       - Age: 24 to 37 million years ago.
       - Odd-toed ungulates were widespread.
       - Even-toed ungulates began to radiate.
       - Human-like primates appeared.
     - (d) **Miocene Epoch**
       - Age: 6 to 24 million years ago.
       - Human-like apes appeared.
       - The *Merychippus* evolved.
       - Modern dogs, elephants and cats appeared.

(e) **Pliocene Epoch**
- Age: 2 to 6 million years ago.
- Hominids, i.e., human-like forms evolved.

2. **Quaternary Period**
- The Quaternary period includes only the last 1.8 million years.
- It is divided into the following two epochs:

(a) **The Pleistocene Epoch**
- Age: 0.6 to 2 million years ago.
- Humans and the modern horse *Equus* evolved in this period.
- Hunting by humans, along with climatic change, served to kill off most of the planet's mega fauna.

(b) **The Holocene Epoch**
- It began 0.6 million years ago.
- The age of humans.

ADAPTATION

- Adaptation may be defined as a trait that enables an animal to become suited to its environment.
- Adaptations are the result of evolution and are one of the basic phenomena of biology.
- Adaptations generally take place because genes mutate or are changed by accident.
- Adaptation is a slow process caused by natural selection.
- Animals must adapt to their environment or face extinction.
- Adaptation may be structural or behavioural.

Structural (Physical) Adaptations

- Body adaptations are known as structural adaptations that help animals to survive in their environment.
- Structural adaptations are specially found in the bones and muscles that are adapted for running, swimming, climbing, flying, gliding, jumping, etc.

Behavioural Adaptations

- The adaptation that helps an organism to enhance either survival or reproduction is known as behavioural adaptation.
- Behavioural adaptation may be instinctive (occur naturally) or learnt (must be taught).
- Birds calls and migration are behavioural adaptations.

ZOOGEOGRAPHICAL REGIONS

- Zoogeography is the study of patterns of distribution of animals in nature.

- Wallace (1876) has divided the whole world into six regions known as zoogeographical regions or realms. These six regions are as follows:

I. Neotropical Region

- The Neotropical region is one of the six major zoogeographical regions of the world.
- This region includes South and Central America, the West Indies and the Southern Mexico.
- This is the most biologically diverse region on the earth.
- The region has tropical climate except the southern region, which has south temperate zones.
- There are many diverse habitats in this region that are tropical.
- It has been divided into the following subregions:
  (a) Chilian subregion
  (b) Brazilian subregion
  (c) Mexican subregion
  (d) Antillean subregion
- 43 families and subfamilies of fishes are endemic to this region. The gymnotiformes, loricarridae and characidae families of fishes are unique to this region. Family cyprinidae is completely absent.
- Amphibians are represented by both anura and urodela.
- The region is inhabited by 15 families of lizards. Agamidae are completely absent. The characteristic genera of snake are *Ungalia*, *Elaps*, *Dromicus*, *Epicrates*, etc. Crocodiles, alligators, turtles and tortoises are common.
- 31 bird families are endemic to the Neotropical region. They include rhea, tinamous, curassows and toucans. Bird families originally unique to the Neotropical region include hummingbirds (family trochilidae) and wrens (family troglodytidae).
- The unique mammals of this region are American opossum, shrew opossum, anteaters, sloths, armadillos, new world monkeys, cavimorpha rodents, etc.

II. Palaearctic Region

- The Palearctic region occupies 39 per cent of the mainland. It is the largest region.
- It is a modest zone as far as the number of kinds of animal is concerned.
- It has no tropical zone.
- It includes Europe and middle North Asia.
- It has been divided into the following subregions:
  (a) European subregion
  (b) Mediterranean subregion
  (c) Siberian subregion
  (d) Manchurian subregion
- This region includes 136 families of vertebrate animals, 100 genera of mammals and 174 genera of birds. There are 35 and 37 unique genera of mammals and birds.
- One bird family Pynellidae is endemic to this region. The Holoarctic has four other endemic bird families, viz., Tetraoninae, Alcidae, Gavidae and Bombycillidae. Parrots are absent in this region.
- The Palearctic region has no endemic mammals but several families are endemic, viz., Ailuridae, Calomyscidae, Prolagidae, etc. The camel family has discontinuous distribution.

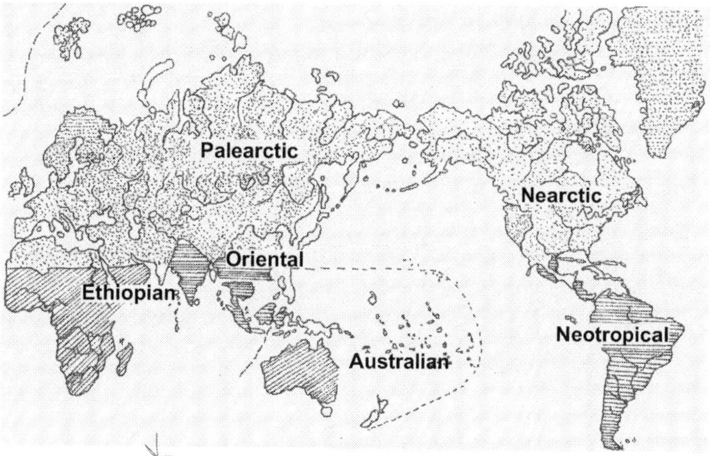

Fig. 14 The six zoogeographical regions of the earth

III. Ethiopian Region

- The Ethiopian region includes Africa south of Sahara, Madagascar and South Arabia.
- This region has poor diversity of animals.
- It resembles the Neotropical and Oriental regions in its fish and amphibious fauna, while in bird fauna, it shows much similarity with the Oriental region.
- This region has the following four subregions:
  (a) East African subregion
  (b) West African subregion
  (c) South African subregion
  (d) Malagasy subregion
- This region represents 174 families of vertebrate animals, 140 genera of mammals and 294 genera of birds.
- The characteristic fauna of this region are *Protopterus*, secretary birds, giraffes, lions, zebras, baboons, gorillas, chimpanzees, hippopotamuses, etc.
- *Rana*, *Bufo* and tailed amphibians, iguanids, talpidae, ursidae, cervidae and wrens are absent.

IV. Australian Region

- The Australian region includes Australia, New Zealand, New Guinea, Tasmania and the Oceanic Islands of the Pacific.
- This region has no land connection with the rest of the regions.
- This region is characterised by both tropical and temperate climates.
- The flora and fauna of this region are peculiar due to their isolation from other regions of the world.
- The Australian region has been divided into the following subregions:
  (a) Austro–Malayan subregion
  (b) Australian subregion
  (c) Polynesian subregion

(d) New Zealand subregion

- This region includes 18 families of mammals, 71 of birds, 31 of reptiles, 11 of amphibians and 11 of freshwater fishes.
- The region is characterised by an abundance of endemic genera and families and a small number of placental mammals. Marsupials and monotremes are found only in this region.
- Out of 670 species of birds inhabiting the Australian region, about 450 are endemic.
- The diagnostic fauna are *Echidna*, duck-billed platypus, kangaroo, marsupial moles, honeyeaters, kiwi, cassowaries, paradise birds, crowned pigeons, *Sphenodon*, *Lioplema*, narrow mouthed toads, *Neoceratodus*, etc.
- The crotalidae and viperidae (reptiles) as well as tailed amphibians are totally absent.

V. Oriental Region

- The Oriental region includes the whole of India, Sri Lanka, Burma, South China, Thailand peninsula, Malaya and Malayan islands (Sumatra, Philippines, Java, Bali and Borneo).
- This entire region is located within the tropics and is covered by luxuriant tropical forests.
- This region is known for its varied physical features.
- This region is subdivided into the following sub-regions:
  (a) Indian subregion
  (b) Ceylonese subregion
  (c) Indo-Chinese subregion and
  (d) Indo-Malayan subregion
- The Oriental region includes 164 families of all vertebrates, 118 genera of mammals and 340 genera of birds. There are 12 unique families of vertebrates, 55 unique genera of mammals and 165 unique genera of birds.
- Primitive fishes are absent in this region.
- Some important animals of this region are the Indian elephants, tigers, rhinoceroses, antelopes, gibbons, flying lemurs, sunbirds, honeyguids, woodpeckers, peacocks, broadbill birds, geckos, iguanas, *gavialis,* etc.

VI. Nearctic Region

- It consists of the geographical territories of North America and Greenland.
- The Neotropical region has a mixture of fauna of the Palaearctic and Neotropical regions.
- It has varied types of climates as well as an extensive mountain range.
- This region is subdivided into the following subregions:
  (a) Californian subregion
  (b) Rocky mountain subregion
  (c) Alleghany subregion and
  (d) Canadian subregion
- This region has 24 genera of mammals, 50 genera of birds, 21 of reptiles, 15 of amphibians and 29 of freshwater fishes.
- The characteristic fauna of this region are *Amia calva, Necturus*, axolotl, rattlesnakes, *Heloderma*, *Phrynosoma*, blue jay turkeys, raccoons, opossums, etc.

Wallace Line

- The Wallace Line is an imaginary line that separates the zoogeographical region of Asia and Australasia.
- It is so named after the name of its originator Alfred Russel Wallace.
- It runs through the Malaya Archipelago, between Borneo and the Celebes and Bali and Lombok.
- To the west of line are found organisms related to Asiatic species, while to its east are found organisms related to Australian species.

Weber's Line

- Weber's Line is an alternative to the Wallace Line.
- It is the new line that separates the Oriental and Australian regions.
- It is situated further east than the Wallace Line.

MIMICRY

- The resemblance of one organism to another or to natural objects in its surroundings, for protection from predators, is known as mimicry.
- The organism that mimics is known as the mimic and the organism that is copied is known as the model.
- The concept of mimicry was given by Henry W Bates (1962) based on his studies of butterflies in the forests of Brazil.
- Mimicry takes place when a group of organisms (mimics), tend to share common perceived characteristics with another group (model).
- Mimicry is a form of co-evolution.
- In Batesian mimicry, the palatable and unprotected species of animals resembles an unpalatable and relatively protected species. This type of mimicry is beneficial only to the mimic.
- In Mullerian mimicry, two or more unpalatable species resemble each other. In this type of mimicry, both the mimic and the model are benefited.
- Mimicry may be protective, aggressive and conscious.
  1. **Protective Mimicry** – It occurs in organisms to protect themselves from predators. It is of the following two types:
     (a) **Concealing Mimicry** – This is the most common type of mimicry. The animal hides itself according to its environment (e.g., the stick insect shows close similarities with a twig).
     (b) **Warning Mimicry** – In warning mimicry, the organism resembles forms which are poisonous or harmless (e.g., snakes of genus *Elaps* mimic the poisonous coral snake).
  2. **Aggressive Mimicry** – In aggressive mimicry, the animal either conceals itself for attack (e.g., spiders living on flowers mimic the flower so that they become invisible to their prey) or allures the prey (certain spiders mimic the flowers of orchids and allure the insects).
  3. **Conscious Mimicry** – Certain animals simulate death whenever in danger to protect themselves (e.g., the opossum becomes unconscious and simulates death, whenever in danger).

EVOLUTION

EVOLUTION

Short-Answer Questions

1. When did the earth originate?
   Answer: The earth originated about 4.8 billion years ago.
2. When did life first appear on the earth?
   Answer: It is estimated that life first appeared on the earth about 3.5 billion years ago.
3. Who wrote the book *Origin of Life*?
   Answer: The *Origin of Life* was written by A I Oparin (1936).
4. Where did life first appear?
   Answer: Life first appeared in the oceans.
5. Define coacervates.
   Answer: Coacervets are complex forms of organic compounds that are believed to be the precursors of the cells.
6. What is the theory of spontaneous generation?
   Answer: According to the theory of spontaneous generation, certain complex organisms are generated by the decaying of certain organic substances.
7. What was the nature of atmosphere in the early stages?
   Answer: In the early stages, the atmosphere was reducing in content.
8. Define the autotrophic hypothesis of the origin of life.
   Answer: The autotrophic hypothesis of the origin of life states that the first organisms on the earth were producers (having the ability to synthesise their own food) like plants and chemosynthetic.
9. What are black smokers and when were they first discovered?
   Answer: Black smokers are a type of hydrothermal vent found on the floor of oceans, which were first discovered in 1977 around the Galapagos Islands.
10. Name the most accepted theory on the origin of life.
    Answer: The heterotrophic theory
11. Name the gases present in the atmosphere of the primitive earth.
    Answer: Methane, hydrogen, ammonia and water vapour were present in the atmosphere of the primitive earth.
12. What is the heterotrophic theory of the origin of life?
    Answer: According to the heterotrophic theory of the origin of life, the first organisms were heterotrophic, i.e., they were not able to produce their own food and were formed as a result of aggregation of complex organic molecules called coacervates.

13. When did oxygen become abundant on the earth?
Answer: Oxygen became abundant on the earth after the emergence of photosynthetic organisms (about 1.5 billion years after the emergence of life on earth).
14. What are the oldest fossils of the earth?
Answer: Stromatolites are the oldest fossils of the earth.
15. Define stromatolites.
Answer: Stromatolites are laminated rock formed by the growth of blue-green algae (cyanobacteria)
16. Define thrombolites.
Answer: Thrombolites are cousins of stromatolites.
17. What is RNA world hypothesis?
Answer: RNA world hypothesis suggests that life began as a simple self-replicating molecule called the RNA that through natural selection evolved into the present form of life.
18. Name the substances used by Miller in his experiment.
Answer: Methane, hydrogen and ammonia were used by Miller in his experiment in the ratio of 2:1:2.
19. Name the substances obtained by Miller from his experiment.
Answer: Miller obtained amino acids, simple sugars and organic compounds from his experiment.
20. In the primitive age, how did ultraviolet rays freely reach the earth?
Answer: Ultraviolet rays were able to reach the earth because there was no ozone layer in the atmosphere of the primitive earth.
21. Define evolution.
Answer: Evolution is the slow process of change from simple to complex.
22. Who coined the term 'evolution'?
Answer: Herbert Spencer coined the term 'evolution'.
23. Define homologous and analogous organs.
Answer: Organs having a similar basic structure but different appearances and functions are known as the homologous organs, while organs having different basic structures but similar appearances and function are called analogous organs.
24. What are connecting links? Give two examples.
Answer: Connecting links are those organisms that posses characteristics of two different groups. For example,
    (a) *Peripatus* – Connecting link between Annelida and Arthropoda
    (b) *Neoplina* – Connecting link between Annelida and Mollusca.
25. What is the zoological importance of *Echidna*?
Answer: *Echidna* is a mammal that connects two groups, viz., cold-blooded reptiles and warm-blooded mammals. It has characters of both reptiles and mammals.
26. What are missing links?
Answer: Extinct connecting links are known as missing links. For example,
    (a) *Seymouria* – Connecting link between amphibians and reptiles
    (b) *Archaeopteryx* – Connecting link between reptiles and birds
27. Define vestigial organs.
Answer: Vestigial organs are those organs that are nonfunctional and useless but were functional in ancestral forms. For example, vermiform appendix and coccyx of humans.
28. What is atavism? Give examples.

Answer: The reappearance of ancestral characteristics is known as atavism. It is also known as reversion (e.g., the occurrence of tail in human infants).

29. Name the scientist who first provided the evolutionary theory.
Answer: Jean Baptiste de Lamarck was the first to provide the evolutionary theory.

30. What is the assumption of Lamarck's theory?
Answer: Lamarck's theory of evolution states that changes acquired during the lifetime of an individual are transmitted to the next generation. It is based on new needs, use and disuse of organs and inheritance of acquired characteristics.

31. Who was the first to give experimental proof against the theory of inheritance of acquired characteristics?
Answer: Weismann (1890) was the first to provide experimental proof against the theory of inheritance of acquired characteristics.

32. Name the scientists who first categorically explained organic evolution.
Answer: Charles Darwin was the first to categorically explain organic evolution in detail

33. Who gave the concept that evolution proceeded through natural selection?
Answer: Charles Darwin (1859) gave the concept that evolution proceeded through natural selection.

34. Give the criteria in order for natural selection to act on a trait.
Answer: In order for natural selection to act on a trait:
    (a) The trait must possess heritable variation.
    (b) Must confer an advantage in the competition for resources.

35. What is the name of the ship on which Darwin travelled?
Answer: Darwin travelled on a ship called HMS Beagle.

36. Who said that population increases geometrically, while food and shelter increase arithmetically?
Answer: Malthus

37. What is the concept of natural selection?
Answer: Natural selection states that nature creates struggle and only those animals that win this competition are selected by nature.

38. Give examples of natural selection.
Answer: Industrial melanism is the best example of natural selection. The moth *Biston betularia* is of two colours, viz., grey and black. Before industrialisation, grey colour moths were abundant but after industrialisation their number decreased and black colour moths became more abundant as they had protective colouration. Thus, before industrialisation, nature selected grey colour moths and after it, black colour ones.

39. Define the synthetic theory of evolution.
Answer: It is the latest and most accepted theory of evolution. According to this theory, the evolution of new species occurs due to the interaction of genetic variations and natural selection. According to the synthetic theory of evolution, the following factors are responsible for evolution:
    (a) Gene mutation
    (b) Changes in the structure and number of chromosomes
    (c) Genetic recombination
    (d) Reproductive isolation
    (e) Natural selection.

40. Define natural selection.
Answer: Natural selection is a means by which useful variations in a population tend to be preserved while useless variations are not preserved.

41. How does natural selection bring change in gene frequencies?
Answer: Natural selection tends to preserve useful variations in populations and passes it from generation to generation, leading to change in gene frequencies.
42. Does natural selection not act on individuals?
Answer: Natural selection does not act on individuals because individual organisms cannot become better adapted to their environment as they are unable to change their genes. This is the reason that natural selection acts on populations and not on individuals.
43. Does natural selection produce evolutionary change?
Answer: Yes, because it changes the genetic composition of the populations.
44. What are Darwin's finches?
Answer: Darwin's finches are a group of sparrow-sized birds that inhabit the Galapagos Islands. These were observed by Darwin, which provide evidence of speciation. They are classic examples of evolution by natural selection. These birds inspired Charles Darwin to formulate the theory of the 'origin of species'.
45. What is name of the island that is related with Darwin?
Answer: The Galapagos Islands
46. Define sexual selection.
Answer: The selection that acts on traits relating to attracting mates is termed as sexual selection.
47. Define fossils. How do fossils provide evidence in favour of evolution?
Answer: Fossils are remains of dead animals and plants that lived in the remote past. They provide evidence that present animals and plants have originated from the pre-existing ones through a process of the continuous evolution. For example, the fossil of the bird *Archaeopteryx* looks like a bird. It had both reptilian characters (such as teeth and tail) as well as avian characters (such as feathered wings). Thus, it is a connecting link between reptiles and birds suggesting that birds may have evolved from reptiles.
48. Name the branch of biology, the study of which provides clear proof of organic evolution.
Answer: Palaeontology (study of fossils) is the branch of biology that provides clear proof of organic evolution.
49. What are coprolites?
Answer: Coprolites are fossilised faeces.
50. In which type of rocks are fossils found?
Answer: Fossils are only found in sedimentary rocks.
51. What is biogenetic law? Who gave this law?
Answer: Biogenetic law states that 'ontogeny recapitulates phylogeny'. This law was given by Ernst Haeckel.
52. Name the type of evolution that results in analogous structures.
Answer: Convergent evolution results in analogous structures.
53. What is divergent evolution? Give an example.
Answer: A pattern of evolution in which two closely related species gradually become increasingly dissimilar in different habitats of the same area is called divergent evolution. Darwin's finches are a very good example of divergent evolution.
54. Define co-evolution.
Answer: The combined change in two or more species, which are in close interaction, usually dependent upon each other is known as co-evolution. Plants and animals that pollinate them are a common example of co-evolution.

55. Define gene flow.
Answer: The movement of genes from one population to another is known as gene flow.

56. Define genetic drift. How might the process of genetic drift increase?
Answer: Genetic drift is the sudden change in gene frequency in a population. The process of genetic drift may increase by the bottleneck effect or founder effect, as genetic drift works more properly in small populations.

57. What term is given to the sudden reduction in the number of individuals in a population due to natural disasters like earthquakes, droughts, floods, habit destruction, etc.?
Answer: The bottleneck effect

58. Name the major factors of evolution.
Answer: The major factors of evolution are as follows:
(a) Natural selection (b) Mutation (c) Genetic drift
They can cause change in genotypes and phenotypes over time.

59. Who proposed the mutation theory?
Answer: The mutation theory was proposed by Hugo de Vries (1901).

60. Define phylogeny.
Answer: The evolutionary history of an organism is called phylogeny.

61. Define isolating mechanism.
Answer: Any mechanism (such as geographical barrier or biological differences) that prevents mating between individuals of population is called an isolating mechanism. Isolating mechanism is one of the most important factors in speciation. The term 'isolating mechanism' was coined by T Dobzhansky (1930).

62. Name the primary mechanisms of reproductive isolation.
Answer: Sexual isolation and hybrid sterility are the primary mechanisms of reproductive isolation between species.

63. Define species and speciation.
Answer: Species is the group of interbreeding individual. The formation of new species is known as speciation.

64. How does the formation of species occur?
Answer: The formation of new species occurs through allopatric speciation and sympatric speciation.

65. What are allopatric and sympatric speciation?
Answer: Speciation within the same geographical region is called allopatric speciation. Most of the speciation occurs through allopatric speciation. A famous allopatric speciation is Darwin's Galapagos finches. Sympatric speciation is the splitting of one species into two within the same geographical region. Sympatric speciation may occur if a mutation results in an immediate reproductive barrier in a segment of the species. The most common example of this is polyploidy in plants.

66. Mention the components of speciation.
Answer: Following are three components of speciation:
(a) Prevention of gene flow.
(b) Establishment of reproductive isolation, preventing interbreeding.
(c) Accumulation of morphological and behavioural differences.

67. Which law states that gene frequencies and genotype ratios in a randomly breeding population remain constant from generation to generation?
Answer: The Hardy–Weinberg law

68. When does the Hardy–Weinberg law fail to apply?
Answer: The Hardy-Weinberg law fails to apply when there is:
(a) Mutation
(b) Genetic drift
(c) Nonrandom mating
(d) Natural selection

69. Is evolution by genetic drift adaptive?
Answer: No, evolution by genetic drift is aimless and not adaptive.

70. Which region was the primary centre of the evolution of the horse?
Answer: The primary centre of the evolution of the horse was North America, especially its Great Plains region.

71. When did the evolution of horse begin?
Answer: The evolution of the horse began in the Eocene epoch.

72. What is the name of the earliest animal that bears horse-like anatomy?
Answer: *Hyracotherium* (*Eohippus*) was the earliest animal that had horse-like anatomy.

73. Name the continents where fossils of *Equus* are not found.
Answer: Fossils of *Equus* are found in every continent except Antarctica and Australia.

74. Name the most common Equidae of North America during the Pliocene epoch.
Answer: *Dinohippus* was the most common Equidae of North America during the late Pliocene epoch.

75. Name the first three-toed grazer fossil of the horse.
Answer: *Merychippus* was the first three-toed grazer.

76. Name the oldest species of the modern horse and the place from where it was discovered.
Answer: *Equus stenonis* is the oldest species of the modern horse whose fossil was discovered from Italy.

77. Name the first one-toed horse.
Answer: *Pliohippus* was the first one-toed horse.

78. Name the vestigial organs of the horse.
Answer: Splint bones are the vestigial organ of the horse.

79. Who first claimed the discovery of fragments of Piltdown humans?
Answer: It was Dawson (1908) who first claimed the discovery of fragments of Piltdown humans.

80. When did the evolution of humans begin?
Answer: The evolution of humans and apes began in the Oligocene epoch and a common stock has been recognised called *Propliopithecus.*

81. Name the banding pattern of chromosomes that are similar in humans and chimpanzees.
Answer: The banding pattern of chromosome 3 and 6 are similar in humans and chimpanzees.

82. Name the closest relative of human beings.
Answer: The chimpanzee is the closest relative of human beings.

83. What do you know about WT-1500?
Answer: Kamoya Kimeu (1977) discovered a fossil skull of a boy near Lake Turkana in Northern Kenya, which belonged to about a 16 lakh year-old ancestor of a hominid. This fossil has been named as WT-1500.

84. What do you know about *Dryopithecus*?
Answer: *Dryopithecus* is a genus of extinct ape having characteristics of old world monkeys, great apes and humans. *Dryopithecus* was found as fossil in the Miocene and Pliocene epoch deposits (23 to 2.6 million years old). It apparently originated in Africa. It had the following characteristics:

(a) Semi-erect gait
(b) Suspensory forelimbs
(c) Large canines and five cusped molars (Y-5 arrangement)
(d) It was a tree-dwelling animal living on berries and fruits.
(e) It was 60 cm long.

85. Name the earliest known catarrhine.
Answer: The earliest known catarrhine is *Kamoyapithecus* from the uppermost Oligocene.

86. Why was *homo habilis* called the tool man?
Answer: Because tools were found with its fossil remains

87. Name the place from which the fossil of *Ramapithecus* was discovered.
Answer: The fossil of *Ramapithecus* was discovered from the Shivalik Hills in India.

88. Mention one difference between *homo sapiens* and *homo erectus.*
Answer: The size of the brain of *homo erectus* was smaller than *homo sapiens.*

89. Name the fossil that is considered the last and direct ancestor of present humans.
Answer: The Cro-Magnon man is considered last and direct ancestor of modern-day humans.

90. Mention the ape-like characteristics of *Australopithecus.*
Answer: (a) Its teeth were larger than those of modern humans
(b) The face was prognathous
(c) Eyebrow ridges projected over the eyes
(d) Cranial capacity 450- 600 cc

91. Give the cranial capacity of:
(a) *Homo sapiens neanderthalensis*
(b) Java man
(c) Peking man
(d) Cro-Magnon man
Answer: (a) *Homo sapiens neanderthalensis* – About 1200 cc to 1600 cc.
(b) Java man – 725 cc to 90 cc
(c) Peking man – 915 cc to 1225cc
(d) Cro-Magnon man – 1600 cc

92. What is geological time scale?
Answer: Geological time scale refers to the various stages in the history of life on earth.

93. What is geological period?
Answer: Geological period is the basic unit of the geological time in which a single type of rock system is formed. Some periods are divided into epochs.

94. What is the name of the largest division of the geological era?
Answer: The eon is the largest division of the geological era.

95. Name the era that is known as the 'Age of Invisible Life'.
Answer: The Archaeozoic era is known as the 'Age of Invisible Life'.

96. Define Cambrian explosion.
Answer: During the Cambrian period, major groups of animals first appeared in the fossil record in a relatively short time, which is termed as Cambrian explosion. Cambrian explosion is the outcome of changes in environmental factors. The changes in environmental factors lead to change in selective pressures, leading to diversification on a large scale.

97. Name the period known as the 'Age of Fishes'.
Answer: The Devonian period is known as the 'Age of Fishes'.

98. When did the first jaw vertebrate appear?
Answer: The first jaw vertebrate appeared during the Silurian Period.

99. Which era is known as the 'Age of Mammals'?
Answer: The Coenozoic era is known as the 'Age of Mammals'.
100. What is the name of the present era?
Answer: The Coenozoic era
101. What is the name of the period in which we are living?
Answer: We are living in the Quaternary period.
102. When did the first birds and mammals appear?
Answer: The first birds and mammals appeared during the Mesozoic era.
103. Name the era that is known as the 'Age of Dinosaurs'.
Answer: The Mesozoic era is known as the 'Age of Dinosaurs'.
104. What is the present geological era?
Answer: The present geological era is the Coenozic era (Holocene epoch of the Quaternary period).
105. Name the periods of the Coenozoic era.
Answer: The Cenozoic era is divided into the following two major periods:
(a) Tertiary period (b) Quaternary period
106. Name the first part of the Phanerozoic eon.
Answer: The Palaeozoic era is the first part of the Phanerozoic eon.
107. When did the eutherian mammals make their first appearance?
Answer: The eutherian mammals made their first appearance in the cretaceous period of the Mesozoic era.
108. What is zoogeography? Name the six zoogeographical regions of the world.
Answer: The study of patterns of distribution of animals on the earth is known as zoogeography. Wallace has divided the whole world into six zoogeographical regions, which are as follows:
(a) Neotropical region (b) Paleartic region
(c) Oriental region (d) Australian region
(e) Ethiopian region (f) Nearctic region.
109. Name the zoogeographical region to which the following animals belong:
Answer: (a) *Sphenodon* (b) *Neoceratodus* (c) *Liopolema* Australian region
110. Name the family of fish which is completely absent in the Neotropical region.
Answer: The cyprinidae family of fish is completely absent from the Neotropical region.
111. What is the Holoarctic region?
Answer: The Nearctic and Palaearctic regions have very similar and closely related faunas and are therefore often combined as the Holoarctic region.
112. Name the zoogeographical region having the maximum number of avian fauna.
Answer: The Neotropical region has the maximum number of bird species.
113. Who is regarded as the 'Father of Animal Geography'?
Answer: Alfred Wallace is regarded as the 'Father of Animal Geography'.
114. State the theory of continental drift? Who first proposed this theory?
Answer: The theory of continental drift states that the continents of today once formed a single mass, named Pangaea by Wegner. The theory of continental drift was first proposed by the German geologist and meteorologist, Alfred Wegner, in 1915.
115. Name the book in which Wegner published his theory.
Answer: Wegner published his theory in the book *On the Origin of Continents and Oceans.*

116. Name the imaginary line that separates the Oriental and Australian zoogeographical regions.
Answer: Wallace's Line

117. Give example of the following:
(a) Protective mimicry
(b) Warning mimicry
(c) Simulation of death

Answer: (a) Protective mimicry – The leaf insect (*Phyllium*) and stick insect (*Carausis*)
(b) Warning mimicry – Certain nonpoisonous snakes of the family elopidae resemble cobras
(c) Simulation of death – The American opossum (*Didelphis virginiana*)

Long-Answer Questions

1. Give an account of the origin of life.
2. Define organic evolution. Describe briefly evidences in favour of organic evolution.
3. What is evolution? Give an account of biochemical, physiological and connecting link evidences in favour of evolution.
4. Describe the following with suitable examples and explanatory notes to show how they support the theory of organic evolution:
   (a) Connecting links
   (b) Fossils
   (c) Homologous and analogous organs
   (d) Vestigial organs
5. Give an account of Lamarckism. Why it is also called the theory of inheritance of acquired characters?
6. What is natural selection? Describe Darwin's theory of the origin of species.
7. Give an account of the synthetic theory of evolution.
8. What do you mean by isolating mechanism? Describe different types of isolating mechanisms. Add a note on the role of isolation in evolution.
9. What is isolation? Discuss isolation as a factor of evolution.
10. What is the Hardy–Weinberg law? What are the conditions that disturb genetic equilibrium causing evolution?
11. Write short notes on the following:
   (a) Speciation
   (b) Connecting links
   (c) Biogenetic law
   (d) Acquired characters
   (e) Coacervates
   (f) Continental drift
   (g) Genetic drift
   (h) Zoogeographical realms
   (i) Cro-Magnon man
   (j) Wallace's Line
   (k) Industrial melanism
12. Distinguish between the following:
   (a) Homologous and analogous organs
   (b) Convergent and divergent evolution
   (c) Allopatric and sympatric speciation
   (d) Microevolution and macroevolution
   (e) Geographical isolation and reproductive isolation
   (f) Cursorial and fossorial adaptations

(g) Germplasm and somatoplasm
(h) Adaptive divergence and adaptive convergence
(i) Horizontal and vertical distribution
(j) Macroevolution and microevolution

13. Trace the evolutionary history of the horse.
14. Give an account of the fossil history of humans.
15. What are fossils? Describe the different types of fossils. Add a note on the significance of fossils.
16. What is fossilisation? Describe the process of fossilisation. How is age of fossils determined?
17. What do you mean by geological time scale? Give an account of geological periods indicating the characteristic animals of those periods.
18. Define geological period. Describe the characteristic features of the Mesozoic era.
19. What are zoogeographical regions? Name the major zoogeographical regions of the world with their four characteristic fauna.
20. Describe geographical limits, physical and climatic features and zoological characteristics of the Australian region.
21. Give an account of subregions and fauna of the Oriental region.
22. Describe the faunal peculiarities of the Ethiopian region.
23. Describe the subregions of the Neotropical region. Add a note on its avian and mammalian fauna.
24. What is mimicry? Give an account of mimicry in animals with its significance.

ORIGIN OF LIFE

Multiple-Choice Questions

1. The primitive composition of the atmosphere on the earth was chiefly composed of:
(a) Water vapour (b) Argon (c) Nitrogen (d) Hydrogen
2. The first scientist to hypothesise how life was formed is:
(a) Oparin (b) Miller (c) Richter (d) Cuvier
3. Which one of the following scientists proposed the Bing Bang hypothesis?
(a) Father Suarez (b) Louis Pasteur (c) Abbe Lemaitre (d) Archbishop Ussher
4. Which one of the following theory is not related to the origin of life?
(a) Theory of inheritance of acquired characters (b) Theory of biogenesis
(c) Cosmozoic theory (d) Theory of abiogenesis
5. The spontaneous generation of flies from rotting was first disproved by:
(a) Louis Pasteur (b) Francesco Redi (c) Arrhenius (d) Archbishop Ussher
6. Louis Pasteur is famous for:
(a) Germ theory of diseases (b) Cosmozoic theory
(c) Theory of catastrophism (d) Big Bang hypothesis
7. Who among the following scientists first pointed out that 'organisms develop from pre-existing organisms'?
(a) Edward Jenner (b) Louis Pasture (c) Francesco Redi (d) A V Leeuwenhoek
8. Who among the following scientists proposed the modern theory of origin of life?
(a) Oparin (b) Haldane (c) Louis Pasteur (d) Miller
9. 'Life originates from pre-existing life'. This was scientifically explained by:
(a) A I Oparin (b) G J Mendel (c) Louis Pasteur (d) Charles Darwin
10. Which one of the following theory was greatly supported by Father Suarez?
(a) Cosmozoic theory (b) Theory of biogenesis
(c) Theory of spontaneous creation (d) Theory of special creation
11. Who among the following scientists rejected the 'theory of spontaneous generation'?
(a) Francesco Redi (b) Abbe Lazzaro Spallanzani
(c) Louis Pasteur (d) All of these
12. Which one of the following scientists published the book *Origin of Life?*
(a) Harold Urey (b) A I Oparin (c) Stanley Miller (d) J B S Haldane
13. The first compound molecules of the primitive earth were _______ and ______.
(a) Water, Methane (b) Methane, Hydrogen Cyanide
(c) Water, Ammonia (d) Nitrites, Carbides
14. Miller demonstrated that ultraviolet radiation or electrical discharge of heat or a combination of these can produce a complex organic compound from a mixture of:
(a) Methane, ammonia, nitrogen and hydrogen

(b) Water vapour, ammonia, methane and hydrogen
(c) Hydrogen cyanide, water, methane and ammonia
(d) Methane, ammonia, carbon-dioxide and hydrogen

15. Which one of the following chemicals was absent at the time of 'origin of life'?
(a) Oxygen (b) Hydrogen (c) Nitrogen (d) Ammonia

16. Haldane suggested that the primitive atmosphere contained:
(a) Only carbon dioxide (b) Only ammonia and water vapour
(c) No oxygen (d) All of these

17. Oparin's concept regarding the 'origin of organic compound' is based on the:
(a) Origin of the earth proposed by Sir James Jean (b) Abiogenesis theory
(c) Biogenesis theory (d) Cosmozoic theory

18. According to Oparin, the first carbon compounds were:
(a) Hydrocarbons (b) CO_2 and CO
(c) Glucose (d) Hydrocarbons and glucose

19. Which one of the following compounds were synthesised by Miller?
(a) Proteins (b) Amino acids (c) Vitamins (d) Carbohydrates

20. Which one of the following scientists used 'swan neck flask' in his experiment?
(a) Louis Pasteur (b) Francesco Redi (c) Stanley Miller (d) Sydney Fox

21. Which one of the following scientists used the term 'microsphere'?
(a) J B S Haldane (b) Sydney Fox (c) Harold Urey (d) Abelson Oro

22. The ratio of methane, ammonia and hydrogen in Miller's experiment was:
(a) 1:1:1 (b) 2:1:1 (c) 3:1:2 (d) 2:1:2

23. Life originated on:
(a) Air (b) Land (c) Water (d) Mountains

24. Which one of the following elements was deficient in primitive earth?
(a) Hydrogen (b) Oxygen (c) Carbon (d) Nitrogen

25. The chemical theory of origin of life was given by:
(a) Louis Pasteur (b) Oparin and Haldane (c) Stanley Miller (d) Sydney Fox

26. The chemical evolution stage was completed with the evolution of:
(a) Proteins (b) Nucleoproteins
(c) Proteins, nucleoproteins and nucleic acid (d) Carbohydrates, proteins and nucleic acid

27. According to Oparin, the origin of the earth took place about:
(a) 2 billion years ago (b) 3·5 billion years ago
(c) 4·5 billion years ago (d) 5·5 billion years ago

28. At the time of origin, the temperature of the earth was:
(a) 500°C to 600°C (b) 5000°C to 6000°C
(c) 1000°C to 2000°C (d) 10,000°C to 1500°C

29. Coacervates are:
(a) Carbohydrate droplets held together by electrostatic forces
(b) Polypeptide molecules held together by cohesive forces
(c) Protein droplets held together by electrostatic forces
(d) Lipoprotein molecules held together by electrostatic forces

30. In Miller's experiment, the resultant liquid obtained after the experiment showed the presence of:
    (a) Nucleic acids (b) Nucleoprotein
    (c) Glycine, alanine and β-alanine (d) Ammonia, methane and water vapour
31. Which one of the following statements about 'protobiont' is incorrect?
    (a) They were aerobes. (b) They were heterotroph.
    (c) They exhibited mutation. (d) They were spherical in shape and double layered.
32. Cyanobacteria:
    (a) Evolved the ability to split water into hydrogen atoms and oxygen
    (b) Created atmospheric oxygen
    (c) Led to the evolution of aerobic metabolism, sexual recombination and multicellularity
    (d) All of these
33. The first living organisms that originated on the earth were:
    (a) Autotrophs (b) Heterotrphs and aerobic
    (c) Heterotrophs and anaerobic (d) Autotrophs and aerobic
34. Which one of the following organisms evolved first on the earth?
    (a) Viruses (b) Bacteria (c) Algae (d) Coacervates
35. Which one of the following has not been proposed as a type proto-cell?
    (a) Endosymbionts (b) Micells (c) Coacervates (d) Protobionts
36. The first oxygen-producing organisms were:
    (a) Certain Protozoa (b) Green algae (c) Cyanobacteria (d) Protovirus
37. The mode of nutrition in the first formed Prokaryotes was:
    (a) Autotrophic (b) Chemo-autotrophic (c) Heterotrophic (d) Obligate aerobes
38. Oparin named the larger and denser colloidal system as:
    (a) Microspheres (b) Protovirus (c) Procells (d) Coacervates
39. The principle of sterilisation is based on the experiments of:
    (a) Oparin (b) Pasteur (c) Miller (d) Redi
40. Oceanic water containing a rich mixture of organic compounds was named by Haldane as:
    (a) Microspheres (b) Hot dilute soup of organic substances
    (c) Prebiotic soup (d) Both (b) and (c)
41. Which one of the following statements is incorrect?
    (a) Hydrogen atoms were more numerous and most reactive in the primitive atmosphere.
    (b) According to the cosmozoic theory, protoplasm reached the earth in the form of spores or germs.
    (c) The nucleoproteins appeared in the primitive ocean as a self-duplicating system.
    (d) The cell wall of bacteria and the cell wall of plants are homologous.
42. Which one of the following statements is correct?
    (a) The early source of energy was probably chlorophyll.
    (b) Early sources of energy were ultraviolet rays and lightening.
    (c) SO_2 played an important role in prebiotic evolution.
    (d) There is life on the moon.
43. Which one of the following about ozone is correct?
    (a) Accumulation of free O_2 in the atmosphere causes formation of the ozone layer in the atmosphere.
    (b) Formation of the ozone layer starts absorbing ultraviolet rays of the sun.
    (c) Absorption of ultraviolet rays by the ozone layer permits migration of oceanic organisms on land.
    (d) All of these

44. Fossils of microspheres and coacervates like microstructures have been found in:
(a) Germany and Africa
(b) Isua rocks of Greenland and Bavaria in Germany
(c) Isua rocks of Greenland and some African rocks
(d) Sedimentary rocks of France and certain rocks of Australia
45. It is presumed that protobionts were formed in the primitive seawater about:
(a) 1 billion years ago
(b) 2·4 billion years ago
(c) 3·8 billion years ago
(d) 5·6 billion years ago
46. Which one of the following gases is absent in the present atmosphere of the earth?
(a) Hydrogen (b) Argon (c) Nitrogen (d) Carbon dioxide
47. Which one of the following groups of living organisms evolved from primitive Eukaryotes?
(a) Protists (b) Plants and animals (c) Fungi (d) All of these
48. Which one of the following groups of living organisms evolved first?
(a) Cyanobacteria (b) Bacteria (c) Chemoheterotrophs (d) Chemoautotrophs
49. Which one of the following statements is correct?
(a) Water played an essential role in the evolution of life.
(b) The end product of Miller's experiment was amino acid.
(c) Protoplasm is the physical basis of life.
(d) All of these
50. In the atmosphere, ozone layer is found in:
(a) Mesosphere (b) Troposphere (c) Stratosphere (d) Lithosphere
51. A fossil of blue-green algae like prokaryote called *Archaeospheroides barbertonensis* has been recently reported from:
(a) Australia (b) Germany (c) Africa (d) New Zealand
52. Nuclear membrane is absent in:
(a) Protista (b) Protozoa (c) Prokaryotes (d) None of these
53. The first sign of life was shown by:
(a) Hydrocarbons (b) Nucleoproteins (c) Proteins (d) Microspheres
54. The first free-living eukaryotic cell-like organism originated in the primitive ocean about:
(a) 1.5 billion years ago
(b) 2 billion years ago
(c) 3.5 billion years ago
(d) 4.6 billion years ago
55. Which one of the following chemicals is necessary for the evolution of life?
(a) Methane (b) Ammonia (c) Water (d) Carbon dioxide
56. Viroids are the virus-like particles discovered by T O Diener. Which one of the following statements about viroids is correct?
(a) It consists of nucleic acid.
(b) It lacks protein coating.
(c) It appears to be an intermediate between living protoplasm and nonliving complex organic compounds.
(d) All
57. The origin of Cyanobacteria dates back to about:
(a) 10 million years ago
(b) 250 million years ago
(c) 3,000 million years ago
(d) 36,000 million years ago
58. According to the Big Bang theory, our planet earth originated due to:
(a) Explosion of condensed gaseous body
(b) Chemical evolution
(c) Collision of the sun with some other planet
(d) Gradual condensation of interstellar dust
59. 'Life begets life' was concluded on the experimental observations of:
(a) John Needham (b) Louis Pasteur (c) Francesco Redi (d) Von Helmont

60. In the primordial atmosphere of the earth, carbon was present in the form of:
(a) Dicarbon and cynogen (b) Methane and Metal carbides
(c) Carbon monoxide and carbon dioxide (d) All of these

61. The liquid obtained by Urey and Miller in their experiment showed the absence of:
(a) Aldehyde and hydrogen cyanide (b) Amino acids
(c) Purines (d) Purines and pyrimidines

62. Which one the following is considered as the precursor of the first organism?
(a) Coacervates (b) Ylem
(c) Hot dilute soup (d) Microspheres

63. The 'theory of abiogenesis' was proposed by:
(a) Kant (b) Pasteur (c) Aristotle (d) Needham

64. Which one of the following chemicals was used for catalysing reactions by the first formed organism?
(a) DNA (b) RNA (c) Histones (d) Protamines

65. The reducing atmosphere of the primitive earth was changed to oxidising by:
(a) Viruses (b) Chemoautotrophs (c) Hetertrophs (d) Photo-autotrophs

66. Oxygen gas in its free state appeared on the earth about:
(a) 1.2 billion years ago (b) 2.3 billion years ago
(c) 3.4 billion years ago (d) 5 billion years ago

67. The most abundant chemical on the earth when life originated was:
(a) Hydrogen (b) Carbon dioxide (c) Oxygen (d) Nitrogen

68. The molecular basis of origin and evolution of life was experimentally proved by:
(a) Oparin and Haldane (b) Stanley B Prusiner
(c) Miller and Chang (d) Urey and Miller

69. Polymer formation takes place by:
(a) Dehydration synthesis (b) Hydration synthesis
(c) Hydrolysis (d) Dehydrogenation and methylation

70. On the primitive earth, the source of energy was:
(a) Ultraviolet radiation (b) Ionising radiation
(c) Electric discharge (d) All of these

71. During the course of 'origin of life', the combination of hydrocarbons, ammonia and water under the influence of freely available energy reacted to form:
(a) Amino acids (b) Protein (c) Hot dilute soup (d) Nucleic acid

72. Seawater containing carbohydrates, proteins, fats and other organic substances was termed:
(a) Coacervates (b) Hot dilute soup (c) Protovirus (d) Microspheres

73. Coacervates had the ability of:
(a) Nutrition (b) Reproduction (c) Mutation (d) All of these

74. Early cells were characterised by:
(a) Synthesis and growth (b) Development and division
(c) Fermentation (d) All of these

75. Seawater contains dissolved:
(a) Ammonia (b) Methane (c) Minerals and salts (d) All of these

76. The age of prokaryotic microbes was:
(a) Azoic (b) Cambrian (c) Precambrian (d) Permian

77. In terms of origin of life, which one of the following is older?
(a) Euglena (b) Coacervates (c) Cyanobacteria (d) Algae

78. The substance undergoing decomposition inside coacervates was probably:
(a) Amino acids (b) Fatty acids and glucose (c) Glycerol (d) Glucose

79. The first cells which originated on earth were:
(a) Phototrophs (b) Autotrophs (c) Heterotrophs (d) Chemotrophs

80. 'Origin of life' from inorganic substances is now not possible because:
(a) The concentration of atmospheric oxygen is high
(b) The environment is highly polluted
(c) There is dearth of inorganic substances
(d) All of these

81. The basis of life is:
(a) Amino acids (b) Proteins (c) Lipoproteins (d) Nucleic acids

82. The first formed prokaryote contained:
(a) DNA (b) RNA (c) Both DNA and protein (d) Naked DNA

83. The first experimental evidence of the origin of life was given by:
(a) Urey and Miller
(b) Oparin and Haldane
(c) Charles Darwin
(d) Fraenkel Conrat and B Singer

84. Simulation experiments are related with:
(a) Oparin (b) Melvin and Calvin (c) Miller and Urey (d) Redi

85. The age of the universe is about:
(a) 3.9 billion years (b) 5.8 billion years (c) 15 billion years (d) 18 billion years

86. Which one of the following scientists gave the statement 'Abiogenesis first and biogenesis ever since'?
(a) Haldane (b) Oparin (c) Pasture (d) Fox

87. The oldest microfossils discovered so far are that of:
(a) Photosynthetic bacteria
(b) Autotrophic bacteria
(c) *Amoeba*
(d) Red algae

88. The hydrosphere appeared later when the earth cooled down to a temperature below:
(a) 250°C (b) 200°C (c) 100°C (d) 75°C

89. The history of life comprises:
(a) Origin
(b) Evolution
(c) Origin and degeneration
(d) Origin and evolution

90. The age of the oldest surviving terrestrial rock is about:
(a) 1.5 billion years old
(b) 2.5 billion years old
(c) 3.3 billion years old
(d) 4.3 billion years old

91. Which one of the following process is responsible for converting primitive reducing environment into the present-day oxidising environment?
(a) Formation of macromolecules
(b) Photosynthesis
(c) Respiration
(d) Evolution

92. Life originated in water because:
(a) Water acts as a temperature buffer
(b) Being a good solvent, it increases biochemical reaction
(c) It remains in liquid form in a temperature range of 4°C to 90°C
(d) All of these

93. Primitive oceans were:
(a) Acidic (b) Alkaline (c) Neutral (d) Highly acidic

94. In the early earth, the requirement of mutation would have been fulfilled by:
(a) High temperature (b) Chemicals present in the ocean
(c) Reducing environment (d) None of these

95. The appearance of anaerobic photosynthetic bacteria took place about:
(a) 3.5 billion years ago (b) 2.4 billion years ago
(c) 1.2 billion years ago (d) 0.6 billion years ago

96. The free-living eukaryotic cell-like organisms originated in ancient oceans about:
(a) 1.5 billion years ago (b) 3.5 billion years ago
(c) 4.8 billion years ago (d) 5 billion years ago

97. The first chemical compound molecules of primitive earth were probably:
(a) Water and ammonia (b) Acetylene and water
(c) Methane and acetylene (d) Methane and hydrogen cyanide.

98. Which one of the following structure is absent in a prokaryotic cell?
(a) Nuclear membrane (b) Nucleus
(c) Mitochondria (d) All of these

99. Find out the correct sequence of substances which appeared during the origin of life:
(a) Glucose ⟶ Fats ⟶ Proteins ⟶ Nucleic acids
(b) Glucose ⟶ Amino acids ⟶ Ammonia ⟶ Protein ⟶ Nucleic acids
(c) Ammonia ⟶ Amino acids ⟶ Nucleic acids
(d) Ammonia ⟶ Amino acids ⟶ Proteins ⟶ Nucleic acids

100. Match column I with column II and select the correct answer of matching using answer codes:

Column I	Column II
(A) J B S Haldane	1. Ocean
(B) Miller	2. Swan neck flask
(C) Origin of life	3. Hot dilute soup
(D) Louis Pasteur	4. Spark discharge apparatus

Answer codes:

	A	B	C	D
(a)	4	1	2	3
(b)	3	4	2	1
(c)	3	1	2	4
(d)	3	4	1	2

101. The first found fossils 'stromatolites' were of:
(a) Cyanobacteria (b) Bacteria (c) Protozoa (d) Fungi

102. L-amino acids were elected as the building blocks of proteins as:
(a) They can form polypeptides. (b) They were abundant in ocean water.
(c) Purely by chance (d) They had intrinsic advantage over D-amino acids.

103. The most important biological condensing agent of the present time is:
(a) Cyanogen (b) Cyanoacetylene (c) HCN (d) ATP

104. Among all condensing agents formed in the primitive earth conditions, __________became the forerunners of adenosine triphosphate:
(a) HCN (b) Cyanoglutanidine (c) Polyphosphates (d) Carbodiamide

105. Organic compounds first arose by reactions between various inorganic compounds of the primitive atmosphere and geosphere. Such reactions continued for about:
   (a) 100 million years (b) 500 million years
   (c) 2,000 million years (d) 5,000 million years
106. The lipids, polysaccharides, polypeptides and polynucleotides:
   (a) Are macromolecules (b) Are polymers
   (c) Were the precursors of life on earth (d) All of these
107. The proposition of Crick has been supported by all modern geneticists that:
   (a) DNA was the first form of life on earth
   (b) Short chain RNA was the first form of life on earth
   (c) The structure of life might have appeared from lifeless molecules, many times at many places
   (d) The first form of life must have had the capacity of self-reproduction
108. The rich broth of the various organic molecules containing short chains of RNA as the first form of life was named by Miller as the:
   (a) Gowen's gene (b) Gynogenotes (c) Coacervates (d) Protogene
109. RNA was the first formed biomolecule during the origin of life, is supported by one of the following evidences:
   (a) RNA is a short chain (b) It has ribose sugar
   (c) RNA copies from the chain can be formed spontaneously in a protein-free environment
   (d) Glycerol derivatives might have been involved in place of ribose in RNA.
110. Polyanions are polymers like:
   (a) DNA (b) RNA (c) Protein and lipid (d) All of these
111. During the course of origin of life:
   (a) RNA was first evolved and from RNA, DNA was evolved possibly by reverse transcription
   (b) DNA was first evolved
   (c) DNA was first evolved and from DNA, RNA was evolved possibly by transcription
   (d) Both DNA and RNA evolved at the same time
112. Which one of the following property of RNA is relevant to the origin of life?
   (a) Short chain (b) Self-replication and self-catalysis
   (c) Ability to synthesise DNA by reverse transcription (d) None of these
113. After the evolution of RNA from DNA, the RNA became an agent to decode the stored information and subsequently evolved the:
   (a) DNA $\longrightarrow$ RNA $\longrightarrow$ Protein path (b) DNA $\longrightarrow$ Protein $\longrightarrow$ RNA path
   (c) RNA $\longrightarrow$ DNA $\longrightarrow$ Protein path (d) RNA $\longrightarrow$ Protein $\longrightarrow$ DNA path
114. According to Fox (1973), the first form of life is:
   (a) RNA (b) DNA (c) Protein (d) Carbohydrate
115. The root from RNA to DNA is known as:
   (a) Gene conversion (b) Gene mutation (c) Gene transformation (d) Genetic engineering
116. Which one of the following group of organisms is closely related with bacteria?
   (a) Protozoa (b) Viruses (c) Blue-green algae (d) Red algae
117. The process of photosynthesis probably originated in the common ancestors of:
   (a) Chloroplasts, eubacteria and holobacteria (b) Chloroplasts
   (c) Chloroplasts and eubacteria (d) Eubacteria, holobacteria and eocytes

118. Which one of the following evolved most recently?
(a) Prokaryotic cell (b) Multicellularity (c) Eukaryotic cell (d) Photosynthesis

119. Any information cannot be transmitted over a long period of time without some deterioration in quality due to copy error. This is known as:
(a) Allen's rule (b) Jordon's rule (c) Shannon's rule (d) Copy choice theory

120. Shannon's theorem has a selective advantage of:
(a) Genetic diversity (b) Genetic heterogeneity
(c) Gene mutation (d) Genetic redundancy in information transmission

121. The early RNA genome was exposed to damage by:
(a) Solar radiation (b) High temperature and other environmental agencies
(c) Both (a) and (b) (d) None of these

122. The sequence of evolutionary events in speciation:
(a) Begins with race formation and ends with reproductive isolation (b) Begins with reproductive isolation and ends in race formation
(c) Both (a) and (b) (d) None of these

123. According to geological evidence, free oxygen began to accumulate in the atmosphere about:
(a) 7 billion years ago (b) 3·1 billion years ago (c) 2·6 billion years ago (d) 2 billion years ago

124. The energy necessary for the formation of first organic molecule was not provided by the:
(a) Ultraviolet rays (b) Lightlning (c) Rain (d) Volcanic eruptions

125. Which one of the following is the key compound produced in various experiments by different scientists to explain evolution of the hereditary materials?
(a) Aspartic acid (b) Glutamic acid (c) Valine (d) Adenine

126. Which one of the following was lacking in the Miller–Urey original mixture?
(a) Phosphorous (b) Nitrogen (c) Hydrogen (d) Oxygen

127. Eukaryotes originated on the earth about:
(a) 5 million years ago (b) 3.5 million years ago
(c) 2 million years ago (d) 1.5 million years ago

128. The Cosmozoic theory was proposed by:
(a) Miller (b) Richter (c) Cuvier (d) Arrhenius

129. The presence of sodium chloride in animal body indicates that life first originated in:
(a) Land (b) Ocean (c) Freshwater (d) Estuary

130. In the Miller and Urey experiment, the source of energy was:
(a) Ultraviolet light (b) Radioactive substances
(c) An electrical spark (d) Lightning

131. Which one of the following is not applicable to lipid coacervates?
(a) Nucleus (b) Ability to carry out reactions
(c) Ability to grow (d) None of these

132. Who proved that some microorganisms are airborne?
(a) John Needham (b) Louis Pasteur (c) Francesco Redi (d) LazzaroSpallanzan

133. The genetic and enzymatic components of early cells were:
(a) DNA (b) DNA and RNA (c) DNA and protein (d) RNA

Answers to Multiple-Choice Questions

1. (d)	2. (a)	3. (c)	4. (a)	5. (b)	6. (a)	7. (b)	8. (a)
9 (c)	10. (d)	11. (d)	12. (b)	13. (c)	14. (b)	15. (a)	16. (d)
17. (a)	18. (a)	19 (b)	20. (a)	21. (b)	22. (d)	23. (c)	24. (b)
25. (b)	26. (c)	27. (c)	28. (b)	29 (c)	30. (c)	31. (a)	32. (d)
33. (c)	34. (d)	35. (a)	36. (c)	37. (b)	38 (d)	39. (b)	40. (d)
41. (d)	42. (b)	43. (d)	44. (c)	45. (c)	46. (a)	47. (d)	48. (c)
49. (d)	50. (c)	51. (c)	52. (c)	53. (b)	54. (a)	55. (c)	56. (d)
57. (d)	58. (a)	59. (b)	60. (d)	61. (d)	62. (a)	63. (c)	64. (b)
65. (d)	66. (c)	67. (a)	68. (d)	69. (a)	70. (d)	71. (a)	72. (b)
73. (d)	74. (d)	75. (d)	76. (c)	77. (b)	78. (d)	79. (c)	80. (a)
81. (d)	82. (d)	83. (a)	84. (c)	85. (c)	86. (b)	87. (a)	88. (c)
89. (d)	90. (d)	91. (b)	92. (d)	93. (b)	94. (a)	95. (a)	96. (a)
97. (a)	98. (d)	99. (d)	100. (d)	101. (a)	102. (c)	103. (d)	104. (c)
105. (c)	106. (d)	107. (b)	108. (d)	109. (c)	110. (d)	111. (a)	112. (b)
113. (a)	114. (c)	115. (a)	116. (c)	117. (a)	118. (b)	119. (c)	120. (d)
121. (c)	122. (c)	123. (d)	124. (c)	125. (d)	126. (a)	127. (d)	128. (b)
129. (b)	130. (c)	131. (a)	132. (b)	133. (d)			

Fill in the Blanks

1. The first organisms were ____________ and____________.
2. The origin of life was followed by ____________.
3. The Big Bang hypothesis was proposed by ____________.
4. ________________ provided the first solid evidence against spontaneous generation.
5. ____________ and ____________ were probably the first compound molecules of the primitive earth.
6. The solar system evolved about ____________ billion years ago.
7. Life is believed to have originated in ____________.
8. The food of the earliest forms was ____________.
9. The earth is formed from the clouds of ____________.
10. The availability of ____________ molecules as building blocks was essential to the spontaneous origin of life.
11. The primitive atmosphere was ____________ while the present-day atmosphere is ____________.
12. The source of present-day oxygen must have been ____________.
13. The apparatus used by Miller is called ____________ apparatus.
14. Miller's experiment contained gases ____________, ____________, ____________ and water .
15. The earliest prokaryotes were ____________
16. The presence of ____________ is the most important condition for the origin of life.
17. Coacervates are colloidal droplets containing____________.
18. The oldest fossil cells resemble ____________.

19. Sydney W Fox studied the formation of_______________ structures.
20. The Miller and Urey experiment produced trace amounts of ____________ and ____________.
21. Multicellular fossils appeared at the beginning of____________ era .
22. The study of the origin of life and evolution of different forms of life in earth was termed by Mayer as ____________.
23. Atoms, stars and galaxies self-assembled from the fundamental particles produced by ____________.
24. The only molecule, that can encode genetic information and also acts as a biocatalyst is the____________.
25. The ALH84001 meteorite arrived from ____________.

Answers to Fill in the Blanks

1. Heterotphic, anaerobic
2. Evolution
3. Abbe Lemaitre
4. Francesco Redi (1686)
5. Water, ammonia
6. 4.5
7. Oceans
8. Organic compounds
9. Gas
10. Organic
11. Reducing, oxidising
12. Photosynthesis
13. Spark discharge
14. Hydrogen, methane, ammonia,
15. Chaemoautotrophs
16. Water
17. Nucleoprotein
18. Heterotrophic bacteria
19. Peptide
20. Amino acids, melanoids
21. Phanerozoic
22. Evolutionary biology
23. Big Bang
24. RNA
25. Mars

True or False

1. Life always arises from pre-existing life.
2. The primitive atmosphere favoured chemical evolution.
3. The Big Bang hypothesis was given by Kant.
4. Coacervates lacked the ability to mutate.
5. Father Suarez was the greatest supporter of the Theory of spontaneous generation.
6. The first experimental evidence of origin of life was provided by Oparin and Haldane.
7. Scientists have reported the presence of water on Mars.
8. Primitive atmosphere was reducing in nature.
9. According to the symbiotic theory, eukaryotic cells have originated by the symbiotic association of two or more types of prokaryotic cells.
10. Life was present on earth about 5.9 billion years ago.
11. Oceanic water containing a mixture of organic compounds was named by Oparin as 'hot dilute soup'.
12. Water and ammonia were probably the first compounds of the primitive earth.
13. Polyanions are a constructive force.
14. The catastrophe theory of origin of life was proposed by Cuvier (1776).
15. Primitive oceans were acidic.

16. Mercury is the hottest planet.
17. The swan-neck flask was used by Miller for his experimental work.
18. According to Oparin, if a mixture of large protein and polysaccharide is shaken, coacervates are formed.
19. Stanley Miller published the book, *The Origin of Life*.
20. The oldest fossil cells resemble red algae.
21. The early belief of origin of life was disproved by Lederberg.
22. The Oparin model of origin of life provides a mechanism for self-replication, mutation and evolution.
23. Archaic bacteria have survived the early harsh environment of the earth.
24. Nitrogen was lacking in the Miller–Urey's original mixture.
25. Ozone was a source of energy on early earth.

Answers to True or False

1. True	2. True	3. False	4. False	5. False	6. False	7. True	8. True
9. True	10. False	11. False	12. True	13. False	14. True	15. False	16. True
17. False	18. True	19. False	20. False	21. False	22. False	23. True	24. False
25. False							

Give Reasons

1. In primitive air, ultraviolet rays freely reached the earth.
   – Because it had no ozone layer.
2. The atmosphere was reducing for some years even after development of chlorophyll.
   – Because the first autotrophs were anaerobic and utilised hydrogen donors other than water, so there was no liberation of oxygen gas, because of which the atmosphere remained reducing.
3. Recent studies on plant viruses have demonstrated that 52 nucleotides can act as an accurate RNA catalyst.
   – Because very short RNA can be catalytic.
4. Ribozymes are expectedly less versatile than protein enzymes.
   – Because RNA lacks the large variety of functional groups.
5. The atmosphere of the earth at the time of its origin was without oxygen.
   – Because hydrogen combined with all available oxygen to form water.
6. During the course of origin of life, complex organic compounds accumulated in seawater as such.
   – Because of absence of free oxygen and bacteria, these organic molecules were neither oxidised nor putrefied.
7. Early autotrophism is called chemosynthesis.
   – Because during the beginning of autotrophism, the energy for the synthesis of organic molecules was obtained from anaerobic breakdown as it was not possible to trap solar energy in the absence of chlorophyll.

8. Miller's apparatus is also known as 'spark discharge apparatus'.
   - Because Miller used two tungsten electrodes connected to an electric source as a source of energy of electric discharge.
9. There was maximum synthesis of proteins in the primitive ocean.
   - Because amino acid molecules have great tendency of polymerisation.
10. It is believed that pyrophosphates have played an important role in origin of life.
   - Because of the presence of high-energy bonds in them.
11. The evolution of oxygen-consuming aerobic cells was of selective advantage.
   - Because they could extract more energy per molecule of exogenous nutrient such as glucose.
12. In the early stage of biological evolution, chemical evolution continued for some time.
   - Because, the early cells continued to select from the environment those organic molecules, which enhanced their survival.
13. Probably, polyphosphate group of ATP was selected as a biological condensing agent.
   - Because polyphosphates have desirable advantage from a biological point of view as they are kinetically stable and at the same time, thermodynamically unstable.
14. With the growth and proliferation of early cells, the sea gradually became depleted of organic compounds.
   - Because these early cells used the organic materials of the sea for their survival.
15. The very first form of life must have had the capacity of self-replication.
   - Because self-replication (reproduction) is the foremost characteristic of life.

LAMARCKISM, DARWINISM AND ORGANIC EVOLUTION

Multiple-Choice Questions

1. Who gave the first theory of evolution?
(a) Huxley (b) Wallace (c) Lamarck (d) Darwin
2. Who proposed the concept of evolution?
(a) Lamarck (b) Wallace (c) Darwin (d) None of these
3. The theory of inheritance of acquired characteristics is related with:
(a) Lamarck (b) Darwin (c) Louis Pasteur (d) Weismann
4. The theory of use and disuse of organs was given by:
(a) Jean Baptistie de Lamarck (b) Alfred Russel Wallace
(c) T R Malthus (d) Charles Darwin
5. The main reason for the failure of the Lamarckian theory is:
(a) Its failure to explain overspecialisation of certain organs
(b) Mainly due to the failure of recognition that the phenotype is the by product of gene reproduction
(c) Both (a) and (b) (d) None of these
6. The key point of Darwin's theory of evolution is:
(a) Overproduction (b) Variation (c) Survival of the fittest (d) Natural selection
7. Organic evolution was explained by two British naturalists, who were:
(a) Lamarck and Darwin (b) Darwin and Wallace
(c) Darwin and Huxley (d) Lyell and Darwin
8. In the life history of birds, nitrogenous waste products are ammonia, urea and uric acid. It provides evidence in favour of:
(a) Divergent evolution (b) Convergent evolution
(c) Parallel evolution (d) Organic evolution
9. Darwin and Wallace were influenced by the essay of T R Malthus on:
(a) Principles of population (b) Theory of continuity of germplasm
(c) Saltatory variations (d) Principles of geography
10. Who among the following was the co-advisor of Darwin?
(a) Smith (b) Dodson (c) Wallace (d) Hugo de Vries
11. Sudden, discontinuous and heritable change is called:
(a) Variation (b) Mutation
(c) Inheritance of acquired character (d) Natural selection
12. The theory of pangenesis was proposed by:
(a) August Weismann (b) Charles Darwin (c) Sewall Wright (d) Galton

13. The theory of continuity of germplasm has been given by:
    (a) T R Malthus (b) August Weismann (c) Charles Lyell (d) Herbert Spencer
14. Darwin's theory was unable to explain:
    (a) Natural selection (b) Mutation
    (c) Variation (d) Reproductive isolation
15. The theory of mutation to explain the mechanism of evolution was proposed by:
    (a) August Weismann (b) H M S Beagle (c) Hugo de Vries (d) Sewall Wright
16. The principle of Darwin called Darwinism is based on:
    (a) Use and disuse of organs (b) Development of vestigial organs
    (c) Effect of nature on organisms (d) None of these
17. Darwin's finches is the:
    (a) Name of the ship on which Darwin travelled (b) Main points of theory of natural selection
    (c) 20 birds studied by Darwin on the Galapagos Islands (d) Extinct animals studied by Darwin
18. The ultimate source of organic evolution is:
    (a) Use and disuse of organ (b) Natural selection
    (c) Mutation (d) Isolation
19. According to Darwin:
    (a) The organs that are not used become vestigial and gradually disappear
    (b) Organisms have a tendency to produce more offsprings
    (c) Nature selects individuals having better adaptability
    (d) Both (b) and (c)
20. Neo Lamarckism states that only those characters are heritable which:
    (a) Arise due to change in environment (b) Affect germplasm
    (c) Are useful (d) All of these
21. Which one of the following is an incorrect match?
    (a) Darwin – Theory of natural selection (b) Lamarck – Use and disuse of organs
    (c) Genetic drift – Sum total of genes in the population of a particular area
    (d) Hugo de Vries – Mutation theory
22. Which one of the following statements about Lamarckism is incorrect?
    (a) Use and disuse of organs cause variations
    (b) All characteristics acquired during the lifetime of an organism are transmitted to the next generation
    (c) Inheritance of acquired characteristics has been experimentally proved by many workers
    (d) Lack of use of a particular organ causes its degeneration and ultimately, disappearance
23. Neo Lamarckism has laid emphasis on:
    (a) Use and disuse of organs (b) Internal vital force
    (c) Both (a) and (b) (d) None of these
24. According to Lamarck, variations in the organisms arise due to:
    (a) Use and disuse of organs (b) Change in germplasm
    (c) Both (a) and (b) (d) Change in the structure of chromosome
25. Which scientist pointed out that Lamarckism is self-contradictory?
    (a) Waddington (b) Payne (c) Ray Lankester (d) Lysenko
26. The first evidence against inheritance of acquired character was proved by:
    (a) Spencer (b) Weismann (c) Malthus (d) Ray Lankester

27. The work of which scientist lends support to Lamarck?
    (a) Weismann (b) Kammerer (c) Wallace (d) Sewall Wright
28. Like Darwin, Wallace also laid emphasis on __________ as the main factor in evolution:
    (a) Struggle for existence (b) Survival of the fittest
    (c) Variation (d) Natural selection
29. Darwin's theory cannot explain:
    (a) Overproduction (b) Overspecialisation of certain structures
    (c) Neutral flowers and sterility of hybrids (d) Both (b) and (c)
30. Who proposed the theory of 'continental drift'?
    (a) Wegener (b) Wallace (c) Moody (d) Huxley
31. Lamarck's theory of evolution stresses upon:
    (a) Internal vital force (b) Use and disuse of organs
    (c) Appetency (d) All of these
32. Charles Darwin is associated with:
    (a) Theory of mutation (b) Population migration
    (c) Natural selection (d) Recapitulation theory
33. Which one of the following statements is incorrect?
    (a) Mutations are the initiating force in evolution. (b) Mutations result in certain fault in gene duplication.
    (c) Only useful mutations are inheritable. (d) Mutation is responsible for speciation.
34. The Galapagos Islands are associated with the name of:
    (a) Darwin (b) Wallace (c) Spencer (d) Huxley
35. Two scientists – Castle and Philips – transplanted ovaries of a black guinea pig in another white guinea pig whose ovaries were replaced and found that all the progenies were black. It lends support to:
    (a) Inheritance of acquired characteristics (b) Theory of natural selection
    (c) Theory of germplasm (d) Mutation theory
36. According to modern concept, evolution involves:
    (a) Genetic variations (b) Isolation (c) Natural selection (d) All of these
37. Who has written the – the man who mused – Birds can fly, so why can't I?
    (a) Mc Dougall (b) Dodson (c) Crew (d) Karl Marx
38. Industrial melanism in *Biston betularia* was explained by:
    (a) Bernard Ketlewell (b) J B S Haldane (c) Sewall Wright (d) Stebbins
39. In mosquitoes, resistance against DDT is an example of:
    (a) Natural selection (b) Artificial selection (c) Mutation (d) Survival of the fittest
40. The concept of Homology and analogy was given by:
    (a) Buffon (b) Richard Owen (c) Julian Huxley (d) A Weismann
41. Adaptive similarities in animals living in the same habitat is known as:
    (a) Parallel evolution (b) Convergent evolution (c) Divergent evolution (d) None of these
42. Homologous organs provide evidence of:
    (a) Parallel evolution (b) Divergent evolution (c) Convergent evolution (d) None of these
43. Hands and legs of humans are:
    (a) Vestigial organs (b) Homologous organs
    (c) Analogous organs (d) Adaptive convergence

44. The presence of analogous organs indicates:
    (a) Similar structure and function of the organs
    (b) Similar adaptation by unrelated groups
    (c) Different adaptation by related groups
    (d) Presence of nonfunctional organs
45. Analogous organs present:
    (a) Divergent evolution
    (b) Convergent evolution
    (c) Parallel evolution
    (d) Adaptive radiation
46. Stings of honeybees and scorpions shows:
    (a) Parallel evolution
    (b) Divergent evolution
    (c) Convergent evolution
    (d) Phylogenetic homology
47. Which one of the following favours the theory of natural selection?
    (a) Limitation of resources
    (b) Overproduction
    (c) Abundance of variations
    (d) All of these
48. Which one of the following is responsible for speciation?
    (a) Natural selection (b) Mutation (c) Reproductive isolation (d) Random mating
49. Principles of geology were propounded by:
    (a) Malthus (b) Lerner (c) Charles Lyell (d) Wallace
50. The modern synthetic theory of evolution was proposed by:
    (a) Huxley (b) Stebbins (c) Sewall Wright (d) Darlington
51. The pioneer workers in the field of organic evolution are:
    (a) Muller, Lamarck, Darwin and Stebbins
    (b) Darwin, Malthus, de Vries and Lerner
    (c) Lamarck, Darwin, de Vries and Wallace
    (d) Darwin, Huxley, Darlington and Simpson
52. 'Food increases geometrically while population increases arithmetically' was proposed by:
    (a) Herbert Spencer (b) T R Malthus (c) Charles Darwin (d) August Weismann
53. The source of raw material for evolution is:
    (a) Mutation
    (b) Natural selection
    (c) Environment
    (d) Use and disuse of organs
54. Evolution is irreversible, this is known as:
    (a) Sewall Wright effect
    (b) Allen's rule
    (c) Cope's law
    (d) Dollo's law
55. The improved race of pigeons developed due to:
    (a) Effect of environment
    (b) Parthenogenesis
    (c) Artificial selection
    (d) Mutation
56. Darwin raised many types of ____________ by artificial selection:
    (a) Fowls (b) Pigeons (c) Ornamental plants (d) Dogs
57. According to Neo Darwinism, new species originate due to:
    (a) Natural selection
    (b) Use and disuse of organs
    (c) Mutations
    (d) All of these
58. Darwin could not differentiate between:
    (a) Useful and harmful variations
    (b) Continuous and discontinuous variations
    (c) Use and disuse of organs
    (d) Changes in environmental conditions
59. The theory of continuity of germplasm:
    (a) Supports Lamarckism
    (b) Criticised Lamarckism
    (c) Supports the theory of natural selection
    (d) Criticised the mutation theory

60. The main centre for the evolution of dinosaurs was:
(a) Australia
(b) South America
(c) North America
(d) The Islands of Madagascar

61. Prototherians have evolved from:
(a) Lung fishes (b) Amphibians (c) Reptiles (d) Aves

62. Prototheria provides evidence in favour of organic evolution, which is:
(a) Biogeographical (b) Biochemical (c) Paleontological (d) Embryological

63. Darwin, during his voyage around the world, studied the flora and fauna of:
(a) Greenland
(b) Channel Islands
(c) Galapagos Islands
(d) Queen Elizabeth Islands

64. Darwin's theory is not applicable to:
(a) Vestigial organs (b) Arrival of the fittest (c) Degeneration of organs (d) All of these

65. Which one of the following theories is not associated with Darwin?
(a) Theory of pangenesis
(b) Theory of sexual selection
(c) Natural selection theory
(d) Theory of continuity of germplasm

66. Darwin pointed out that like artificial selection, an analogous process operates in nature, known as:
(a) Sexual selection
(b) Induced breeding
(c) Natural selection
(d) Struggle for existence

67. Darwin pointed out that:
(a) Continuous variations are heritable
(b) Continuous variations are partially inheritable
(c) Continuous variations are not heritable
(d) None of these

68. Continuous variations are also known as:
(a) Sports (b) Mutations (c) Saltations (d) Fluctuations

69. Darwin's finches provided an evidence of evolution, which is:
(a) Paleontological (b) Embryological (c) Biogeographical (d) Biochemical

70. Which one of the following is incorrect?
(a) All mutations are inheritable.
(b) Majority of mutations are retrogressive or negative.
(c) Mutations are the source of variations.
(d) The mutation theory clearly explains the role of nature.

71. Mutations have given rise to new varieties such as:
(a) Ancon sheep (b) Hornless cattle (c) Hairless cats (d) All of these

72. Mutation cannot explain:
(a) Connecting link (b) Living fossil (c) Mimicry (d) Industrial melanism

73. The formation and disappearance of Ancon sheep presents a very good example of:
(a) Continuous variation
(b) Discontinuous variation
(c) Natural selection
(d) Artificial selection

74. Hugo de Vries performed his experiment on:
(a) *Oenothera lamarckiana*
(b) *Drosophila melanogaster*
(c) *Salmonella typhi*
(d) *Pisum sativum*

75. Genetic drift is also known as:
(a) Founder effect
(b) Sewall Wright effect
(c) Bottleneck effect
(d) Gene migration

76. Saltation theory was given by:
(a) Hugo de Vries (b) J B D Kettlwell (c) Julian Huxley (d) T H Morgan
77. Who termed the recapitulation theory, the law of biogenetics?
(a) Hugo de Vries (b) Wallace (c) Haeckel (d) Huxley
78. Which one of the following statements is incorrect?
(a) Useful mutations are selected by nature.
(b) Microevolution leads to change in the genetic equilibrium in a Mendelian population.
(c) Microevolution results in the splitting of a population of a species into several subgroups.
(d) Macroevolution operates about species level.
79. Substantive variation brings changes in:
(a) Size (b) Weight (c) Colour (d) All of these
80. Meristic variation brings about changes in:
(a) Skin colour (b) Height
(c) In a certain number living beings (d) All of these
81. Which one of the following is a correct match?
(a) Darwin – Inheritance of acquired characters (b) Wallace – Origin of species
(c) Hugo de Vries – Theory of pangenesis (d) Recapitulation theory – Biogenetic law
82. The Ordovician period is known as the:
(a) Age of Giant Molluscs (b) Age of Reptiles
(c) Age of Birds (d) Age of Mammals
83. Sudden appearance of an ancestral characteristics is known as:
(a) Reversion (b) Variation (c) Mutation (d) Vestigial character
84. Which one of the following is a vestigial organ in humans?
(a) Vermiform appendix (b) Coccyx
(c) Ear muscles (d) All of these
85. Which one of the following is a set of vestigial organs?
(a) Hindlimbs of boas, ear pinna of whales and eye muscles of humans
(b) Muscles of ear pinna of humans, hindlimbs of pythons and feathers of birds
(c) Splint bones in horses, eyes of Proteus and ear pinna of whales
(d) Tail fin of tadpole larvae, splint bones of horses and pinna of dogs
86. Who is popularly known as the 'Father of Palaeontology'?
(a) Charles R Richet (b) Leonard da Vinci
(c) Charles Lyell (d) Boltwood
87. The founder of modern palaentology is:
(a) George Cuvier (b) Birbal Sahni (c) K V Frisch (d) Leonard da Vinci
88. The largest animal to have ever walked on the earth is the:
(a) *Brontosaurus* (b) *Tyrannosaurus* (c) *Archosaurus* (d) *Lepidosaurus*
89. The earliest known mammal-like reptile was:
(a) *Lepidosaurus* (b) *Cynognathus* (c) *Archosaurus* (d) *Sauropods*
90. Which one of the following is a living fossil?
(a) *Sphenodon* (b) Prototheria (c) *Latimeria* (d) All
91. Which one of the following is a living fossil and connecting link between Annelida and Arthropoda?
(a) *Limulus* (b) *Peripatus* (c) *Polygordius* (d) *Sacculina*

92. Which one of the following is an extinct reptile and is a connecting link between Amphibia and Reptilia?
(a) *Dodo* (b) *Seymouria* (c) *Lycaenops* (d) *Hesperonis*
93. The connecting link between reptiles and birds is:
(a) *Sphenodon* (b) *Phrynosoma* (c) *Lingula* (d) *Archaeopteryx*
94. *Neoplina* is a connecting link between:
(a) Protozoa and Porifera (b) Annelida and Mollusca
(c) Nonchordata and Chordata (d) Pisces and Amphibia
95. *Balanoglossus* is a connecting link between:
(a) The living and the nonliving (b) Plants and animals
(c) Annelida and Arthopoda (d) Nonchordata and Chordata
96. Lungs fishes are considered a connecting link between:
(a) Cyclostomes and fishes (b) Cartilagenous and bony fishes
(c) Fishes and amphibians (d) None of these
97. *Chimaera* is a connecting link between:
(a) Nonchordata and Chordata (b) Annelida and Arthropoda
(c) Catilaginous fishes and bony fishes (d) Fishes and amphibians
98. Which one of the following is an extinct animal?
(a) Mammoth (b) Polar bear (c) Silverfish (d) Anaconda
99. Presence of facial hair in some people presents an example of:
(a) Gene interaction (b) Atavism
(c) Reverse evolution (d) Ontogeny repeats phylogeny
100. Which one of the following represents Atavism?
(a) Cervical fistula of humans (b) Starfish with six arms
(c) Coccyx of humans (d) Tuft of hair on the pinna of humans
101. Lion Boy of Russia is an example of:
(a) Convergent evolution (b) Divergent evolution
(c) Parallel evolution (d) Atavism
102. The presence of gill clefts in vertebrate embryos provides an evidence of:
(a) Vestigial organ (b) Homologous organs
(c) Organic evolution (d) Use and disuse of organs
103. The era which includes the maximum periods is:
(a) Cenozoic (b) Mesozoic (c) Palaeozoic (d) Proterozoic
104. The earliest Prokaryotes and Eukaryotes were seen during the:
(a) Archaeozoic era (b) Proterozoic era (c) Palaeozoic era (d) Mesozoic era
105. Which one of the following about the Cenozoic era is correct?
(a) The Cenozoic era includes only epochs. (b) The Cenozoic era is known as the age of birds and mammals.
(c) It is divided into four epochs. (d) All of these
106. Which one of the following has the complete fossil record to reveal the evolutionary history?
(a) Dinosaurs (b) Dodos (c) Horses (d) Whales
107. The correct sequence of evolution of the horse is:
(a) *Eohippus* → *Mesohippus* → *Pliohippus* → *Equus* → *Merychippus*
(b) *Eohippus* → *Merichippus* → *Mesohippus* → *Pliohippus* → *Equus*

(c) *Eohippus* → *Mesohippus* → *Merichippus* → *Pliohippus* → *Equus*
(d) *Eohippus* → *Pliohippus* → *Mesohippus* → *Merichyppus* → *Equus*

108. Development of running habit in deer, which is two-toed and one-toed horse, represents:
(a) Parallel evolution (b) Divergent evolution (c) Convergent evolution (d) Biogenetic evolution

109. Blood is an example of multiple alleles. In humans, four types of blood groups, viz., A, B, AB and O are found while apes have blood group (A) and (B). In monkeys, there are no such blood groups. This clearly indicates that:
(a) Humans, apes and monkeys have originated from a common ancestor
(b) Humans and apes have originated from a common ancestor
(c) Humans are more closely related to apes than to monkeys
(d) All of these

110. In a mammalian embryo, the kidney is initially pronephric, then mesonephric and finally metanepheric. It provides support to:
(a) The theory of natural selection (b) Use and disuse of organ
(c) Biogenetic law (d) Cope's law

111. It has been found that individuals of many species inhabiting cold waters are larger than those inhabiting warmer waters. This conforms to:
(a) Allen's rule (b) Jordan's rule (c) Gloger's rule (d) Bergman's rule

112. The presence of vestigial organs in animals suggests that:
(a) Animals having vestigial organs are descendents of such ancestors in which these organs were fully developed and functional
(b) Animals having vestigial organs have a common ancestor
(c) Presence of vestigial organs provide evidence of organic evolution
(d) Both (a) and (c)

113. The concept of adaptive divergence was given by:
(a) Simpson (b) Romer (c) Osborn (d) Huxley

114. Proteins found in humans and apes are similar. It shows:
(a) Homology (b) Analogy (c) Parallel evolution (d) Bio-evolution

115. Fins of fishes and flippers of whales are:
(a) Homologous organs (b) Analogous organs
(c) Vestigial organs (d) Specialised organs

116. Which one of the following is not known, either living or dead, in the main line of chordate evolution?
(a) Echinoderms and hemichordates (b) Urochordates
(c) Cephalochordates (d) All of these

117. The possession of pharyngeal gill slits and pouches in all the embryo of vertebrates indicates that:
(a) Lungs develop from gills (b) Embryos are not able to utilise oxygen of air
(c) There is less demand of oxygen during early development (d) All vertebrates have evolved from a common ancestor

118. There is a tendency among animals to increase in size during the long course of evolution. This is known as:
(a) Cope's law (b) Biogenetic law (c) Allen's rule (d) Gause's law

119. The first eutherian mammals were shrew-like and arboreal:
(a) Herbivorous (b) Insectivorous (c) Omnivorous (d) Sanguivorous

120. Which one of the following has the complete fossil record to reveal evolutionary history?
(a) Camel (b) Elephant (c) Horse (d) All of these

121. Industrial melanism in peppered moth (*Biston betularia*) provides an example of :
    (a) Directional selection (b) Disruptive selection
    (c) Stabilising selection (d) None of these
122. The formation of a different genotype in a new settlement by a small population is known as:
    (a) Bottleneck effect (b) Founder effect
    (c) Haldane effect (d) Tangler effect
123. Which one of the following is an incorrect match?
    (a) Sickle anaemia – Common in Asia (b) Coacervates – Cell-like structures but not organised structures
    (c) Haldane – Worked on origin of life (d) *Latimera* – Living fossil
124. Which one of the following about *Biston betularia* is correct?
    (a) A connecting link (b) A fossil
    (c) An example of natural selection (d) An example of temporal variation
125. Which one of the following about Hardy–Weinberg law is correct?
    (a) A simple mathematical model showing how genetic equilibrium can be maintained in a gene pool
    (b) Helps in calculation of gene frequencies
    (c) Helps in the calculation of genotypic frequencies
    (d) All of these
126. The natural selection of Darwin does not explain:
    (a) The absence of a connecting link (b) Inheritance of vestigial organs
    (c) Origin and inheritance of variations (d) All of these
127. The relative frequency of alleles in sexually reproducing organism remains constant from generation to generation. This is known as:
    (a) Haldane effect (b) Hardy–Weinberg law
    (c) Law of inheritance (d) Cope's law

Answers to Multiple-Choice Questions

1. (c)	2. (d)	3. (a)	4. (a)	5. (b)	6. (d)	7. (b)	8. (d)
9. (a)	10. (c)	11. (b)	12. (b)	13. (b)	14. (c)	15. (c)	16. (c)
17. (c)	18. (c)	19. (d)	20. (b)	21. (c)	22. (a)	23. (d)	24. (a)
25. (c)	26. (b)	27. (b)	28. (d)	29. (d)	30. (a)	31. (d)	32. (c)
33. (c)	34. (a)	35. (c)	36. (d)	37. (b)	38. (a)	39. (a)	40. (b)
41. (b)	42. (b)	43. (b)	44. (b)	45. (b)	46. (c)	47. (d)	48. (c)
49. (c)	50. (b)	51. (c)	52. (b)	53. (a)	54. (d)	55. (c)	56. (b)
57. (d)	58. (b)	59. (b)	60. (c)	61. (c)	62. (a)	63. (c)	64. (d)
65. (d)	66. (c)	67. (a)	68. (d)	69. (c)	70. (d)	71. (d)	72. (b)
73. (b)	74. (a)	75. (b)	76. (a)	77. (c)	78. (c)	79. (d)	80. (c)
81. (d)	82. (a)	83. (a)	84. (d)	85. (c)	86. (b)	87. (a)	88. (a)
89. (b)	90. (d)	91. (b)	92. (b)	93. (d)	94. (b)	95. (d)	96. (c)
97. (c)	98. (a)	99. (b)	100. (a)	101. (d)	102. (c)	103. (c)	104. (a)
105. (b)	106. (c)	107. (c)	108. (a)	109. (d)	110. (c)	111. (b)	112. (d)
113. (c)	114. (a)	115. (b)	116. (d)	117. (d)	118. (a)	119. (b)	120. (d)
121. (a)	122. (b)	123. (a)	124. (c)	125. (d)	126. (d)	127. (b)	

Fill in the Blanks

1. Paleontology is the study of ________________.
2. The form of mimicry in which one edible species resembles an inedible species is called ______________ mimicry.
3. The Palaearctic region is separated from the Nearctic region by the ________________ and ________________ oceans.
4. A connecting link between reptiles and mammals found in Australia is ________________.
5. Synapsid reptiles gave rise to ________________.
6. ________________, a protein-splitting enzyme, is found in all animals from Protozoa to mammals.
7. The half life of potassium is ________________ years.
8. The first reptiles appeared during the ________________ period.
9. The oldest vertebrates are ________________, while the youngest vertebrates are ________________.
10. The genetic basis of adaptations in bacteria was experimentally confirmed by ________________.
11. Australian marsupial and placental mammals show ________________ evolution.
12. Similarity in organisms with different genotypes indicates ________________ evolution.
13. ________________ is the first requisite for fossilisation.
14. The evolution of pentadactyl limbs occurred during the ________________ period.
15. Adaptive radiation is an example for ________________.
16. The unit of natural selection is the ________________.
17. Genetic drift in the new colony is known as the ________________.
18. According to ________________ law, warm-blooded mammals of hot and humid climates have abundance of melanin pigments.
19. Random change in gene frequency is known as ________________.
20. Functionless homologous organs are known as ________________.
21. A species having two or more varieties is called ________________.
22. *Proterospongia* is a connecting link between ________________ and ________________.
23. The biogenetic law is based on ________________ evidences.
24. The unit of evolution is ________________.
25. A recently extinct bird is ________________.
26. The process of natural selection acts only on ________________ traits.
27. The immediate ancestor of the present-day horse is the ________________.
28. ________________ was a missing link between reptiles and mammals.
29. The age of humans is ________________.

Answers to Fill in the Blanks

1. Fossils
2. Batesian
3. Atlantic, Pacific

4. Duck-billed platypus
5. Mammals
6. Trypsin
7. 1.3 to 109
8. Carboniferous
9. Fish, mammals
10. J Lederberg
11. Convergent
12. Convergent
13. Immediate burial
14. Devonian
15. Diversifying selection
16. Individual
17. Founder effect
18. Gloger's
19. Genetic drift
20. Vestigial organ
21. Polytypic
22. Protozoa, Porifera
23. Embryological
24. Population
25. Dodo
26. Inherited
27. *Pliohippus*
28. *Cynognathus*
29. Holocene

True or False

1. C_{14} is used for fossil dating.
2. Vestigial organs illustrate convergent evolution.
3. Sickle cell anaemia helps in the prevention of haemophilia.
4. Reproductive isolation brings about sympatric speciation.
5. *Heloderma* is a living fossil.
6. The wooly mammoth is one of the unaltered fossils.
7. Stings of honeybees and scorpions are homologous structures.
8. The concept of adaptive radiation was developed by H F Osborn in 1898.
9. The synthetic theory is the most modern theory of evolution.
10. Reproductive isolation is essential for speciation.
11. Mammals became dominant on the earth after the extinction of dinosaurs.
12. Gills of fish and gills of prawn are analogous organs.
13. *Archaeopteryx* is a connecting link between reptiles and birds.
14. *Peripatus* is a living fossil.
15. Industrial melanism shows natural selection.

Answers to True or False

1. True
2. False
3. False
4. True
5. False
6. True
7. False
8. True
9. True
10. True
11. True
12. True
13. True
14. True
15. True

Give Reasons

1. There is a gradual complexity in the organisation of fossils.
   – Because fossils from the lowest strata of the earth have the simplest organisation and became more and more complex in the strata lying above them.

2. Antigen-antibody reactions are used as tools to determine the genetic relationship.
   - Because antibodies against the blood of an animal react quickly with the blood of closely related forms and less vigorously with the blood of distantly related forms.
3. Discontinuous variations are heritable.
   - Because they arise from mutation.
4. Somatogenic variations are useless for evolution.
   - Because they are nonheritable.
5. The natural force is a creative force in evolution.
   - Because it selects efficient gene combinations.
6. Serological tests have been invaluable in taxonomy.
   - Because they help to establish natural relationships among animals.
7. *Sphenodon* is called a living fossil.
   - Because it is a relic species and is the sole survivor of order. Rhynchocephalia have many primitive characters which are almost identical to those of Mesozoic fossil reptiles.
8. Bats are found in oceanic islands.
   - Because they are capable of crossing large oceans.
9. The distribution of mammals has been used as a determinant to divide the earth into different zoological realms.
   - Because, being warm-blooded, mammals are capable of occupying a wide range of habitats and being the most recently evolved groups, they had comparatively less time to branch out from the centre of origin.
10. The use of insecticides causes resistance among the insect population.
    - Because resistance against insecticides is necessary for the survival of insect populations, if insecticide is being regularly used.
11. Llama has haemoglobin with an unusually high affinity for oxygen.
    - Because it helps the llama to breath in the rarified mountain air.
12. In the beginning, the evolution of mammals was very slow.
    - Because of the unavailability of habitats, as most habitats were already occupied by reptiles.
13. Lemurs are found both in South India and South Africa.
    - Because both South India and South Africa were once connected by land bridges.
14. Radio carbon dating cannot be used to determine the age of fossils older than 70,000 years.
    - Because the amount of radioactive C_{14} decreases steadily with time.
15. Lack of trees limits the distribution of primates.
    - Because arboreal primates require trees for safety, food and progression.
16. Germinal mutations are important for organic evolution.
    - Because they are inheritable.

BASICS OF ORGANIC EVOLUTION

Multiple-Choice Questions

1. Adaptive radiation is an example of:
   (a) Convergent evolution (b) Divergent evolution
   (c) Co-evolution (d) Parallel evolution
2. The joint changes of two or more species in a close interaction is:
   (a) Co-evolution (b) Parallel evolution (c) Convergent evolution (d) Divergent evolution
3. An organism adapts to the environment through:
   (a) Divergent evolution (b) Convergent evolution
   (c) Co-evolution (d) All of these
4. Which one of the following evolution commonly occurs in a predator-prey relationship?
   (a) Parallel evolution (b) Divergent evolution (c) Convergent evolution (d) Co-evolution
5. Which one of the following is incorrect?
   (a) A species may evolve together. (b) Species may evolve apart.
   (c) Species may evolve with similar structure and appearance. (d) None of these
6. Which one of the following is the major difficulty involved in convergent evolution?
   (a) Homologous structures (b) Analogous structures
   (c) Vestigial structures (d) All of these
7. Parallel evolution occurs between two unrelated species that:
   (a) Occupy the similar niches in a given habitat (b) Do not occupy the similar niches in a given habitat
   (c) Both (a) and (b) (d) Share common traits
8. Consider the following statements about divergent evolution:
   (A) Such type of evolution generally occurs when closely related species diversify to new habitats
   (B) On a large scale, it is responsible for the formation of the present diversity of life on the earth from the first living cells
   (C) Macroevolution results in evolutionary divergence
   (D) Macroevolution results in parallel special adaptations in divergent groups

   The correct statements are:
   (a) All (b) A, B and C (c) B and C (d) C and D
9. Which one of the following is applicable to macroevolution?
   (a) Genetic divergence (b) Production of adaptive types
   (c) Population fragmentation (d) All of these
10. The natural selection becomes operative when the correlation between reproductive success and a trait is:
    (a) Positive (b) Negative (c) Positive or negative (d) Zero
11. Which one of the following prevents many communities from reaching competitive equilibrium?
    (a) Abiotic factors (b) Predation (c) Disturbance (d) All of the above

12. Which one of the following results in the splitting of parental population into two or more new populations?
    (a) Co-evolution (b) Parallel evolution
    (c) Convergent evolution (d) Divergent evolution
13. The varied group of plants and animals, either related or unrelated, provide an example of:
    (a) Divergent evolution (b) Convergent evolution
    (c) Parallel evolution (d) Co-evolution
14. Macroevolution results in the formation of:
    (a) Genera (b) Families (c) Orders (d) All of these
15. Genetic equilibrium of a population is disturbed by:
    (a) Genetic recombination (b) Genetic drift
    (c) Natural selection (d) All of these
16. The establishment of a true link of co-evolution between species requires:
    (a) Experimentation (b) Genetic analysis
    (c) A large number of observations (d) All of these
17. Which one of the following types of species interaction provides the ideal ground for co-evolution?
    (a) Plant – Herbivore (b) Predator – Prey (c) Parasite – Host (d) All of these
18. The development of the same biological traits in unrelated lineages is:
    (a) Convergent evolution (b) Divergent evolution
    (c) Parallel evolution (d) Co-evolution
19. Microevolutionary forces, which are operative for a shorter period, result in the production of:
    (a) Parallel evolution (b) Sequential evolution (c) Convergent evolution (d) Co-evolution
20. The Red Queen Hypothesis has been proposed by:
    (a) J D Bermal (1967) (b) Van Valen (1973)
    (c) M Goodman (1982) (d) L H Thompson (1985)
21. Asiatic lady beetle (*Harmonia axyridis*) presents an actual instance of:
    (a) Microevolution (b) Mesoevolution
    (c) Mega evolution (d) Sequential evolution
22. Antifreeze proteins are found both in Arctic and Antarctic fishes. However, the genetic pathways that produce these proteins are different in these fishes. This shows:
    (a) Co-evolution (b) Parallel evolution (c) Convergent evolution (d) Divergent evolution
23. Consider the following statements:
    (A) Mesoevolutionary changes require time lesser than human life
    (B) Mesoevolution causes alteration in numerous genetic units
    (C) Mega evolution is not followed by mesoevolution
    (D) Microevolution may be sequential or divergent

    The correct statements are:
    (a) All of these (b) A, B and C (c) A and C (d) B and D
24. Macroevolutionary changes are:
    (a) Repeatable (b) Reversible
    (c) Both repeatable and reversible (d) Neither repeatable nor reversible
25. Which one of the following is regarded as the engine of adaptive evolution?
    (a) Hardy–Weinberg equilibrium (b) Natural selection
    (c) Genetic drift (d) Mutation

26. Dobzhansky coined the term 'mesoevolution', after obtaining the results from a cross between geographically isolated populations of:
    (a) *Drosophila melanogaster* (b) *D. pseudoobseura*
    (c) *D. bipectinata* (d) *D. malerkotliana*
27. Which one of the following is a general hypothesis about macroevolution?
    (a) Ecological replacements (b) Red queen hypothesis
    (c) Community paradigm (d) All of these
28. Microevolutionary changes are:
    (a) Repeatable (b) Reversible (c) Reproducible (d) All of these
29. The presence of beak and wings in extinct Pteurosaurs and birds presents an example of:
    (a) Parallel evolution (b) Divergent evolution (c) Covergent evolution (d) Sequential evolution
30. Environmental conditions that involve similar developmental or structural adaptations, even in the species having different descent, may lead to:
    (a) Convergent evolution (b) Divergent evolution
    (c) Co-evolution (d) Parallel evolution
31. Galapagos finches studied by Darwin present a very good example of:
    (a) Adaptive radiation (b) Parallel evolution (c) Co-evolution (d) Microevolution
32. Which one of the following is applicable to the evolution of humans and apes from a common ancestor?
    (a) Divergent evolution (b) Convergent evolution
    (c) Parallel evolution (d) Co-evolution
33. Which one of the following is incorrect?
    (a) Macroevolution results in evolutionary divergence (b) Macroevolution is directional\
    (c) Macroevolution is caused due to macromutations (d) None of these
34. Which of the following theories states that species evolve because of biotic interactions?
    (a) Synthetic theory (b) Community paradigm hypothesis
    (c) Red queen hypothesis (d) Punctuated equilibrium theory
35. Divergent evolution is the process by which related lineages develop different __________ over time.
    (a) Behavioural traits (b) Biological traits (c) Genetic traits (d) All of these
36. Which one of the following is incorrect?
    (a) Hindlimb dominance is an analogous trait in lemurs and humans
    (b) The differences in length between the finger bones of birds and bats is an example of divergent evolution
    (c) Macroevolution documents the process of large-scale changes at or above the species level over long periods
    (d) Mammals and cephalopods have developed eyes in a similar manner
37. In insects, the shape and the size of the body as well as the size of the proboscis is different according to the structure of flowers. This is an example of:
    (a) Parallel evolution (b) Co-evolution
    (c) Convergent evolution (d) Divergent evolution
38. Through which one of the following bacteria may a develop resistance to antibiotics?
    (a) Microevolution (b) Mesoevolution (c) Mega evolution (d) None of these

39. Which one of the following may cause macroevolution?
    (a) Entry of a small population in a new niche
    (b) Change in developmental programme
    (c) Both (a) and (b)
    (d) Change in gene frequency.
40. Golschmidt is associated with:
    (a) Microevolution and macroevolution
    (b) Genetic drift
    (c) Isolating mechanism
    (d) Concept of species
41. Which one of the following is incorrect?
    (a) Mutation – Provide raw material for macroevolution
    (b) Flattening of canine – An example of macroevolution
    (c) Genetic drift – Random fixation of gene pool
    (d) Hox genes – Essential for patterning and growth of tetrapod limbs

Answers to Multiple-Choice Questions

1. (b)	2. (a)	3. (d)	4. (d)	5. (d)	6. (b)	7. (b)	8. (a)
9. (d)	10. (c)	11. (d)	12. (d)	13. (a)	14. (d)	15. (d)	16. (d)
17. (d)	18. (a)	19. (b)	20. (b)	21. (a)	22. (c)	23. (d)	24. (d)
25. (b)	26. (b)	27. (d)	28. (d)	29. (a)	30. (a)	31. (a)	32. (a)
33. (d)	34. (c)	35. (d)	36. (d)	37. (b)	38. (a)	39. (c)	40. (a)
41. (b)							

Fill in the Blanks

1. Slow and gradual changes in the genetic composition of an individual are termed as __________.
2. The process by which one species evolves into two or more species is called __________.
3. The mechanism of evolutionary change was first articulated by __________.
4. Vertebrate limb is a common example of __________ evolution.
5. The four common patterns of evolution are __________ evolution, __________ evolution, __________ evolution and __________.
6. The three levels of evolutionary processes are __________, __________ and __________ evolution.
7. The continuous and gradual change in an interbreeding population is called __________.
8. __________ acts on variations produced in a gene pool by mutation and recombination.
9. Macroevolution is also known as __________.
10. Simpson coined the term __________ for large-scale evolution.
11. When the correlation between reproductive success and a trait is __________, natural selection disappears.
12. The Protozoan that inhabits the gut of termite is a classic example of __________ co-evolution.
13. The process of evolution in which a trait held by a common ancestor evolves into variations over time is known as __________ evolution.
14. __________ evolution is the most commonly known type of evolution.

15. The evolution of reptiles that are completely terrestrial from amphibians is an example of __________.
16. Traits evolved as a result of convergent evolution are called __________ structures.
17. __________ evolution is the opposite of convergent evolution.
18. Bird and bat wings are an example of __________ evolution.
19. The evolution that occurs due to relationships between animals is termed as __________.
20. The statement 'Nothing in biology makes sense, except in the light of evolution' was given by __________.
21. According to the theory of __________, fossil records accurately reflect evolutionary change.
22. A long period without evolutionary change is known as ____________.

Answers to Fill in the Blanks

1. Evolution	2. Adaptive radiation	3. Charles Darwin
4. Divergent	5. Divergent, convergent, parallel, co-evolution	
6. Micro, macro, mega	7. Microevolution	8. Natural selection
9. Adaptive radiation	10. Mega evolution	11. Zero
12. Mutualistic	13. Divergent	14. Divergent
15. Mega evolution	16. Analogous	17. Divergent
18. Convergent	19. Co-evolution	20. T Dobzhansky (1973)
21. Punctuated equilibrium	22. Stasis	

True or False

1. Evolution is unrepeatable and irreversible.
2. Allele fixation is the unit process of evolution.
3. Mega evolutionary changes are very rare.
4. Sequential evolution results in the formation of new populations.
5. The isolation of species from one another accelerates evolution.
6. Microevolution is the process of evolution within populations.
7. Divergent evolution is nondirectional.
8. Competition is applicable to microevolution.
9. Co-evolution has a much shorter impact on community structure.
10. A monotypic species shows races or clines geographically.
11. Co-evolution plays a role in community evolution.
12. Similar environments producing similar adaptations result in parallel evolution.
13. Convergent and parallel evolution are very similar.
14. The fossil evidence in favour of mega evolution is very common.
15. The evolution of the horse from *Hyracotherium* is an example of macroevolution.

16. Microevolution is the process of evolution within populations that does include adaptive and neutral evolution.
17. A community may not remain in equilibrium for a long time.
18. Co-evolution generates reciprocal adaptive responses.
19. The origin of birds and mammals from reptiles presents a classical example of parallel evolution.
20. Natural selection and inheritance are the central agencies driving biological evolution.
21. Natural selection acts on phenotypes.
22. Divergent evolution may occur between two genes within a species.

Answers to True or False

1. True	2. True	3. True	4. False	5. True	6. True	7. False	8. True
9. False	10. False	11. True	12. True	13. True	14. False	15. True	16. False
17. True	18. True	19. False	20. True	21. True	22. True		

Give Reasons

1. In co-evolution, each species affects the evolution of others
   - Because each species in a co-evolutionary relationship exerts selective pressures on the other, thus affecting each other's evolution.
2. There is not a single population exhibiting sequential evolution.
   - Because all populations diverge during the course of evolution and split up into new populations.
3. Evolutionary changes generally follow a common pathway.
   - Because of similar environmental pressures.
4. The wings of all flying animals are very similar.
   - Because the same principle of aerodynamics apply for all.
5. Intraspecific competition can be a selective factor in evolution.
   - Because of the variation of reproductive success of individual organisms.
6. The losing of leaves by a tree is not a case of evolution.
   - Because it lacks descent through genetic inheritance.
7. Development of wings in animals is an example of convergent evolution.
   - Because there was no common ancestor for all winged animals. Due to their individual environments, these animals developed wings as an adaptation for their survival.
8. Whales and fishes have the same characteristics.
   - Because both had to evolve methods of moving through the same medium, i.e., water.
9. A community may not remain in equilibrium for long.
   - Because it may be disturbed by random environmental changes.
10. Most quantitative traits have a genetic basis.
    - Because they are determined by many genes.

GENETICS OF SPECIATION

Multiple-Choice Questions

1. Which one of the following is incorrect?
   (a) Speciation has occurred (b) Speciation is occurring
   (c) Speciation will occur (d) None of these
2. Which one of the following is a casual factor of speciation?
   (a) Genetic drift and mutation (b) Geographic isolation and migration
   (c) Habitat geometry (d) All of these
3. In which one of the following speciations does a population split into two geographically isolated populations?
   (a) Natural speciation (b) Allopatric (c) Parapatric (d) Sympatric
4. Which one of the following is a primary component of speciation?
   (a) Genetic isolation (b) Diversification (c) Both (a) and (b) (d) Genetic drift
5. When one or more species arise without geographical segregation of population, it is known as:
   (a) Allopatric speciation (b) Sympatric speciation
   (c) Parapatric speciation (d) Peripatric speciation
6. Natural selection may contribute to the reproductive isolation of incipient species by enhancing varieties to develop barriers to hybridisation. This is known as:
   (a) Wallace effect (b) Bottleneck effect (c) Founder effect (d) Trafalgar effect
7. Disruptive selection was proposed by:
   (a) C C Li (1955) (b) C P Haskins (1957) (c) J M Smith (1962) (d) G L Stebbins (1968)
8. Ring species provides evidence that speciation has occurred:
   (a) Allopatrically (b) Sympatrically (c) Parapatrically (d) None of these
9. Which one of the following is the main difference between different models of speciation?
   (a) Accumulation of genetic differences (b) Evolvement of reproductive isolation
   (c) Way of interruption of gene flow (d) All of these
10. Which one of the following is applicable to genetic drift?
    (a) Sewall Wright effect (b) Wallace effect
    (c) Bottleneck effect (d) Fountain effect
11. In majority of animal groups, the mode of speciation is:
    (a) Peripatric (b) Sympatric (c) Allopatric (d) Parapatric
12. Which one of the following is a post-zygotic isolating mechanism?
    (a) Hybrid sterility (b) Hybrid inviability (c) Hybrid breakdown (d) All of these
13. Consider the following statements:
    (A) Levels and patterns of biodiversity depend on the process of speciation
    (B) Speciation can be observed directly from beginning to end

(C) Changes in protein may provide basis of new species
(D) Speciation may result from changes in gene regulation

The correct statements are:
(a) All of these (b) A, B and C (c) A, C and D (d) A and D

14. Which one of the following plays a more important role in the speciation of plants?
(a) Ethological isolation (b) Mechanical isolation
(c) Geographical isolation (d) Ecological isolation

15. In isolation, changes in the gene pool may occur through:
(a) Natural selection (b) Genetic drift
(c) Founder effect (d) A combination of all of these

16. Three-spined sticklebacks present a good example of:
(a) Allopatric speciation (b) Sympatric speciation
(c) Peripatric speciation (d) Hybrid inviability

17. Which one of the following forms a ring species around the Himalayas?
(a) Greenish Warbler (*Phylloscopus trochiloides*) (b) *Larus* gulls
(c) *Ensatina* salamanders (d) *Drosophila bipectinata*

18. Which one of the following is probably the most common method of sympatric speciation?
(a) Hybridisation (b) Structural changes (c) Polyploidy (d) Founder effect

19. The pattern of species richness depends on:
(a) Changes in species (b) Extinction (c) Speciation (d) All of these

20. Which one of the following is a geographic mode of speciation?
(a) Sympatry (b) Parapatry (c) Allopatry (d) All of these

21. Which one of the following speciations can occur if some individuals begin to mate at a different time of a day or year?
(a) Allopatric (b) Sympatric (c) Both (a) and (b) (d) None of these

22. Genus X has five species which differ in the time of their peak breeding activity. This is due to:
(a) Ecological isolation (b) Behavioural isolation
(c) Mechanical isolation (d) Temporal isolation

23. Which one of the following about *Raphnobrassica* is incorrect?
(a) Tetraploid (b) Fertile
(c) True breeding (d) Completely isolated reproductively from its pro genitors

24. Chromosomes that lack a centromere:
(a) Undergo irregular disjunction at mitosis (b) Unable to be included in the daughter nuclei
(c) Ultimately lost (d) All of these

25. Which one of the following about natural selection is correct?
(a) Natural selection can produce new genes.
(b) Natural selection can produce new gene combination(s) directly.
(c) Natural selection is responsible for moulding the genetic variations present in the population.
(d) All of these

26. Consider the following statements:
(A) Among polyploids, tetraploids have the diploid chromosome complement doubled
(B) Habitat preference may promote speciation
(C) Whenever disruptive or diversifying selection is active, sympatric speciation may occur

(D) Geographical isolation is essential for speciation

The correct statements are:

(a) All of these (b) B, C and D (c) A and B (d) B and C

27. In parpatric speciation, reproductive isolation is:
(a) Temporal (b) Temporal or behavioural
(c) Geographical (d) All of these

28. Which one of the following tends to alter gene frequencies in large populations?
(a) Mutation and genetic recombination (b) Natural selection
(c) Gene migration (d) All of these

29. A large change in gene frequencies by _________ is a characteristic of small populations:
(a) Natural selection (b) Genetic drift (c) Mutation (d) Gene migration

30. Natural selection is the main evolutionary agent in:
(a) Large population (b) Medium population (c) Small population (d) All of the above

31. Sewall Wright has modified his genetic drift theory into:
(a) Sampling error theory (b) Shifting balance theory
(c) Natural selection theory (d) Punctuated equilibrium theory

32. Changes in chromosome numbers (except polyploidy) arise by:
(a) Deletion (b) Duplication (c) Inversion (d) Translocation

33. Which one of the following about biological concept of species is incorrect?
(a) Distinct unit at any given time (b) Groups of interbreeding natural populations
(c) Evolutionary capacity to change continuously (d) None of these

34. Reinforced mating discrimination:
(a) Is exhibited as dominant trait (b) Exhibits variability within species
(c) May be influenced by a known set of candidate genes involved in olfaction (d) All of these

35. Reinforcement occurs when natural selection strengthens behavioural discrimination to prevent:
(a) Interspecies mating (b) Intraspecies mating
(c) Both (a) and (b) (d) None of these

36. Which one of the following is an example of ethological isolation?
(a) Courtship pattern (b) Specific mating calls (c) Specific scents (d) All of these

37. Genetic drift results in the:
(a) Elimination of genes in a small population (b) Fixation of genes in a small population
(c) Both (a) and (b) (d) Neither elimination nor fixation of genes

38. A higher chromosome number of male parent represents:
(a) Mechanical isolation (b) Reproductive isolation
(c) Ethological isolation (d) Both (a) and (b)

39. *Raphonobrassica* is a classical example of:
(a) Autooctoploidy (b) Allopolyploidy (c) Autopolyploidy (d) Hexaploidy

40. Consider the following statements:
(A) Peripatric speciation was proposed by Ernst Mayr
(B) In peripatric speciation, one of the populations is much smaller than the other
(C) Genetic drift plays a significant role in peripatric speciation
(D) Peripatric speciation is related to the founder effect and selection bottlenecks occur in it

The incorrect statements are:
(a) A, B and C (b) B and C (c) C and D (d) None of the above

41. Which one of the following is a special case of genetic drift?
(a) Haldane effect (b) Founder effect (c) Confusion effect (d) Anderson effect

42. The first speciation gene discovered in mammals is the:
(a) *sd* gene (b) *RB* gene (c) *abl* gene (d) *Prdm 9* gene

43. The initial barrier to gene flow in allopatric speciation is:
(a) Mechanical (b) Behavioural
(c) Geographical (d) Different time of mating

44. Consider the following statements:
(A) Reduced gene flow probably plays a key role in speciation
(B) A species must be monophyletic and share one or more derived characteristics
(C) In hybridisation, the heterogametic sex is the most likely to have reduced fertility or viability
(D) Different types of barriers are responsible for different types of speciation.
The incorrect statements are:
(a) A and B (b) B and D (c) C and D (d) None of these

45. Which one of the following is an outcome of speciation?
(a) Adaptive radiation (b) Co-evolution (c) Divergent evolution (d) Parallel evolution

46. A post isolating mechanism prevents_________ of hybrid:
(a) Development (b) Survival (c) Reproduction (d) All of these

47. The formation of many species from a single common ancestor is known as:
(a) Variation (b) Adaptive radiation (c) Parallel evolution (d) Co-evolution

48. Which one of the following functions in isolating a population ?
(a) Time (b) Geography (c) Behaviour (d) All of these

49. Which one of the following is the fundamental unit of evolution?
(a) Species (b) Genus (c) Individual (d) Population

50. The chromosomes of the hybrid individuals:
(a) Fail to separate during meiosis (b) Fail to form gametes
(c) Form defective gametes (d) All of these

51. In which one of the following does interspecific sterility result in the production of offsprings with lethal genes in them?
(a) *Nemeophila* (b) *Xiphophorus* (c) *Limulus* (d) *Rana*

52. Autopolyploids are sterile to varying degrees, because:
(a) Generally, plants lack sex chromosomes (b) The majority of plants are monoecious
(c) Unequal segregation of chromosomes in the quadrivalents (d) None of these

53. Polyploidy is common in plants, because:
(a) A majority of plants are monoecious (b) There is a general absence of sex chromosomes
(c) Asexual reproduction in plants results in partially sterile polyploid species to propagate (d) All of these

54. Polyploidy occurs due to:
(a) Errors during meiosis (b) Errors during fertilisation
(c) Both (a) and (b) (d) None of these

Answers to Multiple-Choice Questions

1. (d)	2. (d)	3. (b)	4. (c)	5. (b)	6. (a)	7. (c)	8. (a)
9. (c)	10. (a)	11. (c)	12. (d)	13. (c)	14. (b)	15. (d)	16. (b)
17. (a)	18. (c)	19. (d)	20. (d)	21. (b)	22. (d)	23. (d)	24. (d)
25. (c)	26. (a)	27. (d)	28. (d)	29. (b)	30. (d)	31. (b)	32. (d)
33. (d)	34. (d)	35. (a)	36. (d)	37. (c)	38. (b)	39. (b)	40. (d)
41. (b)	42. (d)	43. (c)	44. (d)	45. (a)	46. (d)	47. (b)	48. (d)
49. (a)	50. (d)	51. (a)	52. (c)	53. (d)	54. (c)		

Fill in the Blanks

1. The population resulting from a speciation is a __________.
2. ____________ is the genesis of new species from an ancestral species.
3. The process of one species evolving into a different species is called ___________.
4. The splitting of one species into two or more species is called __________.
5. The two main types of cladogenesis are _________ speciation and __________ speciation.
6. The two predominant schools of thought regarding the speed of speciation are_____________ and ___________.
7. Reproductive isolation may operate __________ or __________.
8. Phylogenetic trees depicting incidence of sympatric speciation may be either ___________ or ______________ .
9. The process by which natural selection increases ____________ is called reinforcement.
10. Isolation caused by physical barriers between two populations leads to _____________ speciation.
11. __________ is the key process linking microevolution and macroevolution.
12. _______ species consists of a chain of interbreeding populations that surrounds some physical barrier.
13. Speciation of individuals isolated as a peripheral population is called_______________ speciation.
14. A failure to elicit mating behaviour is called _____________.
15. Beginning of new species is created within two or three generating through _____________.
16. Polyploids that arise by doubling of the chromosomes are called ____________.
17. Hybridisation that introduces some genes of one species into the genotype of another species is called ______________ hybridisation.
18. _____________ and hybrid sterility are the main processes of reproductive isolation between species.
19. The evolution of reproductive isolation between populations due to ecologically based divergent natural selection is called __________ speciation
20. Island radiation suggests that speciation is caused by ___________.
21. Biological factors that prevent gene flow are called _____________.
22. Speciation can result from ____________ separation or _________ specialisation.
23. Evolution of ____________ between the incipient species is the key to speciation.
24. Repeated independent evolution of the same reproductive isolating mechanism is known as _____________ speciation.

25. *Raphnus brassica* is a hybrid between __________ and ___________.

Answers to Fill in the Blanks

1. Species
2. Speciation
3. Anagenesis
4. Cladogenesis
5. Allopatric, sympatric
6. Gradualists, punctuationalists
7. Prezygotically, postzygotically
8. Dichotomous, polytomous
9. Reproductive isolation
10. Allopatric
11. Speciation
12. Ring
13. Peripatric
14. Sexual selection
15. Polyploidy
16. Autopolyploids
17. Introgressive
18. Sexual selection
19. Ecological
20. Founder effect
21. Isolating mechanisms
22. Geographic, ecological
23. Genetic differences
24. Parallel
25. *Raphnus sativus, Brassica oleracea*

True or False

1. A species is a group of individuals that interbreed in nature.
2. Speciation is a gradual process.
3. All new species develop from pre-existing species.
4. A peripheral isolate cannot be formed in different environments.
5. Speciation is the outcome of adaptation.
6. In sympatric speciation, the populations are not geographically isolated from each other.
7. Speciation takes place as a byproduct of genetic change which accumulates between two previously interbreeding populations of the same species.
8. Two populations that lose genetic contact will, very unlikely, evolve in the same way.
9. The formation of race appears to be a characteristic shared by all species.
10. Speciation does not involve only one mechanism.
11. Jumping of genes from one chromosome to another may result in the formation of a new species.
12. Speciation has been studied since the time of Darwin.
13. Microevolution is the first step towards speciation.
14. Different geographic regions may have different selective pressures.
15. Speciation is not related to adaptive radiation.
16. Sympatric speciation is rare.
17. The reproduction of autotetraploids is not normal.
18. Genetic drift causes fixation of new alleles.
19. Genetic drift increases genetic variability.
20. The bottleneck effect may cause elimination of certain alleles.
21. The process of speciation occurs only in allopatric populations
22. Reproductive isolating mechanism is always complete.
23. There is ample evidence for a casual association between pre-zygotic and post-zygotic isolation.

24. Intrinsic post-zygotic isolation may arise from genome doubling or polyploidy.
25. Sometimes, speciation may result from the accumulation of complementary genes.

Answers to True or False

1. True	2. True	3. True	4. False	5. True	6. True	7. True	8. True
9. False	10. True	11. True	12. True	13. True	14. True	15. False	16. True
17. False	18. True	19. False	20. True	21. True	22. True	23. False	24. True
25. True							

Give Reasons

1. Study of speciation is important.
   – Because management of biodiversity needs a proper understanding of the nature of diversity, which comprises the distribution and abundance of individual species, their assembly in the ecosystem as well as genetic variation within species. These facts are in turn, based on our knowledge of the processes which creates diversity in speciation.
2. Reproductive isolation can occur.
   – Because populations become physically separated from each other.
3. Sympatric speciation is more common in plants.
   – Because plants tend to develop multiple homologous sets of chromosomes, resulting in polyploidy.
4. When the interbreeding between two populations is interrupted, they become genetically differentiated.
   – Because the force of gene flow can no longer maintain its similarity.
5. The understanding of evolutionary mechanisms as well as factors that control speciation rates is critical.
   – Because besides generating species diversity, they also have a strong impact on the nature of species.
6. Hybrids are sterile.
   – Because of abnormal development of gonads or breakdown of meiosis before completion.
7. Reproductive isolation is of much biological significance.
   – Because it prohibits irreversibility of the evolutionary divergence at the species level due to establishment of reproductive isolation.
8. When interbreeding of one population with all other populations is stopped, its evolutionary history diverges.
   – Because the genetic changes cannot cross the reproductive barrier.
9. In small populations, generally, there is occurrence of genetic drift.
   – Because of sampling errors.
10. Isolated populations become more and more different.
    – Because the factor which is advantageous in one ecosystem may not be advantageous in another.
11. Cladogenesis takes place.
    – Because reproductive isolating mechanism prevents two subpopulations from interbreeding.
12. Autopolyploids and allopolyploids lack a clear dividing line.
    – Because of the presence of intermediate properties in hybrids.

EVOLUTION OF HORSE

Multiple-Choice Questions

1. Horse belongs to order:
   (a) Perissodactyla (b) Artiodactyla (c) Proboscidea (d) Edentata
2. The evolution of the horse began during the epoch:
   (a) Miocene (b) Eocene (c) Oligocene (d) Early Miocene
3. The evolutionary history of which one of the following is well documented in paleontology?
   (a) Elephants (b) Humans (c) Horses (d) None of these
4. *Parahippus* appeared in the:
   (a) Early Miocene epoch (b) Late Miocene epoch
   (c) Oligocene epoch (d) Eocene epoch
5. The primary centre of the evolution of the horse was:
   (a) Asia (b) North America (c) Africa (d) Australia
6. The fossil of Orohippus was recovered from:
   (a) Bridger Beds, New Mexico (b) Allia Bay, Kenya
   (c) Hadar, Ethiopia (d) Georgia
7. In *Eohippus*, which one of the following digits in the hindlimbs was represented by splints?
   (a) First and Third (b) First and fifth (c) Second and fourth (d) Third and fifth
8. In North America, horses became extinct by the close of:
   (a) Pleistocene epoch (b) Pliocene epoch (c) Miocene epoch (d) Oligocene epoch
9. These days, the true wild types of horses are found only in:
   (a) Asia (b) Africa (c) Europe (d) India
10. In *Mesohippus,* dentition was:
    (a) Brachydont (b) Hyposodont (c) Bunodont (d) Lophodont
11. *Hipparion* is a fossil of the:
    (a) Camels (b) Elephants (c) Horses (d) Humans
12. *Merychippus* were:
    (a) One-toed (b) Two-toed (c) Three-toed (d) Four-toed
13. The first equid fossil was found in:
    (a) Hadar, Ethiopia (b) Santa Fe, Argentina (c) Montmartre, Paris (d) Allia Bay, Kenya
14. Which one of the following about *Hyracotherium* is incorrect?
    (a) Earliest animal having horse-like anatomy (b) Omnivorous
    (c) Browsing (d) First and fifth toes touch the ground
15. The most remarkable change between *Hyracotherium* and *Orohippus* was in the:
    (a) Skull (b) Neck (c) Teeth (d) Limbs
16. Which one of the following did not radiate from *Merychippus?*
    (a) *Protohippus* (b) *Pliohippus* (c) *Hipparion* (d) *Parahippus*

17. The main difference between *Merychippus* and *Hipparion* was in the structure of:
(a) Tooth enamel (b) Size of skull (c) Size of neck (d) Toes
18. In which one of the following did the skull have deep facial fossae?
(a) *Pliohippus* (b) *Merychippus* (c) *Parahippus* (d) *Miohippus*
19. *Hippidion* appears to be endemic to:
(a) North America (b) South America (c) Italy (d) Asia
20. Which one of the following appears to be an evolutionary link between the old forest-dwelling horses and modern plains-dwelling grazers?
(a) *Mesohippus* (b) *Miohippus* (c) *Parahippus* (d) *Merychippus*
21. The increase in size between *Hyracotherium* and *Equus* was:
(a) Gradual (b) Constant (c) Progressive (d) None
22. The side toes became vestigial in:
(a) *Mesohippus* (b) *Miohippus* (c) *Pliohippus* (d) None of these
23. Which one of the following was a common grazer of the Pliocene epoch?
(a) *Hipparion* (b) *Nannipus* (c) *Neohipparion* (d) All of these
24. Who published the description of newly discovered horse fossils from North America?
(a) William Clark (1807) (b) O C Marsh (1870)
(c) Joseph Leidy (1948) (d) George G Simpson (1951)
25. The fossil of *Anchitherium clarencei*, a three-toed browsing horse was discovered in:
(a) West Turkana, Kenya (2002) (b) Panama Canal earthworks (2009)
(c) Neanderthal Valley, Germany (2007) (d) Hadar, Ethiopia (2006)
26. The horse was first domesticated in:
(a) Central Asia (b) Africa (c) East Indies (d) France
27. In any group of animals, there is a tendency for descendants to grow progressively larger. This is known as:
(a) Hamilton's rule (b) Cope's law (c) Trafalgar effect (d) Owen's law
28. Which one of the following about hooves is correct?
(a) Resist wear and fracturing (b) Absorb shock
(c) Conduct moisture throughout the hoof wall (d) All of these
29. The number of horses declined in the:
(a) Neolithic age (b) Bronze age (c) Mesolithic age (d) Palaeothic age
30. Toes started turning into hooves in:
(a) *Pliohippus* (b) *Merychippus* (c) *Miohippus* (d) *Equus*
31. In horses, the number of chromosomes is:
(a) 24 (b) 34 (c) 44 (d) 64
32. In wild horses, the haploid number of chromosomes is:
(a) 32 (b) 33 (c) 37 (d) 41
33. Individuals of a herd of equids communicate to other herd members using:
(a) Vocalisations (b) Mouth gestures
(c) Ear movements and tail movements (d) All of these
34. Horses are most closely related to:
(a) Rhinoceroses (b) Cattle (c) Antelopes (d) None of these
35. The earliest form of horse is known as:
(a) *Hyracotherium* (b) *Eohippus* (c) Dawn horse (d) All of these

36. Which one of the following was the first three-toed grazer?
(a) *Merychippus* (b) *Miohippus* (c) *Parahippus* (d) *Pliohippus*
37. Mountain horse is applicable to:
(a) *Mesohippus* (b) *Orohippus* (c) *Parahippus* (d) *Epihhipus*
38. *Parahippus* represents the transitional form between:
(a) *Miohippus* and *Merychippus* (b) *Miohippus* and *Hyohippus*
(c) *Mesohippus* and *Miohippus* (d) *Protohippus* and *Pliohippus*
39. Which one of the following about *Anchitherium* is incorrect?
(a) Arose from *Miohippus* (b) Simple cheek teeth
(c) Flourished during the Miocene epoch (d) None of these
40. *Merychippus* lacked:
(a) Foot pad (b) Spring mechanisms (c) Hooves (d) Hyposodont
41. *Hippidion* developed in:
(a) Central Asia (b) North America (c) South America (d) Africa
42. The first scientist who came up with his idea of evolution of the horse was:
(a) O C March (b) George G Simpson (c) Caspar Wistar (d) Buffon
43. In *Mesohippus*, *Miohippus* and *Parahippus* there were no evolutionary changes except:
(a) Increase in size (b) Deep facial fossae
(c) Decrease in number of toes (d) All of these
44. The dawn horse was discovered by:
(a) Owen (b) Marsh (c) Huxley (d) Simpson
45. The number of ribs in the modern horse is:
(a) 17 pairs (b) 18 pairs (c) 19 pairs (d) 17, 18 or 19 pairs
46. The evolution of the modern horse from *Eohippus* involves a change in the enlargement and elongation of the _________ digit, with a loss of other digits:
(a) 2nd (b) 3rd (c) 4th (d) 5th
47. *Hyracotherium* gradually transitioned into:
(a) *Parahippus* (b) *Epihhipus* (c) *Orohippus* (d) *Miohippus*
48. The diet of *Eohippus* was mainly:
(a) Foliage (b) Fruits (c) Herbs and berries (d) All of these
49. *Mesohippus* had ____________ on each side:
(a) 3 incisors and 3 molars (b) 2 incisors and 6 molars
(c) 3 incisors and 6 molars (d) 4 incisors and 6 molars
50. Which one of the following is not applicable to *Equus*?
(a) Herbivorous (b) Quadruped
(c) Long and arched neck (d) Four toes in front legs.
51. *Merychippus* appeared in the:
(a) Early Miocene epoch (b) Middle Miocene epoch
(c) Late Miocene epoch (d) Early Pliocene epoch
52. The first fossil horse of the Miocene epoch is:
(a) *Miohippus* (b) *Merychippus* (c) *Orohippus* (d) *Pliohippus*
53. Both milk and permanent teeth were high crowned in:
(a) *Mesohippus* (b) *Orohippus* (c) *Merychippus* (d) *Protohippus*

Answers to Multiple-Choice Questions

1. (a)	2. (b)	3. (c)	4. (a)	5. (b)	6. (a)	7. (b)	8. (a)
9. (a)	10. (a)	11. (c)	12. (c)	13. (c)	14. (d)	15. (c)	16. (d)
17. (a)	18. (a)	19. (b)	20. (c)	21. (d)	22. (d)	23. (b)	24. (b)
25. (b)	26. (a)	27. (a)	28. (d)	29. (a)	30. (b)	31. (d)	32. (b)
33. (d)	34. (a)	35. (d)	36. (a)	37. (b)	38. (a)	39. (d)	40. (a)
41. (c)	42. (a)	43. (a)	44. (a)	45. (d)	46. (b)	47. (c)	48. (d)
49. (c)	50. (d)	51. (b)	52. (a)	53. (d)			

Fill in the Blanks

1. The horse belongs to the family _________.
2. Modern-day members of the equidae include _________, _________, _________and _________.
3. Fossils of *Equus* are found on every continent except _________ and _________.
4. *Hyracotherium* was a mammal with _________ toes on the front feet and _________ toes on the rear.
5. *Merychippus* had _________ toes.
6. *Eohippus* is also known as _________ horse.
7. *Eohippus* had _________ incisors.
8. The first one-toed horse was _________.
9. *Eohippus* evolved into _________.
10. Splints are two icicle-shaped bones found on the back of each _________.
11. At the beginning of the _________ epoch, the *Mesohippus* evolved into a form known as *Miohippus*.
12. The *Hipparion* had _________ toes.
13. The oldest species of horse, _________, was discovered in Italy.
14. The *Kalobatippus* managed to relocate to Asia via the_________ land bridge.
15. Horses share a common ancestry with _________ and _________.
16. *Pliohippus* arose from _________ in the middle of Miocene.
17. Przevalski horse was described by _________.
18. All the breeds of horses sprang from the European _________ horse and the Asian _________ horse.
19. Horses and their relatives are referred to as _________ ungulates.
20. In *Parahippus,* dentition was of the _________ type.
21. The first representative *Equus* appeared in the late _________.
22. Both forelimbs as well as hindlimbs of *Mesohippus* exhibited lengthening tendency due to elongation of _________ and _________.

Answers to Fill in the Blanks

1. Equidae
2. Zebras, horses, donkeys and asses

3. Australia, Antarctica	4. Four, three	5. Three
6. Dawn	7. Three	8. *Pliohippus*
9. *Mesohippus*	10. Leg	11. Miocene
12. Three	13. *Equus stenonis*	14. Bering Strait
15. Tapirs, rhinoceros	16. *Callippus*	17. N Przevalski (1879)
18. Wild tarpan, Przevalski wild	19. Odd toes	20. Hyposodont
21. Pliocene	22. Metacarpals, metatarsals	

True or False

1. The evolution of the horse has no inherent direction.
2. The evolution of the horse has been a steady ladder of progress.
3. The *Eohippus* had four toes in its front feet and three toes in its hind ones.
4. Many horse species were usually present at the same time.
5. Horses fluctuated considerably in form and size over time.
6. The front hooves of a horse carry 60 per cent and the rear hooves carry 40 per cent body weight.
7. *Eohippus* had well-developed lateral vision.
8. The predecessors of horse were multi-odd-toed animals.
9. *Dinohippus* was a woodland animal.
10. The feet of *Merychippus* were adapted to walk on comparatively soft floor of the forest.
11. *Orohippus* was a true horse.
12. Facial fossa of *Miohippus* was larger and deeper.
13. The ancestors of the American horse migrated to Asia, later to Europe and Africa.
14. Horses have changed people's lives.
15. The 2nd and 4th toes were reduced in *Pliohippus*.
16. *Miohippus* were two-toed browsers.
17. In *Merychippus,* footpad was present.
18. During the course of evolution of the horse, the branchydont dentition was replaced by hypsodont.
19. *Equus* attained worldwide distribution during the Pleistocene epoch.
20. *Epihhipus* was slightly smaller than the *Orohippus*.
21. Milk teeth were high crowned and cemented in *Merychippus*.

Answers to True or False

1. True	2. False	3. True	4. True	5. True	6. True	7. False	8. True
9. True	10. False	11. False	12. True	13. True	14. True	15. True	16. False
17. False	18. True	19. True	20. False	21. True			

Give Reasons

1. The ancestry of the horse is one of the classic examples of evolution.
   – Because:
   Its fossil record is the most complete.
   The changes in skeleton and teeth are obvious and easy to identify and it is easy to understand their significance in animal mechanics and adaptation.
2. In the evolution of the horse, we only have the impressions of straight-line evolution.
   – Because only one genus happens to still be alive, so people are of opinion that the one genus was the target of all the evolution.
3. Horses have better crowned teeth.
   – Because such teeth are better adapted to feeding on grass.
4. Evolution of horses contradicts Cope's law.
   – Because during the course of evolution of horses, there are many instances of evolution from small to large size and back to small size.
5. *Merychippus* had wider molars than its predecessors.
   – Most probably for crunching the hard grass of the steppes.
6. *Pliohippus* is unlikely to be the ancestor of the modern horse.
   – Because its skull had deep fossae, while *Equus* lacked fossae. Also, the teeth of *Pliohippus* were curved, whereas they are straight in modern horses.
7. Horses needed long limbs and hooves instead of fingers.
   – Because the climate was changing, vast steppes appeared instead of jungles, so they needed long limbs and hooves for running long distances in search of food and escaping from predators.
8. North American horses disappeared around 8,000 to 10,000 years ago.
   – Because of climate change, hunting by early natives as well as spread of diseases.
9. The horse's evolution was not graded and smooth.
   – Because different traits evolved at different rates and did not always evolve together and often did so in the reverse direction.

EVOLUTION OF MAN

Multiple-Choice Questions

1 *Homo sapiens* belong to the order:
(a) Primates (b) Prosimians (c) Hominoids (d) Tarsiers

2. Lemurs are:
(a) Crepuscular (b) Nocturnal (c) Diurnal (d) All of these

3. Which one of the following is regarded as the common ancestor of catarrhine and hominoid anthropoids?
(a) *Oligopithecus* (b) *Parapithecus* (c) *Pliopithecus* (d) *Limnopithecus*

4. The cradle of human evolution is:
(a) Central Asia (b) North America (c) South Africa (d) Australia

5. During which one of the following epochs did *Homo sapiens* evolve?
(a) Miocene (b) Pleistocene (c) Pliocene (d) Oligocene

6. Which one of the following was a polymorphic species with respect to facial shape, dentition, palate, etc.?
(a) *Australopithecus africanus* (b) *A. robustus*
(c) *A. aethiopus* (d) *A. afarensis*

7. The Kenya hominoid is:
(a) *Ramapithecus* (b) *Paranthropus* (c) *Pithecanthropus* (d) *Dryopithecus*

8. The cranial capacity of modern (humans) is about:
(a) 400 to 800 cc (b) 1000 to 1200cc (c) 1200 to 1600cc (d) 1600 to 2000 cc

9. *Pithecanthropus* was discovered in:
(a) Java (b) Africa (c) China (d) Germany

10. Taung baby is applicable to:
(a) Java ape man (b) *Homo habilis* (c) *Dryopithecus* (d) *Australopithecus*

11. The first *Homo* is:
(a) *Homo ergaster* (b) *Homo habilis* (c) *Homo erectus* (d) *Homo rhodesiensis*

12. *Australopithecus anamensis* was named in 1995 and was found in:
(a) Allia Bay in Kenya (b) Hadar in Ethiopia
(c) West Turkana, Kenya (d) Georgia

13. Goosebumps occur in humans, when they are:
(a) Cold (b) Frightened (c) Angry or in awe (d) All of these

14. *H. heidelbergenesis* lived in:
(a) Africa (b) China (c) Europe (d) All of these

15. Which one of the following is not a diagnostics of archaic sapiens?
(a) Facial prognathism (b) Heavy supra-orbital ridges
(c) Cranial capacity from 500 cc to 900 cc (d) Low, slopping forehead

16. Consider the following statements about Neanderthals:
    (A) Used cave shelters (B) Cared for the injured
    (C) Strongest humanoids (D) Distinct DNA from sapiens
    The incorrect statements are:
    (a) A, B and C (b) A and B (c) C and D (d) None of these
17. Graves were fairly common in:
    (a) Neanderthal man (b) Cro-Magnon man (c) *Homo erectus* (d) *Paranthropus*
18. The skeleton *Homo ergaster* was discovered by:
    (a) Tax and Callender (1960) (b) Margulis (1970)
    (c) Newberne and Rogers (1981) (d) Walker and Kimeu (1984)
19. Which one of the following was lacking in *Dryopithecus*?
    (a) Brow ridges (b) Five-cusped molars (c) Broadened jaws (d) Large canines
20. Handy man is applicable to:
    (a) Neanderthal man (b) *Homo habilis* (c) *Australopithecus* (d) Peking man
21. *Homo habilis* was:
    (a) Omnivorous (b) Sanguivorous (c) Herbivorous (d) Carnivorous
22. Apes are characterised by:
    (a) Simian shelf (b) Erect posture (c) Shortening of jaws (d) Prehensile tail
23. In which one of the following did the chin first appear?
    (a) Peking man (b) Neanderthal man (c) Java man (d) Cro-Magnon man
24. Which one of the following chromosomes formed in the ancestral population is common to chimpanzees and humans?
    (a) 5 (b) 7 (c) 9 (d) 11
25. Which one of the following is incorrect?
    (a) All chimpanzee chromosomes possess heterochromatic telomeres
    (b) All gorilla chromosomes have heterochromatic telomeres
    (c) All human chromosomes have heterochromatic telomeres
    (d) None of these
26. *Homo erectus* inhabited:
    (a) Europe (b) Asia (c) Africa (d) North America
27. *Australopithecus africanus* was described by:
    (a) Obsorn (1891) (b) Dubois (1891) (c) Raymond Dart (1925) (d) Mayer (1950)
28. The earliest known catarrhine is:
    (a) *Kamoyapithecus* (b) *Parapithecus* (c) *Aegyptopithecus* (d) *Propliopithecus*
29. Humans and apes show similarities in:
    (a) Rh blood groups (b) MN blood groups (c) ABO blood groups (d) All of these
30. The number of teeth in *Oligopithecus* was:
    (a) 24 (b) 28 (c) 32 (d) 44
31. Which one of the following about new world monkeys is incorrect?
    (a) Opposable pollex and hallux (b) Shorter forelimbs
    (c) Muscles for facial expression (d) Nails on all digits
32. Bipedal locomotion was first exhibited by:
    (a) *Australopithecus* (b) *Dryopithecus* (c) *Ramapithecus* (d) *Pithecanthropus*

33. Fossil Lucy is applicable to:
(a) *Dryopithecus* (b) *Ramapithecus* (c) *Australopithecus* (d) *Parapithecus*

34. Consider the following statements about a fossil:
(A) Found in Shivalik Hills
(B) Earliest human-like primates
(C) Dentition similar to humans
(D) Semicircular dental arch
This fossil belongs to:
(a) *Dryopithecus* (b) *Ramapithecus* (c) *Australopithecus* (d) Java man

35. Lucy is the skeleton remains of:
(a) *Australopithecus africanus* (b) *Australopithecus afarensis*
(c) *Ramapithecus* (d) *Dryopithecus*

36. The characteristic difference between the *Australopithecus afarensis* and *A. africanus* is the:
(a) Height and flatter face (b) Height and cranial capacity
(c) Cranial capacity and dentition (d) Cranial capacity

37. Which one of the following chromosomes provides a strong evidence of common descent of humans and other apes?
(a) 2 (b) 4 (c) 9 (d) 11

38. Which one of the following is the largest ape?
(a) Gorilla (b) Orangutan (c) Gibbon (d) Chimpanzee

39. Tukana Boy was __________ discovery:
(a) an African (b) an American (c) an Asian (d) an European

40. Which one of the following about human chromosomes 2 is incorrect?
(a) Formed by and to end fusion of two aneestral chromosomes
(b) Contains a vestigial telomere
(c) Contains vestigial a centrosome
(d) None of these

41. In humans, chimpanzees, gorillas and orangutans, which one of the following chromosomes have identical banding pattern?
(a) 6, 13, 19, 21, 22 and X (b) 6, 8, 13, 20, 22 and X
(c) 4, 6, 8, 16, 18 and Y (d) 6, 8, 10, 12, 21 and X

42. Which one of the following chromosomes are different in humans, chimpanzees, gorillas and orangutans?
(a) 10 and 12 (b) 12 and 14 (c) 4 and 17 (d) 6 and 18

43. Which of the following chromosomes are similar in humans, chimpanzees and gorillas?
(a) 2, 5, 17, 19 and X (b) 3, 11, 14, 15, 18, 20 and Y
(c) 3, 11, 15, 20 and X (d) 5, 7, 9, 18, 20 and Y

44. Which one of the following is not a salient feature of humans?
(a) Erect walk (b) Narrow pelvis (c) Bipedal gait (d) Large cranium

45. Which one of the following is not a human character of *Australopithecus?*
(a) Prognathous face (b) Bipedal gait
(c) Lack of simian gap (d) Forward position of foramen magnum

46. Canines were large in:
(a) *Dryopithecus* (b) *Ramapithecus* (c) *Homo erectus* (d) *Homo habilis*

47. The first civilised human being was:
(a) *Homo erectus* (b) Cro-Magnon man (c) Neanderthal man (d) Heidelberg man

48. Piltdown man is applicable to:
(a) *Eoanthropus* (b) *Oreopithecus* (c) *Propliopithecus* (d) *Homo erectus*
49. The first animal domesticated by humans:
(a) Dog (b) Cat (c) Horse (d) Goat
50. Which one of the following chromosomes is a good marker of primate phylogeny?
(a) 7 (b) 9 (c) 15 (d) 17
51. Consider the following statements:
(A) Human telomere is much shorter, which is only 10 kilo-bases long
(B) Chimpanzees and other apes have 23 kilo-bases long telomeres
(C) More than 98 per cent of the DNA on chimpanzee chromosome 21 is present in human chromosome 21
(D) Chromosomes 3, 11, 14, 15, 18 and 20 match between gorillas, chimpanzees and humans
The correct statements are:
(a) All (b) A, B and C (c) B and C (d) A and D
52. Which one of the following is lacking in apes?
(a) Prognathous face (b) Small cranium (c) Longer hind limbs (d) None of these
53. The fossil of *Homo sapiens idaltu* was found in:
(a) Germany (b) France (c) Ethiopia (d) Australia
54. The oldest known modern human fossil is:
(a) *Homo erectus* (b) *Homo sapiens idaltu*
(c) *Homo neanderthalensis* (d) *Homo ergaster*
55. Some *Homo erectus* fossil differs from humans due to the:
(a) Effects of inbreeding (b) Dietary restrictions
(c) A harsh environment (d) All of these
56. *Ramapithecus* was more closely related to:
(a) Orangutans (b) Gorillas (c) Gibbons (d) Chimpanzees
57. The fossil of the Peking man was discovered by:
(a) Dubois (1890) (b) Pei (1924) (c) Raymond Dart (1924) (d) Lewis (1924)
58. The characters of *Aegyptopithecus* have more resemblance with:
(a) Apes (b) Humans (c) Monkeys (d) All of the above
59. The cranial capacity of *Australopithecus* was about:
(a) 500 cc (b) 800 cc (c) 900 cc (d) 1050 cc
60. Which one of the following is applicable to *Australopithecus?*
(a) Omnivorous (b) Carnivorous (c) Herbivorous (d) Sanguivorous
61. The human chromosome 2 has formed from fusion of two acrocentric present is the:
(a) Hylobatidae (b) Pongidae (c) Hominidae (d) Cebidae
62. α-globin gene becomes a pseudo gene in:
(a) Primates (b) Hominidae
(c) Pongidae (d) Pongidae and hominidae
63. Humans and pongidae differ by __________ pericentric inversions:
(a) 5 (b) 7 (c) 9 (d) 11
64. Which one of the following is a prosimian?
(a) Loris (b) Lemur (c) Tarsiers (d) All of these

65. Burial of the dead and religious feelings started with the:
(a) Cro-Magnon man (b) Neanderthal man (c) Java man (d) Peking man

66. The closet relative of modern humans is the:
(a) Chimpanzee (b) Gorilla (c) Orangutan (d) Gibbon

67. The first human to stand erect was the:
(a) Peking man (b) Java man (c) Neanderthal man (d) Cro-Magnon man

68. In which one of the following was cranial capacity more than the modern human?
(a) Peking man (b) Java man (c) Neanderthal man (d) Cro-Magnon man

69. The first fossil of the Cro-Magnon man was discovered by:
(a) Fuhlrott (1856) (b) Mc Gregor (1868) (c) Donjohanson (1974) (d) Mayor (1950)

70. Which one of the following about the Heidelberg man is incorrect?
(a) Chin present (b) Teeth of moderate size
(c) Massive jaw (d) Not in direct evolutionary line of modern man

71. The Cro-Magnon man:
(a) First made stone ornaments (b) Made wall paintings in caves
(c) Used animal skin to cover his body (d) All of these

72. The fossil of which one of the following was discovered from the Shivalik Hills in India?
(a) *Dryopithecus* (b) *Ramapithecus* (c) *Australopithecus* (d) *Homo erectus*

73. Which of the following chromosomes of chimpanzee differs from that of humans having the acquisition of juxta centromeric heterochromatin in humans?
(a) 5 (b) 7 (c) 9 (d) 11

74. Consider the following statements about a fossil:
(A) This fossil human was a cave dweller (B) Chin was present with strong jaws
(C) Face orthognathous (D) Domesticated dogs
This fossil human was the:
(a) Peking man (b) Neanderthal man (c) Cro-Magnon man (d) Java man

75. Which one of the following is an incorrect match?
(a) *Dryopithecus* – Miocene (b) *Parapithecus* – Lower Oligocene
(c) *Homo habilis* – Late Pliocene (d) *Homo erectus* – Mid Pleistocene

76. Which one of the following about *Parapithecus* is incorrect?
(a) Monkey ape (b) Had characters of tarsiers and anthropoids
(c) Common ancestor of apes, humans and monkeys
(d) None of these

77. Which one of the following about the Piltdown man is incorrect?
(a) Discovered by Charles Dawson (1912). (b) The jawbone was ape-like.
(c) Teeth had human characteristics. (d) The skull piece was very ape-like.

78. Consider the following statements:
(A) Cytochromes C of dog is about 90 per cent similar to that of humans
(B) Haemoglobin of horse is about 88 per cent similar to that of humans
(C) All mammals have a large degree of similarity at the genetic level
(D) The euchromatic regions of the chimpanzee genome share more than 98 per cent sequence similarity with the human ones.

The correct statements are:

(a) A, B and D (b) B, C and D (c) C and D (d) All of these

79. The comparison of human and chimpanzee chromosome using G and Q banding technique has revealed that the difference between the karyotype of the two species is due to:
(a) Absence of one secondary constriction (b) 1 centric fusion
(c) 5 pericentric inversions and 1 paracentric inversion (d) All of these

80. Which one of the following banding patterns of chromosomes is almost similar in humans and chimpanzees?
(a) 2 and 4 (b) 3 and 6 (c) 4 and 6 (d) 6 and 8

81. Which one of the following chromosome of humans provides strong evidence in favour of common descent of humans and apes?
(a) 2 (b) 7 (c) 9 (d) 21

82. Which one of the following sheds light on the common ancestry of humans and other primates?
(a) Amino acid sequencing (b) DNA hybridisation
(c) Immunological studies (d) All of these

83. The only human species earlier than the *Homo erectus* was:
(a) *Homo sapiens* (b) *Homo habilis* (c) *Homo ergaster* (d) *Homo rhodesian*

84. The genetic variability among modern *Homo sapiens* is greatest in:
(a) Africa (b) America (c) Australia (d) Germany

85. To which category does Lucy belong?
(a) *Australopithecus* (b) *Paranthropus* (c) *Homo erectus* (d) *Neanderthal man*

86. The earliest known catarrhine is:
(a) *Parapithecus* (b) *Kamoyapithecus* (c) *Propliopithecus* (d) *Aegyptopithecus*

87. Insular dwarfism is applicable to:
(a) *Homo sapiens idaltu* (b) *Homo floresiensis*
(c) *Homo erectus* (d) *Homo ergaster*

88. *Homo rhodesiensis* lived in:
(a) Java (b) China (c) Ethiopia (d) Zambia

89. The first *Homo sapiens* appeared about ________ years ago:
(a) 5,000 (b) 10,000 (c) 1,95,000 (d) 2,30,000

90. The human Y chromosome is ______per cent different from that of a chimpanzee:
(a) 2 (b) 10 (c) 15 (d) 30

91. Which one of the following about platyrrhina is incorrect?
(a) Tail long and prehensile (b) Opposable thumbs
(c) Flat nose (d) Clawed digits

92. Kamoya Kimeu is associated with:
(a) Lucy (b) WT-1500
(c) Ternifier man (d) *Australopithecus robustus*

93. Which one of the following is applicable to *Homo sapiens?*
(a) Herbivorous (b) Cannibalism (c) Prognathous face (d) None of these

94. Which one of the following about the Neanderthal man is incorrect?
(a) Used hides of animals (b) Cranial capacity similar to modern humans
(c) Well-developed chin (d) Began primitive social life

95. Which one of the following is associated with the evolution of man?
(a) Migratory habits (b) Development of brain
(c) Shift from herbivorous to omnivorous (d) None of these

96. The modern concept of human evolution is derived mainly from:
(a) Paleontology (b) Anthropology (c) Genetics (d) All of these

97. The first hominid to spend a major period of time in an extremely cold environment was the:
(a) Java man (b) Peking man (c) Neanderthal man (d) Cro-Magnon man

98. *The Descent of Man's* theory that humans and apes share a common ancestor was based on the many:
(a) Shared morphological features of apes and humans
(b) Shared anatomical features of apes and humans
(c) Shared common social behaviours
(d) All of these

99. Which one of the following may have been the first hominid to reach Eurasia?
(a) *Homo ergaster* (b) *H. erectus* (c) *H. habilis* (d) *H. rhodesian*

Answers to Multiple-Choice Questions

1. (a)	2. (d)	3. (a)	4. (c)	5. (b)	6. (a)	7. (a)	8. (c)
9. (a)	10. (d)	11. (b)	12. (a)	13. (d)	14. (d)	15. (c)	16. (d)
17. (a)	18. (d)	19. (a)	20. (b)	21. (d)	22. (a)	23. (d)	24. (b)
25. (c)	26. (b)	27. (c)	28. (a)	29. (d)	30. (c)	31. (b)	32. (a)
33. (c)	34. (b)	35. (b)	36. (b)	37. (a)	38. (a)	39. (a)	40. (d)
41. (a)	42. (c)	43. (b)	44. (b)	45. (a)	46. (a)	47. (b)	48. (a)
49. (a)	50. (a)	51. (a)	52. (c)	53. (c)	54. (b)	55. (d)	56. (a)
57. (b)	58. (a)	59. (a)	60. (a)	61. (b)	62. (d)	63. (c)	64. (d)
65. (b)	66. (a)	67. (b)	68. (d)	69. (b)	70. (a)	71. (d)	72. (b)
73. (c)	74. (c)	75. (c)	76. (d)	77. (d)	78. (d)	79. (d)	80. (b)
81. (a)	82. (d)	83. (b)	84. (a)	85. (a)	86. (b)	87. (b)	88. (d)
89. (c)	90. (d)	91. (b)	92. (b)	93. (d)	94. (c)	95. (b)	96. (d)
97. (c)	98. (b)	99. (a)					

Fill in the Blanks

1. *The Descent of Man* was published by __________.
2. The human family is __________.
3. The most ancient ancestor of humans is __________.
4. The cranial capacity of *Australopithecus* was __________.
5. The Cro-Magnon man used paints composed of __________ and __________.
6. The name *Homo habilis* was suggested by __________.
7. All members of hominidae have 48 chromosomes, except __________.
8. Human chromosome __________ is an end-to-end fusion of two ancestral chromosomes.

9. The long interspersed element-1 (L1) and __________ elements are the most abundant mobile elements in the human genome.
10. The genus *Homo* first appeared in __________ during the early Pleistocene period.
11. Chromosomes__, __, __, __, __ and __ are structurally the same in all great apes.
12. In humans, foramen magnum is ________.
13. __________realm was the centre of human evolution.
14. The fossil of the Cro-Magnon man was found in _________.
15. Tarsiers have only one genus *Tarsius* found in __________.
16. __________ was the earliest fossil of the old world monkey.
17. Lemurs are confined mainly to the island of __________.
18. Old world monkeys are found in __________ and __________.
19. The face of the Cro-Magnon man was perfectly __________.
20. The fossil of the Piltdown man was discovered by __________.
21. *Homo sapiens fossilis* is _____________ man.
22. __________ was the closet relative of humans.
23. *Homo erectus* is the scientific name of __________.
24. One distinguishing characteristic of an ape is its __________ -shaped jaw.
25. The jaw of humans is __________ -shaped.
26. The study of human fossils is called __________ years ago.
27. Archaic forms of *Homo sapiens* first appeared about __________ years ago.
28. The Cro-Magnon man appeared in __________.
29. The Cro-Magnon man differs from the Neanderthal man in having__________.
30. *Oligopithecus* is regarded as common ancestor of __________ and __________.
31. *Homo sapiens* is the only living species of its genus __________.

Answers to Fill in the Blanks

1. Charles Darwin
2. Hominids
3. *Australopithecus afarensis*
4. 380 to 450 cc
5. Manganese and iron oxides
6. Raymond Dart
7. Humans
8. Two
9. *AlU*
10. Africa
11. 6, 13, 19, 21, 22 and X
12. Anterior
13. Ethiopian
14. France
15. East Indies
16. *Oligopithecus*
17. Madagascar
18. Africa, Asia
19. Orthognathous
20. Charles Dawson (1908)
21. Cro-Magnon
22. Cro-Magnon man
23. Neanderthal man
24. U
25. V
26. Palaeoanthropology
27. 5,00,000
28. Upper Pleistocene
29. Small jaws
30. Catarrhine, hominoid anthropoids
31. *Homo*

True or False

1. Human mortality may be determined through the study of evolution.
2. The Taung baby was a bipedal human ancestor.
3. Modern humans evolved in Africa.
4. Modern humans have undergone a cultural evolution.
5. *Homo sapiens fossilis* is living.
6. Apes are tailless primates.
7. Cheek pouches and ischial callosities are present in New World monkeys.
8. The anthropoid ancestry gave rise to apes and humans.
9. *Homo erectus* was the first user of tools.
10. Foramen magnum at the back of skull is an adaptation to bipedalism.
11. Humans and dinosaurs lived during the last ice age.
12. *Homo habilis* evolved during the late Pliocene epoch.
13. The first group of hominids which left Africa was *Homo erectus*.
14. Chromosomes 4 and 17 exhibit different banding patterns in humans and also among different species of apes.
15. Most human and chimpanzee chromosomes have similar banding pattern.
16. *Australopithecus* was bipedal.
17. *Paranthropus* were users of tool.
18. There is complete absence of 48 chromosomes in human ancestry.
19. Chimpanzees have no blood type B.
20. Gorillas lack blood type A.
21. Our big toe is opposable like that of an ape.
22. The Y chromosome is the most rapidly evolving of the human chromosomes.
23. The Peking man is the representative of the ape man.
24. *Homo erectus* is similar to ape in cranial capacity and physical characteristics.
25. The cranial capacity of *Homo erectus* was 1150 cc.
26. Humans have a strong nuchal crest.
27. Humans lack simian gap and simian shelf.
28. In *Australopithecus*, the position of foramen magnum was backward under the base of the skull.
29. *Australopithecus* had characters of both humans and apes.
30. The forehead was well marked in *Homo erectus*.
31. The Cro-Magnon man was taller than the Neanderthal man.

Answers to True or False

1. True	2. True	3. True	4. True	5. False	6. True	7. False	8. True
9. False	10. False	11. False	12. True	13. True	14. True	15. True	16. True
17. False	18. True	19. True	20. True	21. False	22. True	23. True	24. False
25. False	26. True	27. True	28. False	29. True	30. False	31. True	

Give Reasons

1. *Homo habilis*, or the handy man, was so called.
   – Because of the tools found with its remains.
2. *Australopithecus aethiopicus, A. robustus* and *A. boisei* are known as robust australopithecines.
   – Because their skulls in particular are more heavily built.
3. Bipedal locomotion is advantageous.
   – Because due to this, forelimbs become free for other purposes.
4. Neanderthals were efficient hunters.
   – Most probably because in the cold environment, plant foods were scarce and were available only seasonally.
5. *Australopithecus anamensis* may be the common ancestor of *A. afarensis*, *A. platyops* and *A. garhi.*
   – Because of its early emergence and geographic location.
6. Mitochondrial DNA provides a magnified view of the diversity present in the human gene pool.
   – Because mutations accumulate in the mitochondrial DNA many times faster than the nucleus. Further, mitochondrial DNA is inherited maternally and does not recombine. Also, there are 1016 mitochondrial DNA molecules within a typical human and generally are identical to one another.

GEOLOGICAL TIME SCALE

Multiple-Choice Questions

1. The first geologic time scale was proposed by:
   (a) A Holmes (1913) (b) Adam Sedgwick (1835)
   (c) Charles Lapworth (1879) (d) C S Elton (1927)
2. The geological time scale has been broadly divided into _______ major divisions:
   (a) 4 (b) 5 (c) 6 (d) 7
3. The major divisions of time scale are called:
   (a) Periods (b) Eras (c) Epoch (d) Years
4. Consider the following statements:
   (A) The rocks of this era are mainly of the igneous type
   (B) The rocks of this era are devoid of fossils
   (C) During this era, the earth was formed
   (D) This era is characterised by complete absence of living organisms
   This era is known as the:
   (a) Proterozoic era (b) Azoic era (c) Archaeozoic era (d) Palaeozoic era
5. The first land vertebrate appeared during the:
   (a) Cambrian period (b) Ordovician period (c) Jurassic period (d) Devonian period
6. Which one of the following is known as the 'age of fishes'?
   (a) Devonian period (b) Tertiary period (c) Silurian period (d) Triassic period
7. The melting of glaciers began in the:
   (a) Ordovician period (b) Cambrian period (c) Silurian period (d) Devonian period
8. Placoderms became extinct by the end of the:
   (a) Silurian period (b) Mississippian period
   (c) Devonian period (d) Carboniferous period
9. Which one of the following periods marks the end of the Palaeozoic era?
   (a) Tertiary period (b) Permian period (c) Quaternary period (d) Devonian period
10. Dinosaurs with avian pelvic girdle were:
    (a) Ornithischia (b) Saurischia (c) *Brachiosaurus* (d) *Tyrannosaurus*
11. Cotylosaurs appeared in the:
    (a) Carboniferous period (b) Permian period
    (c) Tertiary period (d) Devonian period
12. Cotylosaurs became extinct during the:
    (a) Permian period (b) Triassic period (c) Devonian period (d) Silurian period
13. Which one of the following periods is known as the 'Age of Man'?
    (a) Tertiary period (b) Triassic period (c) Jurassic period (d) Quaternary period

14. Which one of the following about the Ordovician period is incorrect?
    (a) The first coral appeared in the Ordovician period.
    (b) Brachiopods became dominant.
    (c) Crinoids became dominant.
    (d) The climate was cold.
15. The first mass extinction occurred during the:
    (a) Permian-Triassic (b) Triassic-Silurian (c) Jurassic (d) Cretaceous
16. The main centre of evolution of the dinosaur was:
    (a) South Africa (b) North America (c) Central Asia (d) South America
17. Dinosaurs originated during the:
    (a) Triassic period (b) Permian period (c) Jurassic period (d) Silurian period
18. The first reptiles appeared during the:
    (a) Tertiary period (b) Permian period (c) Carboniferous period (d) Triassic period
19. Which one of the following is divided into epochs and periods?
    (a) Palaeozoic era (b) Mesozoic era (c) Cenozoic era (d) Archeozoic era
20. Dinosaurs were abundant during the:
    (a) Permian period (b) Triassic period (c) Tertiary period (d) Jurassic period
21. *Archaeopteryx* lived during the _______ period:
    (a) Jurassic (b) Permian (c) Devonian (d) Tertiary
22. During which one of the following periods did dinosaurs become extinct?
    (a) Devonian (b) Permian (c) Jurassic (d) Tertiary
23. The era of fishes was:
    (a) Archeozoic (b) Palaeozoic (c) Mesozoic (d) Cenozoic
24. The period during which the first mammal appeared:
    (a) Jurassic (b) Devonian (c) Permian (d) Triassic
25. The first creature whose body was perfectly preserved:
    (a) Trilobites (b) *Icthyostega* (c) *Eryops* (d) *Archaeopteryx*
26. *Eryops* evolved during the:
    (a) Devonian period (b) Carboniferous period (c) Ordovician period (d) Cambrian period
27. Archaic mammals vanished during the:
    (a) Eocene epoch (b) Oligocene epoch (c) Miocene epoch (d) Pliocene epoch
28. Consider the following statements:
    (A) Precambrian is the darkest period of geology
    (B) The key of amphibian evolution is provided by *Icthyostega*
    (C) *Seymouria* was one of the members of cotylosauria
    (D) Triassic was the age of snakes, turtles and saltophosuchus

    The correct statements are:
    (a) All (b) A, B and C (c) A and D (d) B and C
29. Divergent evolution of birds and mammals occurred during the:
    (a) Paleocene (b) Eocene (c) Oligocene (d) Miocene
30. Which one of the following is the 'age of giant molluscs'?
    (a) Ordovician (b) Devonian (c) Triassic (d) Permian
31. The Cambrian era does not include the:
    (a) Cenozoic era (b) Mesozoic era (c) Palaeozoic era (d) Archaeozoic era

32. Which one of the following is not a period of the Palaeozoic era?
(a) Ordovician (b) Permian (c) Triassic (d) Silurian
33. Insects appeared during the _______ period.
(a) Triassic (b) Ordovician (c) Permian (d) Tertiary
34. Which one of the following is the latest era?
(a) Cenozoic (b) Mesozoic (c) Palaeozoic (d) Archaeozoic
35. Trilobites were dominant during the:
(a) Ordovician era (b) Permian era (c) Cambrian era (d) Triassic era
36. The tail of the *Archaeopteryx* was flexible, containing________ vertebrae:
(a) 10 (b) 15 (c) 23 (d) 35
37. Which one of the following was absent in the *Archaeopteryx?*
(a) Sternum (b) Gastral ribs (c) Saurian pelvis (d) Wishbone
38. In which one of the following periods did aquatic Arthropods give rise to air breathing forms like scorpions, spiders, millipedes, etc.?
(a) Ordovician (b) Silurian (c) Devonian (d) Permian
39. Apes originated from monkeys during the________ epoch:
(a) Eocene (b) Oligocene (c) Miocene (d) Pliocene
40. The first mammal fossil was found in:
(a) Triassic (b) Devonian (c) Silurian (d) Ordovician
41. The smallest dinosaur was the:
(a) *Brachiosaurus* (b) *Dimetrodon* (c) *Diplodocus* (d) *Compsognathus*
42. Which one of the following periods ended by the laramide revolution?
(a) Cretaceous (b) Triassic (c) Permian (d) Jurassic
43. Lizards and crocodiles first appeared during the:
(a) Tertiary period (b) Triassic period (c) Silurian period (d) Jurassic period
44. Stegocephalians became extinct during the:
(a) Jurassic period (b) Cretaceous period (c) Triassic period (d) Permian period
45. Which one of the following is a creodonta?
(a) *Coryphodon* (b) *Arctocynon* (c) *Pantolambda* (d) *Unitatherium*
46. The fossil of *Hesperornis* has been found from the :
(a) Cretaceous period (b) Jurassic period (c) Permian period (d) Silurian period
47. *Hesperornis* was a:
(a) Flying reptile (b) Fish (c) Aquatic diving bird (d) Aquatic reptile
48. Ammonites first appeared during the:
(a) Ordovician period (b) Devonian period (c) Permian period (d) Silurian period
49. The only group that exhibited decline during the Devonian period was:
(a) Trilobites (b) Stegocephalians (c) Arthrodires (d) Eurypterids
50. Seaweeds were the only plants of the:
(a) Permian period (b) Ordovician period (c) Silurian period (d) Devonian period
51. The largest division of geologic time is the:
(a) Eon (b) Era (c) Period (d) Epoch
52. Explosion of life forms occurred during the:
(a) Cambrian period (b) Silurian period (c) Ordovician period (d) Permian period

53. Which one of the following is the basic unit of geological time scale in which a single type of rock system is formed?
    (a) Era (b) Period (c) Epoch (d) Eon
54. The period that brought the first reptile is the:
    (a) Permian period (b) Ordovician period (c) Carboniferous period (d) Silurian period
55. Pangaea began to break up during the:
    (a) Permian period (b) Triassic period (c) Tertiary period (d) Silurian period
56. A new period added by geologists in March 2004 to the earth's chronology is the:
    (a) Neocaran (b) Paleocaran (c) Ediacaran (d) Old Cambrian

Answers to Multiple-Choice Questions

1. (a)	2. (c)	3. (b)	4. (b)	5. (d)	6. (a)	7. (b)	8. (c)
9. (b)	10. (a)	11. (b)	12. (b)	13. (d)	14. (d)	15. (a)	16. (b)
17. (a)	18. (c)	19. (c)	20. (d)	21. (a)	22. (b)	23. (b)	24. (d)
25. (a)	26. (b)	27. (a)	28. (a)	29. (b)	30. (a)	31. (d)	32. (c)
33. (c)	34. (a)	35. (c)	36. (c)	37. (a)	38. (b)	39. (b)	40. (a)
41. (d)	42. (a)	43. (d)	44. (c)	45. (b)	46. (a)	47. (c)	48. (b)
49. (a)	50. (b)	51. (a)	52. (a)	53. (b)	54. (c)	55. (b)	56. (c)

Fill in the Blanks

1. The age of the earth has been divided into ______, ______ and ______.
2. The Precambrian time includes ______ and ______.
3. The modern orders of the class of mammalians originated during ______ period.
4. The Cenozoic era has been divided into ______ and ______ periods.
5. The coal-bearing period is ______.
6. Eurypterids were the ancestors of modern ______.
7. Decline of monotremes and marsupials occurred during the ________ epoch.
8. True sharks appeared in freshwater during the ______ period.
9. The best-known arthrodire was ______.
10. The Rocky Mountain Revolution occurred between the ______ and ______ eras.
11. Metatheria evolved during ______.
12. The ______ era was the 'golden age of reptiles'.
13. Almost all types of invertebrates evolved during the ______ era.
14. The three eons are ______, ______ and ______.
15. ______ were the largest four-footed animals that ever lived on land.
16. Ambylopodas were archaic heavy-limbed, slow moving ______ found in swamps.
17. The Mesozoic era has been divided into three, viz., the ______, ______ and ______ periods.

18. The four epochs of the tertiary period are ______, ________, ________ and ______.
19. The origin of primitive teleosts fishes occurred during the ___________ period.
20. The first fossil birds found in the rocks of Jurassic period belonged to genera ______ and ______.
21. Subdivisions of periods are called ______.
22. The ______ and ______ periods together constitute the middle Paleozoic era.
23. ______ was the period of establishment of the first land flora.
24. ______ was the last period of Cenozoic era.
25. The fossil of *Archaeopteryx* was discovered by ___________

Answers to Fill in the Blanks

1. Eras, periods and epochs
2. Archaeozoic, Proterozoic
3. Tertiary
4. Tertiary, quaternary
5. Carboniferous
6. Scorpions
7. Oligocene
8. Devonian
9. *Dinichthys*
10. Mesozoic, Cenozoic
11. Jurassic
12. Mesozoic
13. Palaeozoic
14. Achaean, Proterozoic, Phanerozoic
15. Sauropods
16. Ungulates
17. Triassic, Jurassic, Cretaceous
18. Eocene, Oligocene ,Miocene and Pliocene
19. Jurassic
20. *Archaeopteryx*, *Archaeornis*
21. Epochs
22. Silurian, Devonian
23. Devonian
24. Pleistocene
25. Andreas Wagner (1861)

True or False

1. Era is the largest subdivision in the geological time scale.
2. The Archaeozoic era is the age of Protists.
3. Cotylosaurs are stem reptiles.
4. Cambrian fossils are represented only by marine plants and animals.
5. The modern squid and *Octopus* were represented by Belemnites in the Mesozoic era.
6. Mesozoic mammals were large in size.
7. The Cretaceous period ended by the Cascadian revolution.
8. Grasslands appeared during the tertiary period.
9. Even-toed ungulates began to radiate during the Oligocene epoch.
10. Deer and antelopes were abundant during the Miocene epoch.
11. The Palaeozoic era is known as the 'Cradle of ancient life'.
12. The Proterozoic era ended by the Killarney revolution.
13. The Ordovician period ended by the Green Mountain disturbances in the northeastern part of North America.
14. Pocono formation occurred during the Permian period.
15. Pleiosaurs were aerial reptiles.

16. The climate of the Permian period was warmed and moist.
17. The formation of lead, zinc, gold, silver and gypsum occurred during the Carboniferous period.
18. Placoderms appeared during the Silurian period.
19. Ornithischia were abundant during the Triassic period.
20. Woody plants arose during the Ordovician period.
21. Miocene was a time of mammalian culmination.
22. The rise of nautiloids occurred during the Silurian period.
23. The Archaeozoic era is known as the 'era of invisible life'.
24. The first emergence of mountains occurred during the Ordovician period.
25. The Mesozoic era is characterised by the crustal change of the earth.

Answers to True or False

1. True	2. True	3. True	4. True	5. True	6. False	7. False	8. True
9. True	10. False	11. True	12. True	13. False	14. False	15. False	16. False
17. True	18. True	19. False	20. True	21. True	22. False	23. True	24. False
25. True							

Give Reasons

1. The sedimentation rate technique of estimating the age of fossils is not very accurate.
   - Because sedimentation rates are not always uniform.
2. The Devonian period is known as the 'age of fishes'.
   - Because fossil records indicated the existence of numerous fish forms during this period.
3. The Carboniferous period is called the coal-bearing period.
   - Because of the formation of a huge quantity of coal during this period.
4. The Cenozoic era is called the 'era of mammals'.
   - Because of a marked increase in the diversity and abundance of mammals during this era.

ADAPTATION

Multiple-Choice Questions

1. Consider the following statements:
   (A) Adaptations are produced by natural selection
   (B) Echolocation in bats is an adaptation for catching insects
   (C) Vestigial structures present in animals are adaptation
   (D) An adaptation is a feature that is not common in a population
   The correct statements are:
   (a) All of the above (b) A, B and C (c) B and C (d) C and D
2. Which one of the following about adaptation is incorrect?
   (a) Heritable (b) Functional (c) Adaptive (d) None of the above
3. All changes that are the result of adaptation can be explained by:
   (a) Genetic drift (b) Natural selection
   (c) Mutation (d) Genetic recombination
4. Adaptation includes_________ characteristics that an increase organism's reproductive success in its natural environment:
   (a) Structural (b) Functional (c) Behavioural (d) All of these
5. Adaptation results in:
   (a) Gain of a new feature (b) Loss of an ancestral feature
   (c) Both (a) and (b) (d) Neither gain nor loss of a feature
6. Structural adaptations are:
   (a) The physical characteristics of an organism (b) Not caused by natural selection
   (c) The ability of an organism to behave (d) All of these
7. Which one of the following is a behavioural adaptation?
   (a) Water foul migration (b) Calling behaviour in frogs
   (c) Nesting locations in ground squirrels (d) All of these
8. Match column I with column II and select the correct answer using answer codes:

Column I	Column II
(A) Powerful hind legs	1. Vultures
(B) Long powerful arms	2. Woodpeckers
(C) Pointed beak	3. Kangaroos
(D) Long broad wings	4. Chimpanzees

   Answer codes:

	A	B	C	D
(a)	4	3	2	1
(b)	2	3	1	4
(c)	3	1	4	2
(d)	3	4	2	1

9. Flexible steering tail is found in:
(a) Kangaroos (b) Cheetahs (c) Horses (d) Musk oxens
10. Which one of the following about the eagles is incorrect?
(a) Sharp claws and sharp hooked beak (b) Camouflage colouration
(c) Keen eyesight (d) High flight
11. Which one of the following is an incorrect match?
(a) Opossum – Plays dead (b) Skunk – Camouflage colouration
(c) Spider – Floats in air (d) Moose – Sprays foul smelling scent
12. Seasonal colour change for camouflage occurs in:
(a) Mooses (b) Foxes (c) Rabbits (d) Opossums
13. Which one of the following is an incorrect match?
(a) Bird wing – Structural adaptation (b) Migration – Behavioural adaptation
(c) Photosynthesis – Biochemical adaptation (d) None of the above
14. Foot is good for running in:
(a) Sparrows (b) Pheasants (c) Herons (d) Coots
15. Foul-smelling scent spraying is found in:
(a) Skunks (b) Opossums (c) Mooses (d) Foxes
16. Scopae are found in:
(a) Herons (b) Humming birds (c) Coots (d) Honeybees
17. Head is tapering and ends with a snout in:
(a) Shrews (b) Snakes (c) Moles (d) All of these
18. Which one of the following is found in penguins?
(a) Blubber (b) Countercurrent heat exchange
(c) Uropygial glands (d) All of these
19. Tail is vestigial in:
(a) Lizards and snakes (b) Hedgehogs and moles
(c) Moles and wombats (d) Woodchunks and snakes
20. The free movement of limb segments at joints is lost in a cursorial foot, except at the:
(a) Ball and socket joint of hip (b) Ball and socket joint of hip and shoulder
(c) Ball and socket joint of ankle and wrist joints (d) All of these
21. In which one of the following forms of adaptation are ulna and forelimbs reduced as an adaptation to fast running?
(a) Fossorial adaptation (b) Cursorial adaptation
(c) Arboreal adaptation (d) Volant adaptation
22. External ears are entirely absent in:
(a) Monotremes (b) Fossorial moles
(c) Both (a) and (b) (d) Otter, monotremes and fossorial moles
23. The sole of feet bears adhesive pads in:
(a) Japanese macaques (b) Tree porcupines
(c) Tree coneys (d) All of these
24. Penguins:
(a) Are amphibious (b) Feed only at sea (c) Breed on land (d) All of these

25. Which one of the following is the root concept that grew into Darwin's theory of natural selection?
(a) Genetic drift (b) Mutation (c) Genetic recombination (d) Adaptation

26. Social adaptation is found in:
(a) Lions (b) Wolves (c) Tigers (d) Elephants

27. Which is one of the following about adaptations is correct?
(a) Evolutionary process (b) Basic phenomenon of biology
(c) Occurs over many generations (d) All of these

28. Consider the following statements about Koalas:
(A) Koalas are marsupials
(B) Koalas get most of their water from the succulent *Eucalyptus* leaves
(C) Koalas have thickly padded tails
(D) There is a large gap between the first and second fingers
The correct statements are:
(a) All of these (b) B, C and D (c) A, C and D (d) A and B

29. Which one of the following favours stride rate?
(a) Increase in limb mass (b) Decrease in limb mass
(c) Increase in length of limbs (d) Decrease in length of limbs

30. Teeth are prehensile in:
(a) Bears (b) Seals (c) Musk deer (d) None of these

31. Which one of the following about fossorial mammals is correct?
(a) Small and pentadactyl (b) Relatively defenseless
(c) Plantrigrade (d) All of these

32. Eyes are covered with skin in:
(a) Marsupial moles (b) Cape golden moles (c) Mole rats (d) Hedgehogs

33. Feet are prehensile in:
(a) Primates (b) Marsupials (c) Both (a) and (b) (d) Flying squirrel

34. In all modern birds, there are:
(a) Two equally developed metacarpals (b) Two unequally developed metacarpals
(c) Three equally developed metacarpals (d) Three unequally developed metacarpals

35. Which one of the following is not applicable to *Rhacophorus?*
(a) Webbed feet (b) Adhesive pads (c) Patagium (d) Volant adaptation

36. The hoof has reached the highest degree of perfection in:
(a) Camels (b) Horses (c) Oxen (d) Rhinoceroses

37. The sidewinder is mostly seen during:
(a) Morning (b) Evening (c) Night (d) Afternoon

38. Nocturnal animals have a well-developed sense of:
(a) Sight (b) Hearing (c) Smell (d) All of these

39. The main feature of nocturnality is the:
(a) Adaptation to avoid predation (b) Ecological niche
(c) Both (a) and (b) (d) To avoid high temperature

40. Consider the following statements:
(A) Cats have eyes that can adapt to both night and day levels of illumination
(B) Snakes always keep their eyes open, even when they are asleep

(C) Frogs lack eardrums (D) The skin of polar bear is black

The correct statements are:

(a) All (b) A, B and C (c) A, B and D (d) C and D

41. Polar bears use ________ as rudders:

(a) Front legs (b) Hindlegs (c) Ears (d) Webbed toes

42. Gliding has particularly evolved among animals living in:

(a) Rainforests (b) Deserts (c) Alpine forests (d) Mangroves

43. A fake penis is found in females of:

(a) Polar beers (b) Spotted hyenas (c) Cheetahs (d) Wild dogs

44. Which one of the following always floats on its back at the time of eating?

(a) Dolphins (b) Otters (c) Walruses (d) None of these

45. Knees can be bent backwards in:

(a) Kiwis (b) Emus (c) Flamingoes (d) Ostriches

46. When there is plenty of prey, the polar bear does not eat the ___________ of seal:

(a) Skin (b) Fat (c) Meat (d) Internal organs

47. By flapping its ears, an elephant can lower its body temperature by:

(a) 1°C (b) 3ºC (c) 5ºC (d) 7ºC

48. The only hoofed animal that allogrooms – mutual grooming from one animal to another as well as grooming itself – is the:

(a) Rhinoceroses (b) Impalas (c) Hippopotamuses (d) Elephants

49. Sweet tooth is found in:

(a) Hippopotamuses (b) Giraffes (c) Rhinoceroses (d) Antelopes

50. Which one of the following is a digging organ?

(a) Snout (b) Teeth (c) Forelimb (d) All of these

51. Aquatic mammals lack:

(a) Pinnae (b) Valvular nostrils (c) Foam (d) Melon

52. In cetaceans, lack of hair is compensated by:

(a) Melon (b) Baleen (c) Foam (d) Blubber

53. Bones are filled with oil in:

(a) Cetacea (b) Sirenia (c) Chiroptera (d) Monotremes

54. Which one of the following about amphibious mammals is correct?

(a) Webbed feet (b) Cylindrical tail

(c) Subcutaneous fat (d) Small pinnae

55. Gliding is shown by:

(a) *Draco* (b) *Rhacophorus* (c) *Acrobates* (d) All of these

56. Camels begin to sweat only when their body temperature reaches:

(a) 38°C (b) 40°C (c) 44°C (d) 45°C

57. The process of urohydrosis occurs in:

(a) Deserts rodents (b) Camels

(c) Turkeys (d) Turkeys and black vultures

58. Which one of the following is known to store water in fatty tissues?

(a) Gila monsters (b) Black vultures (c) Kangaroos (d) Opossums

59. The body of a burrowing animal is:
(a) Worm like (b) Spindle shaped (c) Fusiform (d) All of these
60. The second, third and fourth vertebrae are enclosed in:
(a) Lizards (b) Moles (c) Snakes (d) Opossums
61. Cursorial adaptations are found in:
(a) Cassowaries (b) Rheas (c) Emus (d) All
62. In sirenians:
(a) Lungs are unlobed (b) Dentition is greatly reduced
(c) Only two teats are present (d) All of these
63. In fossorial animals:
(a) The shoulder girdle is weak (b) The olecranon process of ulna is very well developed
(c) Forelimbs are weak (d) The humerus is long
64. Second and third toes are syndactlyus in:
(a) *Bradypus* (b) *Koala* (c) *Carpomys* (d) *Chameleon*
65. Brachydont is found in:
(a) Insectivores (b) Herbivores (c) Carnivores (d) Omnivores
66. Match column I with column II and select the correct match using answer codes:

Column I	ColumnII
(A) Aggressive mimicry	1. *Phyllium*
(B) Warning mimicry	2. Spiders
(C) Protective mimicry	3. *Didelphis virginiana*
(D) Simulation of death	4. *Lementis*

Answer codes:

	A	B	C	D
(a)	4	1	3	2
(b)	2	3	4	1
(c)	3	1	4	2
(d)	2	4	1	3

67. Which one of the following lacks digitigrade foot posture?
(a) Deer (b) Antelopes (c) Giraffes (d) All of these
68. Which one of the following is not a brachiator?
(a) Apes (b) Monkeys (c) *Galeopithecus* (d) All of these
69. Skin is hygroscopic in:
(a) *Moloch* (b) *Perognathus* (c) *Dipodomys* (d) *Neotoma*
70. Consider the following statements:
(A) In moles, the snout is highly sensitive
(B) In kangaroos, the tail acts as the third leg
(C) *Amblyopsis* lacks taste buds
(D) In moles, an extra pair of zygapophyses are present in the lumber vertebrae

The correct statements are:
(a) All (b) A, B and C (c) B and C (d) C and D
71. Which one of the following is lacking in aquatic mammals?
(a) Lacrimal glands (b) Sweat glands (c) Sebaceous glands (d) All of these

72. Which one of the following with reference to bats is incorrect?
(a) Knee joints backwards
(b) Tragus present
(c) Echo apparatus
(d) Keeled sternum is lacking

Answers to Multiple-Choice Questions

1. (d)	2. (d)	3. (b)	4. (d)	5. (c)	6. (a)	7. (d)	8. (d)
9. (b)	10. (b)	11. (a)	12. (c)	13. (d)	14. (b)	15. (a)	16. (d)
17. (d)	18. (d)	19. (c)	20. (b)	21. (b)	22. (c)	23. (d)	24. (d)
25. (d)	26. (b)	27. (d)	28. (a)	29. (b)	30. (b)	31. (d)	32. (b)
33. (c)	34. (d)	35. (c)	36. (b)	37. (c)	38. (d)	39. (c)	40. (c)
41. (b)	42. (a)	43. (b)	44. (b)	45. (c)	46. (c)	47. (b)	48. (b)
49. (a)	50. (d)	51. (a)	52. (d)	53. (a)	54. (b)	55. (d)	56. (b)
57. (d)	58. (a)	59. (d)	60. (b)	61. (d)	62. (d)	63. (b)	64. (b)
65. (b)	66. (d)	67. (d)	68. (c)	69. (a)	70. (d)	71. (d)	72. (d)

Fill in the Blanks

1. Adaptations are the result of ____________.
2. The evolutionary process whereby a population becomes better suited to its environment is called ____________
3. The special characteristics of an organism that help it to survive in its natural environment are called ____________ adaptations.
4. Burrowing animals mostly belong to phylum chordata, except ____________.
5. Adaptation of an animal to like something else is called ____________ adaptation.
6. In horses, the ____________ and ulna are represented by splint bones.
7. In invertebrates, the aerial mode of existence is found only in ____________.
8. Adaptations may be ____________, ____________, ____________.
9. Adaptation is based on the fact that populations of an organism change over time as a result of ____________.
10. ____________ are propelling organs of cursorial forms.
11. ____________ is the reverse of albinism.
12. Total loss of colour is known as ____________.
13. Swim bladder is the homologue of the ____________ of the terrestrial animal.
14. Crickets have ears on their ____________ legs.
15. ____________ bears are the largest bears in the world.
16. Polar bears live in the ____________.
17. The smallest flying vertebrates are the ____________ and the ____________.
18. The heaviest flying animal is the ____________.

19. Warning mimicry is also known as ____________ mimicry.
20. Arctic animals are as a rule white, except the ____________ forms.
21. According to Weismann, ____________ is the only factor responsible for causing mimicry.
22. A change in the species over along period of time is known as ____________ .

Answers to Fill in the Blanks

1.	Evolution	2.	Adaptation	3.	Structural
4.	Birds	5.	Mimicry	6.	Fibula
7.	Insects	8.	Structural, physiological, behavioural		
9.	Natural selection	10.	Limbs	11.	Melanism
12.	Albinism	13.	Lungs	14.	Front
15.	Polar	16.	Arctic	17.	Beehummingbird, bumblebeebat
18.	Great bustard	19.	Batesian	20.	Aquatic
21.	Natural selection	22.	Evolution		

True or False

1. The ability to adapt is the most basic fundamental feature of living organisms.
2. The ability to adapt forms the basic differences between living and nonliving matter.
3. Honeybees are highly adapted to functions in the community.
4. Natural selection has been recognised as the means by which evolutionary adaptation occurs.
5. Adaptations are heritable characteristics.
6. Each adaptation has been produced by evolution.
7. All evolutionary changes are the result of adaptation.
8. In sloths, the humerus is very short.
9. Some whales can hold their breath even for an hour.
10. A species becomes extinct when the environment changes and the adaptive characteristics of a species are less efficient to allow its survival.
11. Giraffes have a camouflaged coat.
12. Circulatory system of penguins is adapted to conserve heat.
13. Physical adaptations develop during the lifetime of an individual.
14. Stride length is gained by increasing limb length.
15. Plantigrade animals are generally five-toed.
16. In desert animals, nocturnal behaviour decreases osmoregulation.
17. Cobras are able to kill with a bite as soon as they are born.
18. The claws of polar bears help them to catch seals.

19. All toes of a woodpecker are turned backwards, which assists it in grasping the bark of a tree.
20. The Tasmanian wolf is cursorial.
21. Camels have become secondarily digitigrade.

Answers to True or False

1. True	2. True	3. True	4. True	5. True	6. True	7. False	8. False
9. True	10. True	11. True	12. True	13. False	14. True	15. True	16. False
17. True	18. True	19. False	20. True	21. True			

Give Reasons

1. Adaptation usually occurs.
   - Because a gene mutates or changes by accident.
2. The beaks of birds vary in their shape and size.
   - Because of their different feeding habits.
3. Certain species of female frogs prefer males with long breeding calls.
   - Because males with long calls have better breeding habitats.
4. Adaptations are different from acclimatisation.
   - Because an adaptation takes several generations to develop, while an individual may acclimatise to conditions within his lifetime.
5. Arctic foxes are white:
   - Because it is an adaptation that helps them in hiding from enemies.
6. Desert animals have long limbs and padded feet.
   - Because these help them to run easily on sand for long distances.
7. In penguins, the base of feathers is downy.
   - For allowing entrapment of more air for better insulation.
8. In burrowing animals, the tail is either reduced or vestigial.
   - Because the tail plays no role in burrowing.
9. The majority of fossorial animals undergo winter sleep.
   - Because they lack the ability to maintain constant body temperature.
10. The knees of a camel contain thick leathery patches.
    - Because these patches protect the knees from being burnt when these are placed on the hot desert sand.
11. Camels have two rows of long eyelashes.
    - Because these long lashes protect the eyes from blowing sand and the sun.
12. Behaviours that were once adaptive can no longer be adaptive in the new environment.
    - Because the environment may change from generation to generation.
13. Most desert animals are pale in colour.
    - Because pale colour prevents their bodies from absorbing more heat from the sun.

14. Jackrabbits have large ears.
    – Because they provide greater body surface to dissipate heat.
15. An adaptation must be genetically encoded.
    – Because natural selection cannot act on traits that are not passed on to the offspring.
16. In sirenians, the lungs and diaphragm extend the entire length of vertebral column.
    – Because such features help sirenians in controlling their buoyancy and in maintaining their horizontal position in water.
17. Most adaptations of emperor penguins are related to their sea needs.
    – Because they spend about 75 per cent of their lives in seawater.
18. Carnivores are fast.
    – Because they have flexible trunks.
19. Polar bears walk with their toes pointing inward.
    – Because this protects them from slipping.
20. The tongue of a woodpecker is abnormally long.
    – Because such a long tongue reaches the grubs at the bottom of holes and thus assists in capturing its food.
21. Animals can live in a variety of places in the world.
    – Because they have special adaptations according to the area they inhabit.

COMPARATIVE ANATOMY

INTEGUMENT AND ITS DERIVATIVES

- The integument is the outer protective covering of the body, which includes skin and all the structures derived from it.
- The integumentary system is the largest organ system of the body.
- The integumentary system has a variety of functions. In animals, it may serve to waterproof, cushion and protect the deeper tissues, regulate temperature, excrete waste and is the location of sensory receptor for pain, pressure and temperature.

Structure of Skin

- In all vertebrates, the integument consists of two layers, viz., an outer epidermis and inner dermis.

I. Epidermis

- It is the outer thin layer of the skin and is ectodermal in origin.
- It contains an outermost layer of dead cells called stratum corneum and an inner layer of cubical cells called stratum germinativum or Malpighian layer.
- The stratum corneum first appears in amphibia.
- The cells of stratum germinativum give rise to cells of the epidermis.
- In higher vertebrates, the epidermis is stratified into several layers.

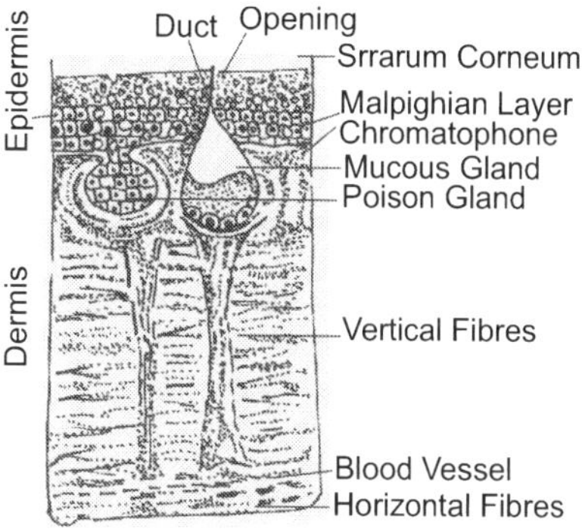

Fig. 1 VS skin of frog

- Generally, the epidermis lacks blood and nerve supply.
- Structures such as hair, nails, feathers, scales, enamel, nails horns and claws are derived from the Malpighian layer.

II. Dermis

- It lies immediately beneath the epidermis and is derived from the mesoderm.
- It is primarily consists of fibrous connective tissues.
- It is richly supplied with blood vessels and lymphatics.

- Pigment cells generally occur in the dermis.

Integument in Different Classes of Vertebrates

I. Cyclostomata

- In cyclostomes, the skin is smooth, moist and devoid of scales.
- The epidermis is many layered and lack keratin.
- The epidermis contains many club-shaped secretary cells.
- The dermis contains an outer thick layer of fibrous connective tissue and an inner layer of pigment cells.
- In between the dermis and body wall musculature, a layer of subcutaneous tissue containing blood vessels' fat, a connective tissue is present.

II. Fishes

- The skin of fishes is rich in mucous glands and is tight fitting.
- The outermost layer of the epidermis, the stratum corneum is absent.
- Poison glands and light-producing organs (photophores) may be present.
- The dermis is made up of fibrous connective tissues that contain a network of blood capillaries.
- A characteristic feature of the skin of fishes is the presence of scales.
- Placoid scale is found in cartilaginous (elasmobranchs) fishes.
- In chondrostei and holostei ganoid, scales are absent while in teleosts cycloid and ctenoid, scales are present.

III. Amphibians

- The skin of amphibians is thin, moist and usually slimy.
- The amphibian skin is without scales, except in gymnophiona.
- Some amphibians have the ability to change body colour.
- The epidermis is many layered, the outer being stratum corneum and the inner stratum germinativum.
- The stratum corneum of epidermis first appears in amphibians.
- In toads, stratum corneum is thicker.
- The dermis is relatively thin and is made up of two layers, viz., the upper stratum spongiosum and the inner stratum compactum.
- The dermis is richly supplied with blood vessels.
- The dermis contains chromatophores, poison glands and mucous glands which are epidermal derivatives.

IV. Reptiles

- In reptiles, the integument is thick, dry and is covered with epidermal scales, spines, shields, scutes, plates, claws, horns, rattles, etc., of dermal origin.
- Epidermal covering is periodically shed (ecdysis) in lizards and snakes.
- Stratum corneum is relatively thick, which makes the skin dry and prevent water loss.
- Epidermal scales are few, except femoral glands in lizards and cloacal scent glands of alligators.

- The dermis is thick having upper and lower layers.
- In snakes and lizards, the upper layer of the dermis contains chromatophores.
- The lower layer of the dermis is made up of connective tissues.

V. Birds

- The integument is thin and loose fitting in order to have maximum freedom of movement for flight.
- The skin is covered with feathers and lacks chromatophores.
- Skin glands are lacking, except the uropygial gland. The uropygial gland is well developed in aquatic birds.
- The epidermis is soft, except on feet and shank, where it is thick and forms epidermal scales.
- The dermis is compact and is devoid of dermal elements.
- The feathers and scales contain pigment cells.

VI. Mammals

- Mammalian skin is elastic, waterproof and thickest among vertebrates.
- Both the epidermis and dermis are highly modified.
- The epidermis is thickest in mammals and has five layers which are stratum corneum, stratum lucidum, stratum granulosum, stratum spinosum and stratum germinativum.
- Stratum corneum contains keratin and is thickest in palms and sole.
- The epidermis lacks mucous glands.
- The dermis in mammals is relatively thicker than any vertebrate, except armadillos.
- The dermis contains hair follicles, sebaceous glands and sweat glands of epidermal origin.
- Dermal scales are absent.

Derivatives of Integument

- Derivatives of skin include epidermal scales, bony plates, teeth, hair, feathers, nails, claws, keratinised scales, etc.
- Depending upon the layer of skin from which derivatives are formed, they are of the following two types:
  1. **Epidermal Derivatives –** Epidermal derivatives are derived from the epidermis and include epidermal glands and hard horny structures.
  2. **Dermal Derivatives –** Dermal derivatives are formed from the dermis and include dermal scales, plates or scutes, antlers, fin rays, etc.

I. Epidermal Glands

- Epidermal scales are formed by the Malpighian layer of the epidermis.
- They are unicellular or multicellular.
- Epidermal glands are formed in the epidermis, but sometimes invade the dermis.
- These epidermal glands are mucous glands (amphibians), poison glands (amphibians), luminescent glands (teleosts fishes), femoral glands (lizards), uropygial glands(birds), sweat glands (mammals), sebaceous glands, scent glands (musk deer, skunks, kangaroos, rats) and mammary glands (characteristic of mammals), etc.

II. Epidermal Scales and Scutes

- The cells of stratum corneum become cornified and form hard horny exoskeletal structures such as hair, feathers, scales, nails, claws, hoofs, beaks, etc.
- In reptiles, there is a continuous layer of scales that prevent loss of water.
- In lizards and snakes, scales are overlapping and are shed periodically (ecdysis).
- Large, thick and rectangular and non-overlapping scutes are found in crocodiles and turtles.
- The horny beak of turtles, rattle of rattlesnakes and horns of horned toad are a modification of the epidermis.
- Claws on all toes are present in reptiles.
- Epidermal scales and claws are also found in the feet of birds.
- The sheath of the beak of birds is the modification of stratum corneum.
- In some mammals, epidermal scales are found on the feet and tails of rats and beavers.

III. Dermal Scales and Scutes

- Dermal scales and scutes are mesodermal in origin and are formed in the dermis.
- Bony dermal scales increase in size and are not shed.
- The dermal derivatives of fishes are scales and fin rays.
- In fishes, five types of scales are found, viz., placoid, ganoid, cosmoid cycloid and ctenoid.
- Bony plates, which are dermal derivatives, are found in the skin of many reptiles, which lie beneath the epidermal scales.
- In turtles, carapace and plastron are found as rigid dermal skeleton below epidermal scutes.
- In mammals, bony plates are found in armadillos and whales.

IV. Digital Cornification

- Digital cornification includes claws, nails, hooves, horns, etc.
- The claws, nails, horns and hooves are formed by the stratum corneum.
- Claws have identical structures in reptiles, birds and mammals.
- Claws are modified to form nails, which are characteristic of primates.
- Claws and nails are similar except shape.
- Horns are found in hoofed mammals.
- Hooves are characteristic of ungulate mammals.

V. Hair

- Hair are characteristic of mammals and are cornified epidermal derivatives.
- Generally, hair cover the entire body but in some cases there may be less hair (e.g., whales, dolphins, etc.).
- Each hair has two parts, called the root and the shaft.
- The root is embedded in the dermis while the shaft projects outward.
- In *Echidna*, porcupine and hedgehog, hair are modified into spines.
- Hair are mainly concerned with insulation of the body.

VI. Feathers

- Feathers are epidermal derivatives of birds.
- Feathers are strong, elastic, waterproof and light in weight.
- Feathers help in flight as well as act as an insulator.
- They are replaced seasonally.

VII. Teeth

- Teeth are derived both from the ectoderm and dermis.
- The enamel of teeth develops from the ectoderm, while the cementing parts like dentine and pulp cavity develop from the dermis.

DIGESTIVE SYSTEM

- Digestive system is a series of organs involved in the digestion of food eaten, absorption of digested food and expulsion of undigested residue.
- Digestive system includes alimentary canal and associated glands.
- The basic plan of the digestive system is similar in all vertebrates. But within this basic plan, the specific components vary from one animal to another.
- The alimentary canal is formed early during embryonic development and is lined with the endoderm, except the mouth and rectum, which have an ectodermal lining.
- The digestive tract of vertebrates include the following parts:

I. Mouth and Oral Cavity

- Mouth is the anterior opening that leads into the oral cavity.
- In cyclostomes, the mouth is a circular opening.
- In gnathostomes, generally the mouth is terminal. However, in elasmobranchs and sturgeons, the mouth is ventral.
- In fishes, the mouth margins of amphibians and most reptiles have small lips.
- In turtles, the lips of birds and a few mammals have been transformed into a horny beak or bill.
- In mammals, the mouth is bounded by muscular lips.
- The mammalian mouth is specialised to serve as suckling and masticatory organ (with muscular cheeks).
- The oral cavity begins at the mouth and ends at the pharynx.
- Fishes have a very short oral cavity, while tetrapods have longer oral cavities.
- In a majority of fishes, the nasal cavities do not open into the oral cavity by internal nares.
- In crossopterygians and amphibians, the internal nares open into the oral cavity,
- In reptiles and birds, the oral cavity is separated from the nasal passage by palatal folds.
- In mammals, the nasal passage has been completely separated from the oral cavity due to the formation of a bony secondary palate,
- Oral cavity contains teeth, tongue and oral glands.

1. **Teeth**
   - Cyclostomes lack true teeth but epidermal teeth are present in the wall of the buccal funnel and on the tongue.
   - Tadpole larvae of anurans have serrated epidermal teeth.
   - True teeth are found in all vertebrates (except cyclostomes, sturgeons, some toads, sirens, turtles and birds).
   - In fishes, amphibians and most reptiles, teeth are acrodont, homodont and polyphyodont.
   - In mammals, teeth are thecodont, heterodont and diphyodont.
   - Teleosts have sharp teeth attached to jaw bones.
   - In mammals, teeth exhibit morphological variation. Four types of teeth are found in mammals, viz., incisors (cutting), canines (piercing and tearing), premolars and molars (mastication).
2. **Tongue**
   - The tongue is a simple crescent-shaped elevation in the floor of the oral cavity.
   - The tongue is nonmuscular in fishes.
   - Some anurans are tongueless while in others, a muscular tongue is present.
   - In snakes and some lizards, the tongue is well developed but in turtles and crocodiles tongue is poorly developed.
   - In birds and mammals, the tongue is well developed.
   - In mammals, taste buds are present on the tongue.
   - In vertebrates, the tongue performs various functions, viz., capturing and gathering of food, swallowing, thermoregulation, manipulating fluids and solids in the oral cavity and assists in human speech.
3. **Oral Glands**
   - In vertebrates, a great variety of oral glands are found that secrete a variety of substances such as saliva, poison (lizards, snakes and mammals) and anticoagulant (vampire bats).

II. Pharynx

- It is the region of foregut between the oral cavity and the oesophagus.
- The pharynx is lined with the endoderm.
- In fishes, the pharynx is the respiratory organ.
- In tetrapods, the pharynx is the cross road between food and respiratory passage.
- In tetrapods, it is the site of openings of auditory (Eustachian) tube.

III. Oesophagus

- The oesophagus is a distensible muscular tube connecting the pharynx and the stomach.
- The oesophagus is very short in neckless vertebrates (fishes and amphibians).
- Sharks and *Latimeria* have a large oesophagus.
- The oesophagus is long in *Polypterus*.
- In amniotes, the oesophagus is fairly long, reaching to the maximum in birds and giraffes.
- In birds, the oesophagus contains a distensible sac called crop, which acts as a temporary site of food storage.

IV. Stomach

- The stomach is a muscular chamber at the end of the oesophagus.
- It serves as storage and macerating site for ingested food and secretes digestive enzymes.
- The shape of the stomach varies from animal to animal and even in the same animal, depending upon whether the stomach is full or empty.
- Stomach is lacking in cyclostomes, chimeras, lung fishes and some teleosts.
- In embryos of vertebrates, the stomach is a straight tube-like structure.
- In certain fishes, long-bodied amphibians, lizards and snakes, the stomach is cigar-shaped.
- In most fishes, the stomach is J-shaped.
- In some deep fishes, the stomach is so distensible, so that it can hold prey larger than themselves.

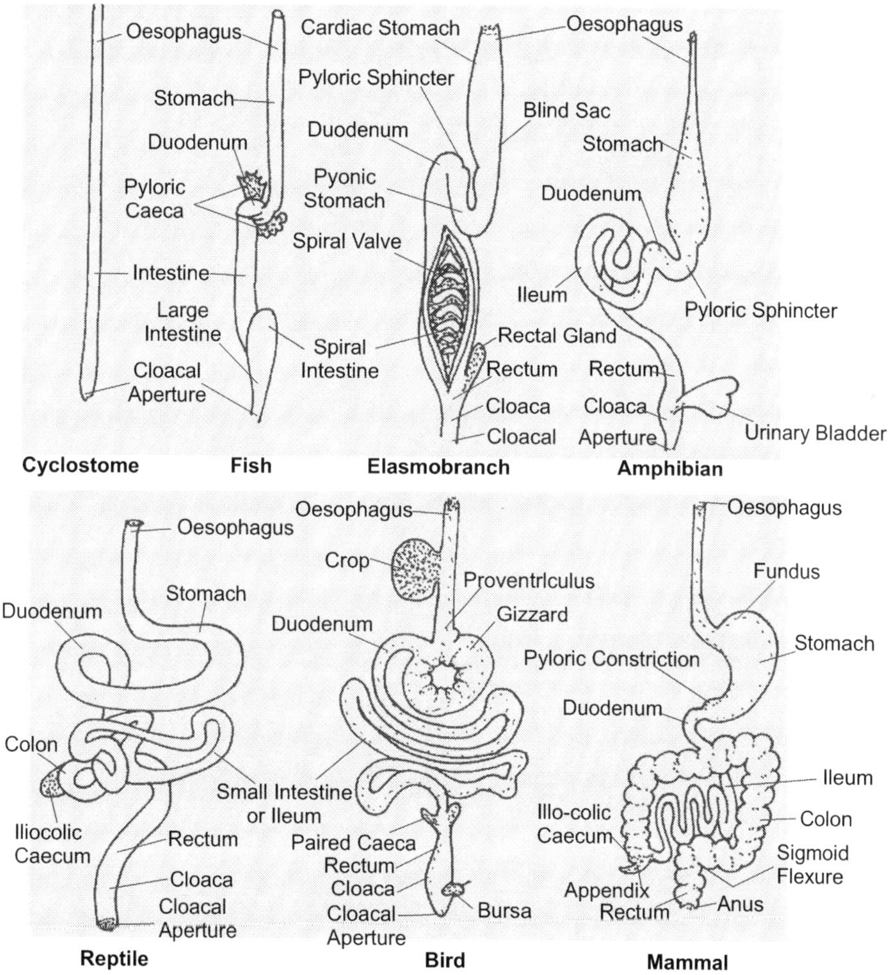

Fig. 2 Digestive tracts of different vertebrates

- In amphibians, the stomach is long and lacks fundus.
- The stomach of *Uromastix* is U-shaped, having thick muscular walls
- In birds, the stomach is short and has no storage capacity. The stomach has a small glandular region called proventriculus and a distal muscular region called gizzard. The gizzard helps in grinding the food.
- Mammals have a well-developed stomach.
- In monotremes, the stomach is in the form of a sac-like structure and is without any gastric glands.

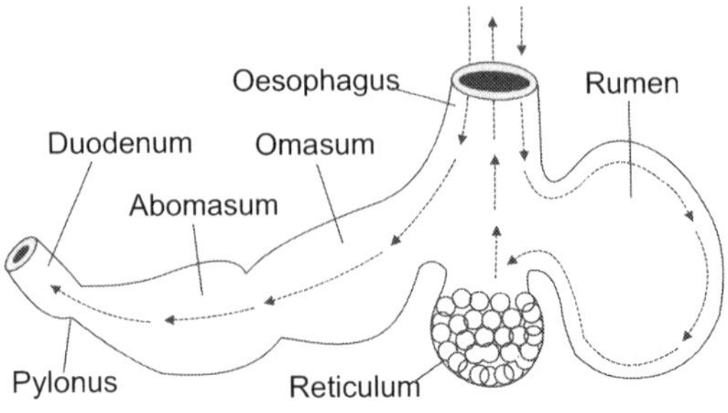

Fig. 3 Stomach of a ruminant

- In ruminants, the stomach has four chambers, viz., the rumen, reticulum, omasum and abomasum. The rumen is a large chamber while the reticulum is a small accessory chamber having a crisscross ridge on its inner surface. In the rumen and reticulum, food is reduced to pulp. In the omasum, food is again triturated. Abomasum is the true stomach in ruminants where enzymatic activity operates.
- In camels, the omasum is absent.

V. Intestine

- The intestine is located between the stomach and the cloaca or anus.
- It is an important site for digestion and absorption.
- The intestines of vertebrates are differentiated to varying degrees into small and large intestines.
- The intestine is relatively straight and short in cartilaginous and primitive bony fishes (lung fish and sturgeon). However, the intestines of cartilaginous fishes have a spiral valve.
- In amphibians, the intestine is differentiated into a coiled small intestine and a short straight large intestine.
- In reptiles and birds, the intestine is differentiated into a coiled small intestine and a relatively short large intestine that empties into cloaca.
- In mammals, the small intestine is long and coiled and differentiated into duodenum, jejunum and ileum. The large intestine is often long (but not as long as the small intestine). A caecum is often present at the junction of the small and large intestine in herbivorous forms.
- The cloaca opens to the exterior by cloacal aperture.
- In many bony fishes and mammals (except monotremes), the urinogenital and anal openings are separate.

Glands Associated with the Digestive System

I. Liver

- The liver is the largest gland of the body and is found in all vertebrates.
- Liver cells manufacture bile, which is stored in the gall bladder.
- The gall bladder is absent in lampreys, many birds and some mammals (perissodactyla, hyracoidea, whales and some rodents).
- In *Myxine,* the liver is bilobed.

- In elasmobranchs, the liver is bilobed while it is two or three lobed in teleosts, amphibians, reptiles and birds.
- In mammals, the liver is multilobed.
- Bile produced by the liver aids in digestion by emulsifying fats.

II. Pancreas

- Pancreas is the second largest digestive glands of vertebrates.
- In lampreys and many bony fishes, the pancreas is diffused while in higher tetrapods, it is a well defined and compact structure.
- The pancreas is both an exocrine as well as endocrine gland.
- Its exocrine part secretes pancreatic juice. Pancreatic juice is alkaline, which neutralises acids coming from the stomach as well as contains digestive enzymes (trypsin, chymotrypsin, amylase and lipase) that help in the digestion of proteins, carbohydrates and fats.
- The endocrine part (islets of Langerhans) secretes hormones (insulin and glucagon) that regulate carbohydrate metabolism.

RESPIRATORY SYSTEM

- Respiration is the process of drawing in of oxygen and giving out carbon dioxide.
- The exchange of oxygen and carbon dioxide between body cells and environment is known as external respiration, while the use of oxygen by the cells for the production of ATP (carbon dioxide is also produced in this process) is known as the internal respiration.
- The body structures that are needed for the exchange of oxygen and carbon dioxide are known as respiratory organs.

Adaptations for External Respiration

- Primary respiratory organs in adult vertebrates are external and internal gills, swim bladders or lungs, skin and buccopharyngeal mucosa.
- Filamentous outgrowths of the posterior trunk and thigh (African hairy frog), lining of the cloaca and lining of oesophagus also act as the sites of gaseous exchange.

Respiratory Organs

I. Cutaneous Respiration

- Cutaneous respiration is the respiration through skin.
- It can occur in water, air or both.
- Cutaneous respiration plays an important role in gaseous exchange in amphibia (especially in family plethodontidae).
- During winter, cutaneous respiration is more important for gaseous exchange.

II. Gills

(a) Agnathans

- There are 6 to 15 pairs of gill pouches.
- Gill pouches are connected to pharynx by afferent branchial ducts and to the exterior by efferent branchial ducts.

(b) Cartilaginous Fishes

- Cartilaginous fishes have five naked gill slits.
- The anterior and posterior walls of the first four gill chambers have a gill surface (demibranch).
- The posterior wall of last (fifth) chamber lacks demibranch.
- Interbranchial septum is present between two demibranchs of the gill arch.
- Gill rakers protrude from gill cartilage and guard the entrance to the gill chamber.
- Two demibranchs, septum, associated cartilage, blood vessels, muscles and nerve form the holobranch.
- Gill filaments are richly supplied with blood capillaries.

(c) Bony Fishes

- Generally, there are five pairs of gill pouches.
- In bony fishes, a bony flap called operculum is present which projects backward over gill chambers.
- Interbranchial septa are very short or absent.
- Spiracles become either closed or lost in adults.

1. Larval Gills

(a) External Gills

- External gills are formed by outgrowth from the external surface of one or more gill arches.
- They are ectodermal in origin and are generally temporary structures found only in larval stages.

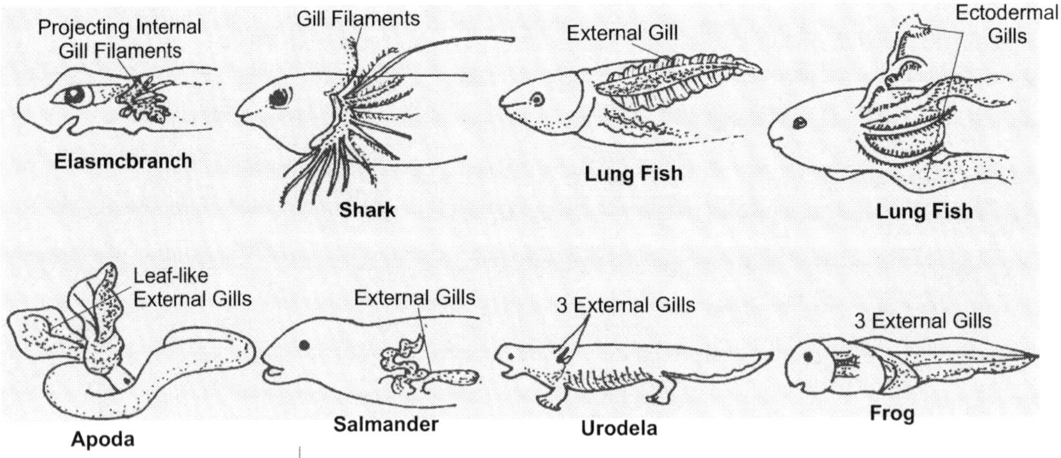

Fig. 4 External gills in larval form (fishes and amphibians)

- They are found in larvae of lampreys, lung fishes and amphibians.
- In amphibians, larval gills are absorbed at the time of metamorphosis but in urodeles, both external gills and gill slits persist in adult.
- The gills may be pectinate, bipinnate, dendritic or leaf-like.
- In the early stages of development of elasmobranchs, filamentous extensions of internal gills project throughout gill slits.

(b) Internal Gills

- In late anuran tadpoles, internal gills are present hidden behind larval operculum.

II. Swim Bladder and Origin of Lungs

- The majority of vertebrates develop an outpocketing of pharynx or oesophagus that becomes one or pair of sacs (swim bladder or lungs) filled with gas derived directly or indirectly from the atmosphere.
- Similarities between swim bladder and lungs indicate that they are the same organs.
- Cyclostomes, cartilaginous fish and few teleosts are vertebrates without swim bladder or lungs.

II. Swim Bladder

Most actinopterygian fishes have a sac-like structure called swim bladder lying between alimentary canal and vertebral column.

- It may be paired or unpaired and varies in length and form in different fishes.
- It may have a pneumatic duct during development that generally connects to the oesophagus. The duct remain open (physostomus) in bowfins and lung fishes, but closes off (physoclistous) in most teleosts.
- It is primarily hydrostatic in function (regulating the fish's specific gravity).
- It gains gas by red body or red gland. The gas is reabsorbed via the oval body on the posterior part of bladder.
- It plays an important role in the following:
  - (a) **Respiration –** The swim bladder of lung fishes has a number of septa (to increase the surface area). Oxygen and carbon dioxide are exchanged between the bladder and blood.
  - (b) **Hearing –** In some freshwater teleosts (catfish, goldfish, carps) hearing is done by way of pressure waves transmitted via the swim bladder and the Weberian ossicles.
  - (c) **Sound production –** Muscles attached to the swim bladder contract to move between sub-chambers of the swim bladder and the resulting vibration creates sounds in fishes.

Lungs and Associated Structures

1. Larynx

- The beginning of larynx has occurred in amphibia.
- The larynx is simplest in *Necturus*.
- In tetrapods besides mammals, the larynx is supported by two pairs of cartilages called arytenoids and cricoids.
- In reptiles, the larynx is more developed than the amphibians.

- In birds, it is small and rudimentary.
- The larynx is highly developed in mammals. Besides paired arytenoids and single cricoids, a single thyroid cartilage is added on the ventral surface of the larynx.
- Amphibians, some lizards and mammals also have vocal cords stretched across the laryngotracheal chamber.

2. Trachea and Syrinx

- Trachea is the part of air duct between the larynx and the lungs.
- Trachea is usually as long as a vertebrate neck (except in a few birds such as cranes).
- In anura, the trachea is extremely short or absent. In *Siren*, *Amphiuma* and apoda a delicate trachea is present.
- Trachea is supported by cartilaginous rings (c rings).
- The lower end of the trachea bifurcates forming two bronchi. Each bronchus is lined with cilia and each bronchus enters the lung.
- At the point where the bronchi meet the trachea, there is a slight enlargement called syrinx (voice box of birds).

3. Lungs

(a) Amphibian Lungs

- Amphibian lungs are two simple sacs, elongated in urodeles and bulbous in anurans.
- The left lung is usually longer except in caecilians, in which it is rudimentary.
- The internal lining of lungs may be smooth (*Necturus*) or may have simple sacculations or pockets (*Amphiuma*).
- The lungs are lined with mucous epithelium and are highly vascular.
- Air exchange occurs via positive pressure ventilation.

(b) Reptilian Lungs

- Lungs of reptiles are abdominal and more complicated.
- In *Sphenodon* and snakes, the lungs remain as simple sacs.
- In lizards, crocodilians and turtles, the lining of lungs is septate with many chambers and sub-chambers.
- The left lung in limbless lizards and in snakes is rudimentary or absent (except in certain forms such as black snakes).
- Air exchange occurs via positive pressure ventilation.

(c) Avian Lungs

- Avian lungs are highly modified due to their aerial mode of life.
- Avian lungs are small, spongy and inelastic and are placed outside the coelom in pleural cavities.
- Avian lungs lack alveoli.
- Air sacs (diverticula of lungs) are extensively distributed throughout most of the body.
- Air flow through lungs is unidirectional.

(d) Mammalian Lungs

- Mammals have a pair of lungs enclosed in a thoracic cavity.
- The lungs are spongy and very elastic.

- The lungs are multichambered and usually divided into lobes.
- Usually the right lobe has more lobes. In humans, the right lung has three lobes and four in rabbits.
- In elephants, whales, sirenians, hyrax and many perissodactyles, the lungs are simple and without lobes.
- In monotremes and rats, only the right lung is lobulated.
- Air flow is bidirectional. Trachea ⟷ Primary bronchus ⟷ Secondary bronchi ⟷ Tertiary bronchi ⟷ Bronchioles ⟷ Alveoli.
- Air exchange occurs via negative pressure ventilation with pressure changing due to contraction and relaxation of diaphragm and intercostal muscles.

HEART

- The heart is a modified blood vessel whose walls are muscular and is found in all animals having a circulatory system.
- The heart contracts periodically to pump blood to different parts of the body.
- The heart is located in a pericardial sac, the walls of which form the pericardium.
- The heart is composed of cardiac muscle (involuntary muscle).
- The heart develops from mesoderm as a straight and cylindrical tube. But during the course of development, the anterior part of the cardiac tube becomes twisted in an S-shaped curve.
- In primitive vertebrates, the heart consists of the following chambers:

  1. **Sinus Venosus**
     - It is a sac-like thin-walled structure having less muscular tissue.
     - It receives venous blood through hepatic and cardinal veins.
  2. **Atrium (Auricle)**
     - It is a thin-walled structure that is distensible.
  3. **Ventricle**
     - It is thick-walled structure that constitutes the main contractile part of the heart.
  4. **Conus Arteriosus**
     - It is the anterior-most thick-walled and tube-like chamber.
     - It contains several sets of valves.
     - In higher vertebrate groups, sinus venosus and conus arteriosus disappear, while auricle and ventricle have a tendency to subdivide.

Heart in Different Groups of Vertebrates

I. Two-chambered Heart

- It is characteristic of most fishes (except lung fish).
- There are two major chambers called atrium and ventricle and two accessory chambers called sinus venosus and conus arteriosus.

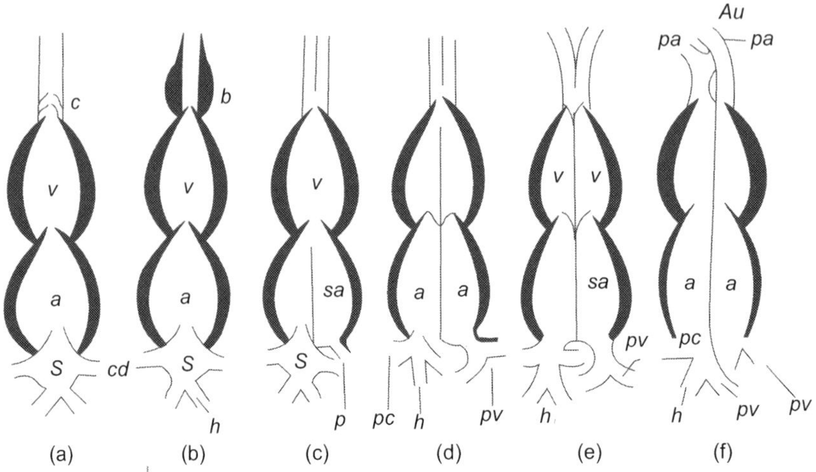

Fig. 5 Different stages in the differentiation of the part of the heart

- Conus arteriosus is much reduced in teleosts, while in some chondrostei (*Polypterus*) and holostei (*Lepidosteus*) conus arteriosus is fairly long with several valves.
- Blood enters into the sinus venosus and then into the auricle.
- From the auricle, blood passes to the ventricle where it is pumped to the conus arteriosus.
- The valve present in the conus arteriosus prevents backward flow of blood.
- In fishes, there is only a single circuit of blood circulation.
  (a) Elasmobranchs
  (b) Teleosts
  (c) Amphibians
  (d) Lower reptiles
  (e) Alligators
  (f) Birds and mammals

  a—atrium; ao—aorta; b—bulbous; arteriosus; c—conus; cd—Cuvierian duct; h—hepatic veins; pa—pulmonary artery; pc—pre and postcaval veins; pv—pulmonary veins; S—sinus venosus; sa—interauricular septum; v—ventricles

II. Three-chambered Heart

- Three-chambered heart is characteristic of lung fishes and amphibians.
- An interarterial septum divides the auricle into two parts. As a result, oxygenated blood remains separate from deoxygenated blood returning from the body.
- Hence blood from the pulmonary veins empties into the left auricle, while the sinus venosus returns blood to the right auricle.
- In slethodentid salamanders, the atria remain undivided.
- There is formation of a partial interventricular septum in lung fishes and or a trebeculae in urodeles within the ventricle. Their function is to separate oxygenated blood from deoxygenated blood.
- Spiral valves in the conus arteriosus likewise direct oxygenated blood into aortic arches supplying the body, while deoxygenated blood is directed into arches supplying gills or lungs.

- In most reptiles, the ventricle is incompletely divided by an incomplete interventricular septum preventing mixing of oxygenated and deoxygenated blood. Thus, the reptilian heart represents a transitional heart having a midway condition between the two-chambered heart of fishes (single circulation) and the four-chambered heart of birds and mammals (double circulation).
- But in crocodilians, the interventricular septum is complete and thus the heart is four-chambered.
- Though in crocodilians the heart is four-chambered (two auricles and two ventricles), there is no complete separation of oxygenated and deoxygenated blood in the ventricle due to presence of foramen of Panizza.
- In urodeles, the conus arteriosus becomes the bulbous arteriosus which maintains a steady flow at blood through the gills.
- In reptiles, the conus and ventral aorta of embryo become spilt in the adult into three distinct trunks called pulmonary, right systemic and left systemic.

III. Four-chambered Heart

- A four-chambered (two auricles and two ventricles) heart is found in birds and mammals.
- The left auricle receives oxygenated blood from lungs, while the right auricle receives deoxygenated blood returning from the body.
- Sinus venosus is absent and it is marked by sinu atrial node.
- In the mammalian embryo, atria communicate through foramen ovale.
- The ventricles of mammals are lined with ridges and columns of muscle, the columnae carneae.
- Both auricles open into their respective ventricles by separate apertures which are guarded by valves.
- The right auriculoventricular valve is tricuspid, while left auriculoventricular valve is bicuspid.
- In birds, only the right systemic arch persists while in mammals, only the left systemic arch is present.
- The right ventricle leads into the pulmonary trunk, while the left ventricle leads into the carotico-systemic trunk.
- From the left auricle blood passes to the left ventricle which pumps it to the body through systemic circulation. Likewise, blood from the right auricle comes to the right ventricle, which pumps it to lungs for re-oxygenation.
- Complete separation of oxygenated and deoxygenated blood has first occurred in birds.
- Thus, in birds and mammals there is a double circulation in which there is no mixing of oxygenated and deoxygenated blood. Such a heart is called pulmonary heart.
- Semilunar valves prevent backward flow of blood from the aorta and pulmonary arteries.

Aortic Arches

- Aortic arches are paired blood vessels that emerge from the ventricle of the heart.
- Aortic arches carry blood from the dorsal blood vessel to the ventral blood vessel.
- The number of aortic arches is different in different adult vertebrates but they are built on the same fundamental plan.
- In ancestral vertebrates, the number of aortic arches was high and variable. Ostracoderms had 10 pairs of aortic arches.
- In all lower vertebrates, aortic arches are bilaterally symmetrical but in amniotes these are asym metrical.

- *Branchiostoma (Amphioxus)* has 60 pairs of aortic arches. Lampreys have 8 pairs of aortic arches and in hagfishes the number of aortic arches 15 pairs.
- In jawed vertebrates, there are 6 pairs of aortic arches found in the embryo of all vertebrates. The first aortic arch is called the mandibular aortic arch, the second is called the hyoid arch and the remaining are called the third, fourth, fifth and sixth aortic arches.

I. Trends of Evolution

- There is a progressive decrease in the number of aortic arches in vertebrate groups during evolution.
- The difference in the number of aortic arches is due to the complexity of blood circulation in the mode of living from aquatic to terrestrial respiration.
- Due to changes in the mode of respiration, certain arches have been completely lost and others have undergone modification.

II. Aortic Arches in Different Groups of Vertebrates

1. Fishes

- In fishes, each aortic arch has been divided into an afferent branchial artery and an efferent branchial artery.
- The afferent artery carries deoxygenated blood from ventral aorta to the gills, while the efferent artery carries oxygenated blood from gills to the radices aorta.
- The afferent and efferent vessels are connected by a capillary loop in the gills.

(a) Elasmobranchs

- In elasmobranchs, there are five pairs of aortic arches.
- The first pair of aortic arches (mandibular arch) has disappeared.

(b) Teleosts

- In most bony fishes, the first and second aortic arches have disappeared, so only four pairs of aortic arches (third, fourth, fifth and sixth) aortic arches persist.
- In teleosts, the first and second gill slits are nonfunctional.

(c) Lung Fishes

- In lung fishes, there are four pairs of aortic arches (third, fourth, fifth and sixth).
- In lung fishes, a pulmonary artery breaks off from the sixth aortic arch carrying blood to the lung or air bladder.

2. Tetrapoda

In tetrapods, aortic arches have undergone great modification due to the following:

(a) Differentiation of neck
(b) Replacement of branchial respiration by pulmonary respiration
(c) Posterior shifting of heart in the thoracic region
(d) Tendency of increasing separation of arterial and venous blood in the heart

3. Amphibians

In amphibians, aortic arches have undergone modification due to the introduction of pulmonary respiration.

(a) Urodels

- Four pairs of aortic arches are usually functional (third to sixth).
- The fifth pair is absent in *Siren* and *Amphiuma*.
- The third pair forms the carotid artery and the fourth pair forms the systemic arches.
- The sixth pair forms the pulmonary arteries, which supplies blood to the skin and lungs.
- The radix or lateral aorta between the third and fourth arches may persist as a vascularised connection called ductus caroticus.
- The ductus arteriosus persists.

(b) Anurans

- In anura, only three pairs of aortic arches (third, fourth and sixth) persist in adults.
- The third aortic arch along with a part of ventral aorta becomes the carotid arch.

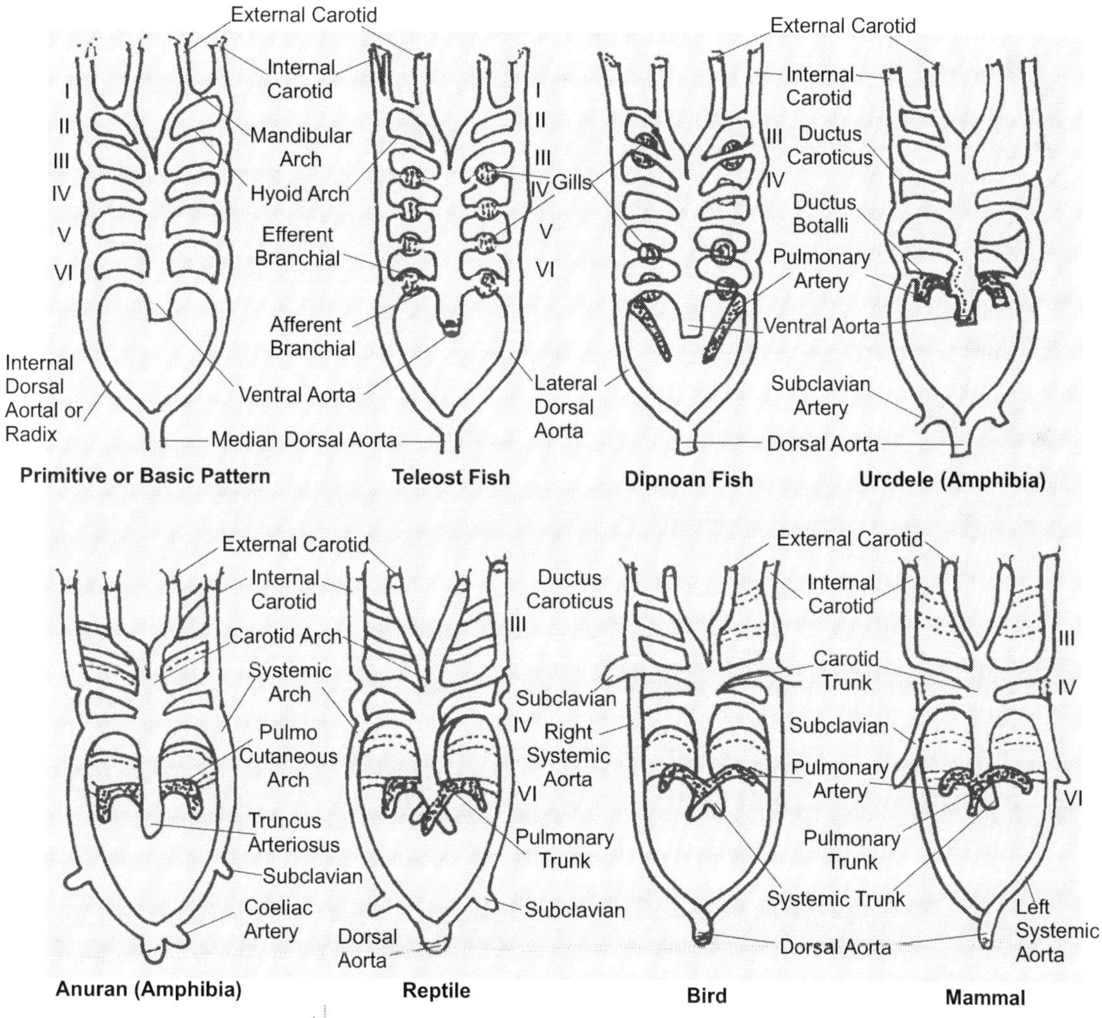

Fig. 6 Modification of aortic arches in defferent vertebrates

- The fourth aortic arch with its lateral dorsal aorta forms the systemic arch.
- Ductus caroticus and ductus arteriosus (ductus botalli) disappear.

(c) Gymnophiona

- In gymnophiona, aortic arches are similar to that of urodels.
- Ductus caroticus and ductus arteriosus also persist in gymnophiona.

4. Reptiles

- In reptiles, the aortic arches become fixed in pattern for a lung system due to the complete disappearance of gill respiration.
- Only three pairs of aortic arches (third, fourth and sixth) persist in adults.
- Ductus caroticus and ductus arteriosus are absent, but in certain snakes and lizards, the ductus caroticus persist and in some turtles and *Sphenodon,* ductus arteriosus persists.

5. Birds

- Only three pairs of aortic arches (third, fourth and sixth) persist.
- The third aortic arch represents the carotid artery while the fourth aortic arch becomes the systemic artery.
- The sixth arch arises from a single pulmonary trunk carrying deoxygenated blood from the right ventricle to the lung.
- Subclavian artery is present on the left side.
- Only the right systemic arch persists.
- Ductus caroticus and ductus arteriosus are absent.

6. Mammals

- In mammals too, three pairs of aortic arches (third, fourth and sixth) persist.
- The subclavian artery is present on the right side.
- Only the left systemic arch persists.
- Ductus caroticus and ductus arteriosus are absent

Thus, there is gradual reduction in the number of aortic arches from lower to higher vertebrate groups. The embryonic condition of aortic arches supports the recapitulation theory of Haeckel.

BRAIN

- The brain is the specialised anterior part of the central nervous system.
- In all vertebrates, the brain is built on the same structural plan but differs in different vertebrates according to their habitats and behaviour.
- It becomes concentrated in the anterior part of the body.
- It develops as an outgrowth from the ectoderm called neural tube.
- The anterior thickened end of the neural tube is called encephalon.
- When the neural tube is fully formed, it becomes divided into three parts by two constrictions called cerebral vesicles.

- These cerebral vesicles become differentiated into forebrain (Prosencephalon), midbrain (mesencephalon) and hindbrain (rhombencephalon).
- The adult brain contains a large number of cavities called ventricles, which are in continuation of central canal of spinal cord.
- The ventricles contain cerebrospinal fluid. The various cavities of brain are shown in Fig. below.

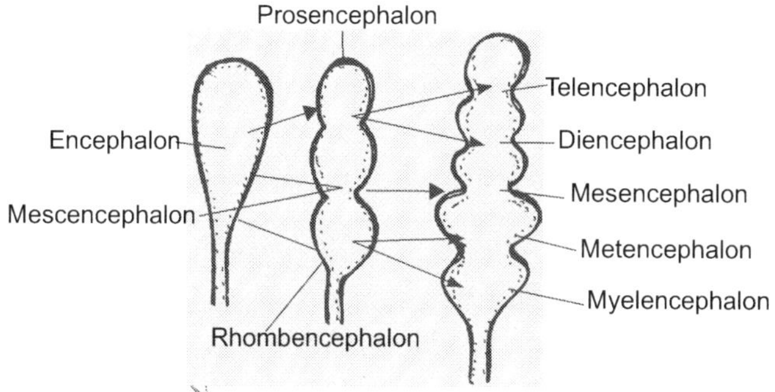

Fig. 7 Origin and development of brain

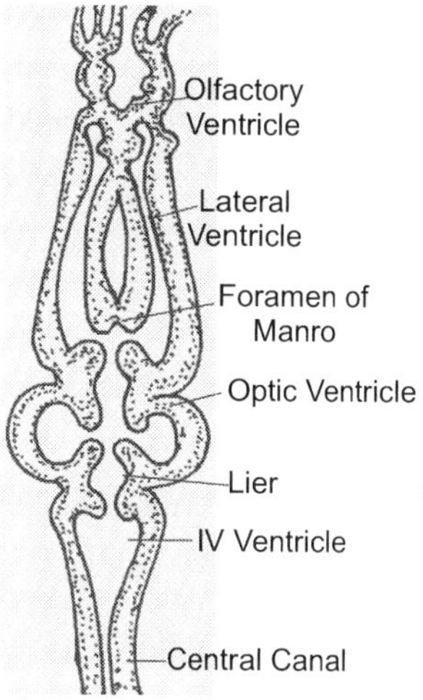

Fig. 8 Different cavities (ventricles) of brain of frog

Comparative Account of Brain in Different Classes of Vertebrates

I. Cyclostomes

- In cyclostomes, the brain is very primitive and subdivisions are not well marked.
- Olfactory lobes are prominent, but cerebral hemispheres are small.
- Pineal organ and parietal body are well developed in *Petromyzon* but in *Bdellostoma,* they are vestigial, while in *Myxine,* they are lacking.
- The two optic lobes are partially differentiated.
- Infundibulum is poorly developed and contains a pituitary body.
- The cerebellum is in the form of a cross-band-like structure.
- The medulla oblongata is well developed.
- Ventricular cavities are well developed in *Petromyzon* but are greatly reduced in *Myxine.*

II. Fishes

1. Elasmobranchs

- The brain is somewhat compact, while in some cases it is long and slender.

- Olfactory lobes are relatively large and are attached to the cerebrum by short and stout olfactory tracts.
- Cerebral hemispheres are large.
- Diencephalon bears pineal apparatus.
- Infundibulum is well marked.
- Parietal organ is absent.
- Optic lobes and pallium are relatively of moderate size.
- The cerebellum is well developed due to active mode of swimming.
- The medulla oblongata is either long (in sharks) or short (in skates).
- In most elasmobranchs, the ventricles are well developed.

2. Teleosts

- Brain is more specialised than the elasmobranchs.
- The prosencephalon is small and remains undivided.
- Olfactory lobes, cerebral hemispheres and diencephalon are small.
- Pineal organ is present.
- Parietal organ is absent in adults.
- Infundibulum is large.
- The optic lobes and cerebellum are well developed.
- The anterior part of the cerebellum grows forwards beneath the midbrain, forming valvula cerebelli.
- The medulla oblongata is well developed.

3. Dipnoi

- Prosencephalon is divided into two, thus comprising two cerebral hemispheres.
- Anteriorly, each cerebral hemisphere passes into an olfactory lobe, which is drawn into an olfactory peduncle that ends distally into an olfactory bulb.
- The cerebral hemispheres are relatively smaller in *Neoceratodus*.
- The diencephalon is relatively small.
- Pineal body is well developed.
- Two optic lobes are indistinctly separate from each other.
- The cerebellum is poorly developed, but the medulla oblongata is quite prominent.
- A valvula cerebelli is present in the cerebellum.

III. Amphibians

- Olfactory lobes merge into the cerebral hemispheres.
- The cerebral hemispheres are distinct and elongated and are separated by a fissure.
- The floor of the cerebral hemispheres is thickened to form corpora striata, which is poorly developed in anura and urodela but is prominent in gymnophiona.
- The cavity of each hemisphere is reduced.
- Diencephalon is broad in anura and narrow in urodela and gymnophiona.
- In adults, the pineal body undergoes regression.
- Parietal organ is lacking.
- Infundibulum is well developed.
- The optic lobes are well developed.

- The cerebellum is poorly developed and is in the form of a transverse band.
- The medulla oblongata is small.
- The ventricles are fairly spacious.

IV. Reptiles

- The reptilian brain shows considerable advancement over the amphibian brain.
- The cerebral hemispheres are comparatively bigger than the amphibians and are elongated and somewhat oval in shape.
- The olfactory lobes are larger than those of amphibians and are fully merged in the cerebral hemispheres.
- The floor of each cerebral hemisphere thickens to form the corpus striatum.
- The two corpora striata are connected together by an anterior commissure.
- The two hippocampal lobes are connected together by hippocampal commissure.
- Due to great development of the cerebral hemispheres, the diencephalon is hardly observable from the dorsal side.
- Crocodiles have the first true cerebral cortex (neo pallium).
- Pineal body and parietal organ are present.
- Parietal organ is well developed in *Sphenodon* and lacertilian, but is rudimentary in other groups.
- The third ventricle is reduced.
- Optic lobes are large and paired. In snakes, each optic lobe is divided into anterior and posterior parts to become corpora quadrigemina.
- The cerebellum is somewhat larger than amphibians and is pear-shaped. In crocodiles, cerebellum is well developed having a median and two lateral parts.
- The medulla oblongata is relatively triangular.

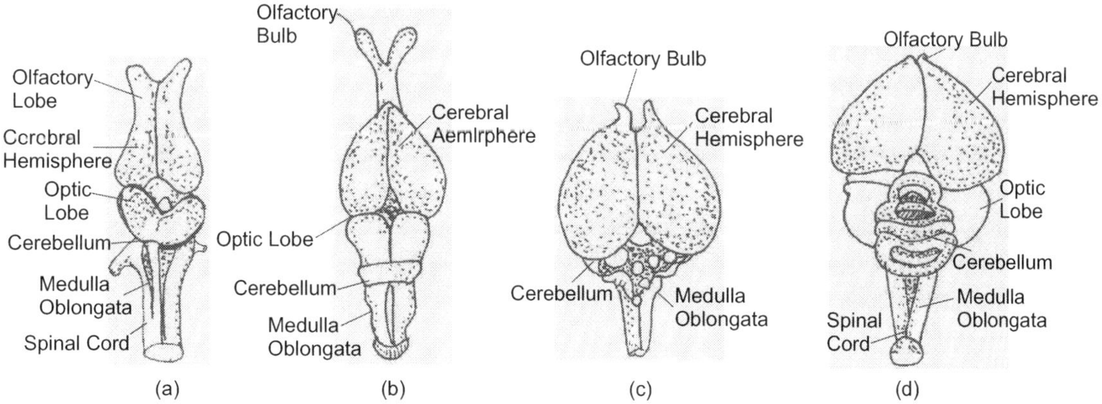

Fig. 9 Brain in different groups of vertebrate
(a) Amphibian, (b) Reptile, (c) Mammal and (d) Brid.

V. Aves

- The brain is relatively short and broad.
- The olfactory lobes are poorly developed due to a poor sense of smell.
- The cerebral hemispheres are large and broad.
- Pallium is thin, but the corpus striatum is highly enlarged due to which the lateral ventricle becomes small and vertical.
- In birds, the corpus striatum is the main integrative centre of the brain.
- The diencephalon is completely concealed by the cerebrum and cerebellum.
- The pineal body is glandular and there is no parietal organ.
- The diocoel (third ventricle) is reduced due to well-developed optic thalami.
- The optic lobes are well developed and somewhat laterally displaced.
- The cerebellum is well developed having several superficial flocculi. It is used to coordinate movement and balance.
- The medulla oblongata is short and is fixed ventrally.

VI. Mammals

- The brain is better developed than any other classes of vertebrates.
- The olfactory lobes are relatively small.
- In prototheria, the cerebral hemispheres are relatively small and smooth. However, in metatheria they are large, but smooth.
- In eutheria, the cerebral hemispheres are greatly enlarged and become divided into lobes having a thick cerebral cortex of grey matter.
- The two cerebral hemispheres are separated by a median fissure. Each hemisphere is divided into an anterior frontal lobe and a postero-lateral temporal lobe by a Sylvian fissure.
- The two cerebral hemispheres are joined together by transverse fibres called corpus callosum (a characteristic feature of the placental brain).
- The diencephalon and midbrain are completely covered by the cerebral hemispheres.
- A stalked pineal body is present.
- The third ventricle is laterally compressed.
- There are four optic lobes.
- The optocoel is absent.
- The cerebellum is large having extensive folding.
- The vermis and flocculi are highly developed. On the ventral surface of the cerebellum, there is a thick band of transverse nerve fibres that form pons varoli. It connects the two halves of the cerebellum.
- The medulla oblongata is comparatively a smaller structure.

JAW SUSPENSION

- The attachment of upper and lower jaws with the brain case is known as jaw suspension.
- Jaw suspension is vital to the feeding system.

- Jaw suspension is of the following types:

I. Autodiastylic

- Autodiastylic jaw suspension was found in early bony fishes.
- The jaws are attached to the cranium by anterior and posterior ligaments.
- The hyoid arch is complete and does not support the jaws.
- The gill cleft in front of the hyoid arch bears a complete gill.

II. Amphistylic

- It is characteristic of primitive sharks such as *Heptanchus*, *Hexanchus*.
- The upper jaw is attached to the cranium by ligaments at orbital and otic processes of the platoquadrate.
- The hyoid arch is attached to chondrocranium as well as lower jaw and is involved in the suspension of both jaws.
- Amphistylic is a double jaw suspension.

III. Hylostylic

- Hylostylic jaw suspension is found in most cartilaginous and all bony fishes.
- The upper jaw lacks direct connection with the brain case (except anterior).
- Both jaws remain attached to the brain case by the hyomandibular of the hyoid arch.

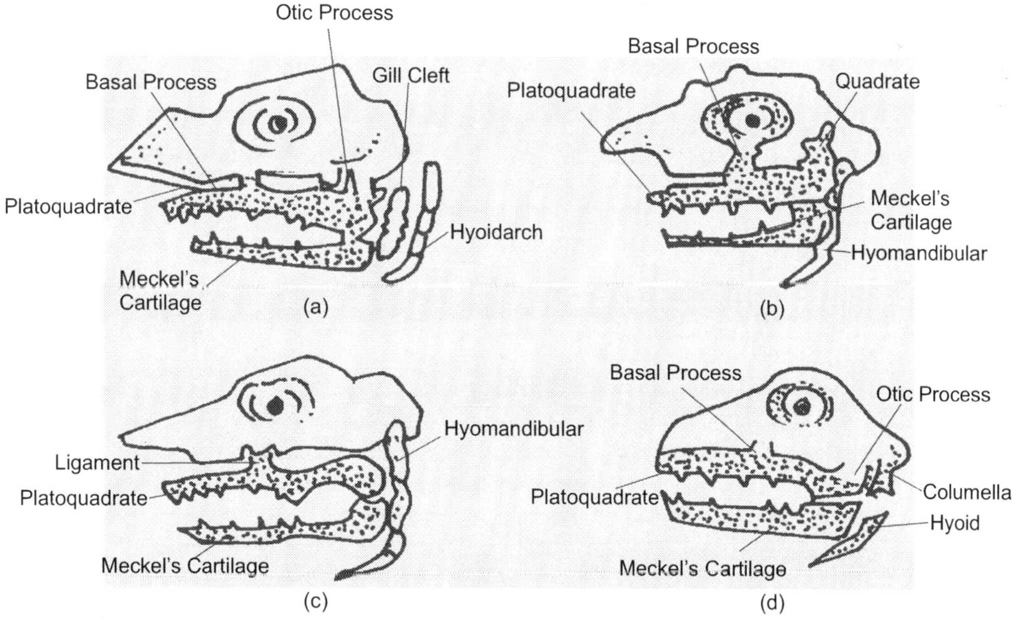

Fig. 10 Types of jaw suspension (a) Autodiastylic, (b) Amphistylic, (c) Hyostylic and (d) Autostylic

- Such type of jaw suspension permits a larger movement of jaws and assists in swallowing of larger prey.

IV. Autostylic

- Autostylic jaw suspension is found in lung fishes and tetrapods.
- The hyomandibular is not involved in jaw suspension.
- In tetrapods, the hyomandibular becomes the columella of the middle ear.
- Autostylic jaw suspension is of the following types:

1. Holostylic

- Holostylic suspension is found in holocephali (chimaeras).
- The hyoid arch is not attached to the skull.
- The upper jaw is fused with the brain case and the lower jaw is suspended from the upper jaw.

2. Monimostylic

- Monimostylic jaw suspension is found in many tetrapods (except mammals).
- The quadrate becomes firmly attached to the skull.
- The hyomandibular becomes modified into the columella.
- The articular of the lower jaw articulates with the quadrate of the upper jaw.

3. Streptostylic

- Streptostylic jaw suspension is found in lizards, snakes and birds.
- The quadrate is movable.
- Articular of the lower jaw articulates with the quadrate bone of the upper jaw.

IV. Craniostylic

- Craniostylic jaw suspension is the characteristic of mammals.
- In mammals, the articular and quadrate have been transformed into malleus and incus respectively, while the hyomandibular becomes modified to form stapes.
- The mammal jaws are completely dermal.
- The maxilla incorporated into the cranium.
- Jaws are suspended directly from squamosal bone.

URINOGENITAL SYSTEM

- Urinogenital system includes both the excretory and the reproductive system.
- Excretory system comprises kidneys and their ducts and is concerned with the removal of nitrogenous waste products.
- The reproductive system includes gonads and their ducts and assists in the continuance of species.
- Kidneys and gonads have unrelated functions, but in most vertebrates they are intimately associated as in most vertebrates, kidney tubules are used for the transport of gametes (particularly in males).
- Therefore, due to their close relationship, these two systems are studied under a common name called the urinogenital system.

Excretory System

I. Kidneys

- Kidneys are compact paired structures and are the chief excretory organs of vertebrates.
- They are lie dorsal to the coelom.
- Each kidney is made up of a large number of coiled tubules called uriniferous tubules or nephrons.
- Uriniferous tubules develop from a specialised part of mesoderm called mesomere in a linear series.
- Each uriniferous tubule consists of three parts, viz., peritoneal funnel, Malpighian body and tubule.

(a) Peritoneal Funnel

- At the free end of the uriniferous tubule there is found a ciliated funnel-like ciliated structure called the peritoneal funnel.
- It opens into the coelom by a wide structure called nephrostome.

(b) Malpighian Body

- Malpighian body comprises a cup-like hollow, double-walled structure called Bowman's capsule and a bunch of arteries called glomerulus.
- Afferent and efferent arteries together form the glomerulus.
- The afferent artery brings blood into the glomerulus while the efferent artery carries blood away from it.
- Glomeruli are lacking (aglomerular kidney) in some fishes.

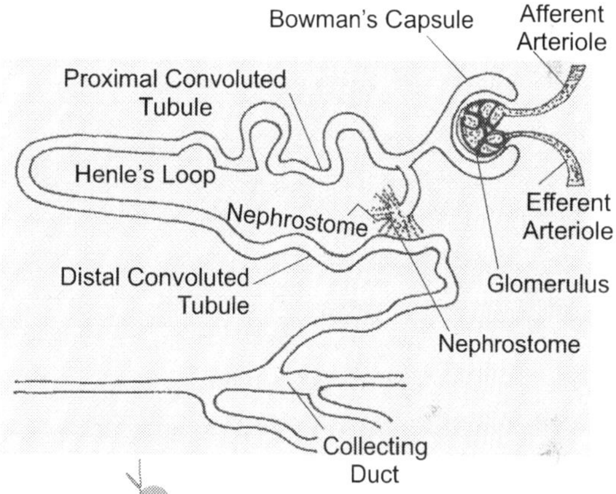

Fig. 11 Basic structure of excretory tubule

(c) Tubule

- It is the coiled tubular part having three parts, viz., the proximal convoluted tubule, Henle's loop and the distal convoluted tubule.
- The distal convoluted tubule opens into the collecting tubule, which leads into the urinary duct.
- The above parts of a typical uriniferous tubule may be lacking in some vertebrate groups.
- In amniotes (reptiles, birds and mammals) nephrostomes are absent.
- In some fishes, Malpighian bodies are lacking.

1. Succession of Kidney

(a) Archinephrous

- It is the earliest vertebrate kidney.
- It extends to the entire length of the body cavity.
- Each kidney had tubules arranged segment-wise that open separately into the body cavity through a nephrostome.
- Glomeruli are external and drain the coelomic fluid.
- The tubules of the kidney open into a duct called archinephric duct that joins the cloaca.
- Such kidney is found in larval *Myxine* and some apodan amphibians.

(b) Pronephros

- It is the first embryonic kidney in all vertebrates.
- They are anteriorly located and are also known as the head kidney.
- Generally, it consists of 1 to 12 uriniferous tubules (e.g., 3 in frogs, 7 in human embryos and 12 in chick embryos).
- The duct that drains the pronephros is called the pronephric duct.
- Generally, the pronephric tubules do not have individual glomerulus. Several glomeruli unite to form a common large glomus.
- Pronephros is functional only during the embryonic stages. But in adult cyclostomes and in many teleosts, it is retained throughout life.

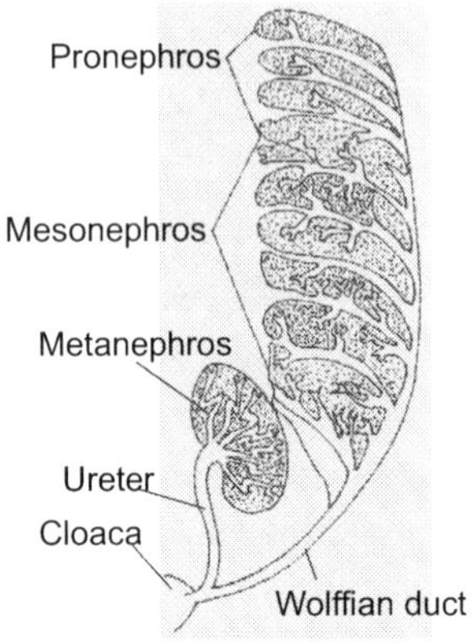

Fig. 12 Diagrammatic representation of different types of kidneys in vertebrates

(c) Mesonephros

- Mesonephros is the functional kidney in fishes and amphibians. But in amniotes (reptiles, birds and mammals) it is functional only in embryos.
- It is formed by corpuscles and tubules that develop caudal to the pronephric region.
- It forms connection with the existing pronephric duct, which is now called the mesonephric duct.
- The segmental arrangement is lost due to branching of tubules in each segment.
- Mesonephric tubules have independent glomeruli.
- In the males of jawed fish, amphibians and some amniotes, the tubules of mesonephros conduct sperm from the testis to the mesonephric duct. This part of mesonephros is called the 'sexual kidney', while the rest is known as the 'uriniferous kidney'.

(d) Metanephros

- Metanephros is the functional kidney of adult amniotes.
- It is formed from the posterior end of the nephrogenic mesoderm.
- During the development of metanephric tubules, all the mesonephric tubules disappear (except in males, which are associated with the testes forming vasa efferentia).
- Each uriniferous tubule develops a Bowman's capsule and glomerulus.
- The segmental arrangement of uriniferous tubules is lost.

II. Urinary Bladder

- Urinary bladder is found in all vertebrates except agnathans, snakes, crocodilians, some lizards and birds (except ostriches).
- In fishes, the urinary bladder is a terminal enlargement of the mesonephric duct.
- In amphibians and all amniotes, urinary bladders arise as evaginations of the ventral wall of the cloaca.
- Ureters usually open dorsally into the cloaca. But in mammals (except monotremes), the ureters directly open into the urinary bladder. The urinary bladder opens towards the outside via a short tube called the urethra.

III. Genital System

- The genital system includes gonads (testis and ovary) and their ducts, called genital ducts.
- Vertebrates are unisexual, having paired testes (in males) and paired ovaries (in females).
- Gonads are located near the kidneys (except mammals, in which testes are extra-abdominal).
- Gonads develop as paired ridges joint medial to mesonephros.
- In some cases, due to fusion or failure of one ridge to differentiate, there is a single testis or ovary (agnathans, female lizard, crocodiles and most female birds).
- Hormones play a key role in the differentiation of early gonads into either testes or ovaries.

IV. Ovaries

- Usually, ovaries are oval structures.
- In some teleosts, ovaries are a hollow structures, while in other teleosts and agnathans, the ovaries are compact and eggs are discharged into coelom.
- In amphibians, ovaries are hollow and eggs are discharged into the coelom.
- In reptiles, birds and monotremes ovaries are solid but develop fluid-filled cavities. Eggs are discharged into coelom.
- In mammals, ovaries are compact without chambers or lacunae.

V. Testes

- Testes are smaller than ovaries because sperms, although numerous, are much smaller than the eggs.
- In mammals, testes are larger than ovaries.
- In many mammals, testes descend permanently into the scrotal sac.
- Each testis is a compact structure and is made up of a large number of coiled tubules called seminiferous tubules.
- Seminiferous tubules are lined with germinal epithelium that divides to form sperms.

VI. Female Genital Ducts

- Female genital ducts typically consist of a pair of gonoducts (oviducts) that extend from the ostia to the cloaca.
- Different segments of ducts perform different functions.
- In cartilaginous fishes, two ostia fuse to form a single ostium.
- In teleosts, ducts are continuous with the cavity of ovary.
- In amphibians, the oviducts are long convoluted and the posterior part may enlarge to form uteri.
- Crocodilians, some lizards and birds have one coiled oviduct lined with glands.
- In placental mammals, embryonic ducts give rise to oviducts, uteri and vagina.
- Oviducts are relatively short, convoluted and lined with cilia. Each oviduct begins at the ostium bordered with fimbriae.
- The condition of uteri varies in different mammals:
  - (a) **Duplex uterus** – In marsupials, there is no fusion of embryonic ducts, so there are two uteri.
  - (b) **Bipartite uterus –** The uteri are partially fused to form two horns and two separate lumens inside it and there is single vagina (e.g., rabbits).

(c) **Bicornuate uterus –** There two uterine horns with single internal cavity and single vagina (e.g., ungulates).

(d) **Simplex uterus –** There are no uterine horns and both uteri fuse completely with a single internal cavity (e.g., primates).

VII. Male Genital Ducts

- In elasmobranchs, the anterior part of the mesonephric duct carries only spermatozoa.
- In teleosts, the mesonephric duct drains the kidney. A separate sperm duct develops.
- In some amphibians, the mesonephric duct carries only sperms and a new accessory urinary duct drains the kidney.
- In reptiles and birds, the urinary and genital ducts are separate (except in the cloacal region).

VIII. Copulatory Organs

- Copulatory organs are lacking in anamniotes.
- A copulatory organ is present in male amniotes.
- In sharks, the pelvic fins become modified into a grooved intromittent organ called claspers.
- All reptiles have hemipenes as copulatory organs (except *Sphenodon*).
- Some birds (ducks, ostriches) have an erectile penis.
- All mammals have an erectile penis (copulatory organ).

COMPARATIVE ANATOMY

Short-Answer Questions

1. What is integument?
   Answer: Integument is the outermost covering of the body, which includes the skin and its derivatives.
2. Name the two parts of the skin.
   Answer: (a) Epidermis which is the outer part and is ectodermal in origin.
   (b) Dermis which is the inner part and is mesodermal in origin.
3. Name the largest organ of the human body.
   Answer: The integument is the largest organ of the human body.
4. What is stratum corneum?
   Answer: Stratum corneum is the outermost layers of dead cells of the epidermis.
5. Name the group of amphibians having dermal scales.
   Answer: Gymnophiona (e.g., *Icthyophis*)
6. Name the epidermal glands found in reptiles.
   Answer: (a) Femoral glands – Lizards
   (b) Inframandibular and cloacal glands – Alligator
7. Name the reptiles that periodically shed their epidermal covering.
   Answer: Lizards and snakes periodically shed their epidermal covering
8. What is the name of the only integumentary gland found in birds?
   Answer: The Uropygial gland
9. Where do we find the thickest epidermis in mammals?
   Answer: In mammals, the epidermis is thickest in the sole.
10. Name three hard epidermal derivatives.
    Answer: Feathers, nails and hair.
11. Name the dermal derivative of fishes.
    Answer: The dermal derivative of fishes is the scales.
12. Name the mammal that has specialised hair as defensive quills.
    Answer: Porcupines
13. Name the amphibians having keratin at the tips of their digits.
    Answer: Only two amphibians are known to have keratin at the tips of their digits are *Xenopus* and spade-foot toads.
14. Rhinoceros horns are not true horns.
    Answer: Because they lack a bony core and are composed of compacted keratin.
15. Name the only class of vertebrates in which sudoriferous and apocrine glands are found.
    Answer: Mammals are the only class of vertebrates in which sudoriferous and apocrine glands exist.

16. Name two structures derived from both the epidermis and dermis.
    Answer: Teeth and scales of fishes
17. Name the only mammal having dermal scales.
    Answer: Armadillo
18. Sebaceous glands are called holocrine.
    Answer: Because their secretion results in the complete disintegration of their cells.
19. Name mammals that lack sweat and sebaceous glands.
    Answer: Elephants lack sweat and sebaceous glands
20. Name the animal in which sweat glands are located:
    (a) Around the lips (b) On the base of the tail
    (c) On the snout (d) On the underside of the paws
    Answer: (a) Around lips – Rabbits (b) On the base of the tail – Deer
    (c) On the snout – Duck-billed platypuses
    (d) On the underside of the paws – Mice, rats and cats
20. Name the birds having the largest uropygial glands.
    Answer: Aquatic birds
21. From which layer of epidermis are nails, claws hooves and horns derived?
    Answer: Nails, claws, hooves and horns are derived from the stratum corneum of the epidermis.
22. What is the function of claws in birds?
    Answer: In birds, claws help in walking and perching.
23. Name the layer of epidermis that divides to form other layers.
    Answer: Stratum germinativum divides mitotically to form other layers of the epidermis.
24. Name the different layers of the mammalian epidermis.
    Answer: The different layers of mammalian epidermis (from inner to outer) are as follows:
    (a) Stratum germinativum (b) Stratum spinosum
    (c) Stratum granulosum (d) Stratum lucidum
    (e) Stratum corneum
25. What is the characteristic of the epidermis of aquatic vertebrates?
    Answer: The epidermis of aquatic vertebrates is thin and is rich in mucous glands.
26. Name vertebrate animals that lack a stomach.
    Answer: Stomach is lacking in cyclostomes, chimeras, lung fishes and some teleosts.
27. Name the mammal that has a sac-like stomach and lacks gastric glands.
    Answer: In monotremes, the stomach is sac-like and without any gastric glands.
28. Name the four chambers of stomach of ruminants.
    Answer: (a) Rumen (b) Reticulum (c) Omasum (d) Abomasum
29. Out of four chambers of the stomach in ruminants, in which chamber does enzymatic activity operate?
    Answer: Enzymatic activity occurs in the abomasum, which is known as the true stomach.
30. Colon is not required for the reabsorption of water in sharks.
    Answer: Because sharks are isotonic to their environment.
31. The saliva of which mammal contains an anticoagulant?
    Answer: Vampire bat
32. Name the vertebrate that stores food in the crop without digesting.
    Answer: Birds may store food in the crop without digesting it, before passing it to stomach for digestion.

33. What are tusks?
Answer: Tusks are excessively developed canines (in the walrus) or incisors (in the elephant)

34. What name is given to:
(a) Teeth with high crowns (b) Teeth with low crowns
(c) Teeth with cusps drawn into ridges
Answer: (a) Hypsodont (b) Bunodont (c) Lophodont

35. Which salivary gland is the largest in herbivores and carnivores, respectively?
Answer: (a) Parotid gland (b) Mandibular gland.

36. What is bile? Where it is produced and stored?
Answer: Bile is a bitter yellowish, blue and green fluid which is secreted by hepatocytes of liver and is stored in the gall bladder.

37. Name the digestive organ that distinguishes birds from other vertebrates.
Answer: Crop is exclusive to birds, which is not found in other vertebrates.

38. Name the parts of a mammalian stomach.
Answer: A mammalian stomach consists of three parts, viz., cardiac, pyloric and fundic glands.

39. What is Bursa entiana?
Answer: At the distal end of pyloric stomach in *scoliodon*, a muscular globular sac is found containing longitudinal folds of pyloric stomach. This is called the Bursa entiana.

40. Name the animal, whose intestine contains spiral valves.
Answer: In cartilaginous fishes, the intestine contains spiral valves.

41. What is the shape of stomach in elasmobranchs?
Answer: In elasmobranchs, the stomach is J-shaped.

42. Name the ruminant that lacks omasum.
Answer: The camel lacks omasum.

43. Herbivores have a large digestive tract while carnivores have a short one.
Answer: Because it takes a longer time to absorb nutrients from the plants but in carnivores, nutrients are easily obtained from the meat they eat. Besides, the caecum is also larger in herbivores, which helps in the digestion of cellulose. But in carnivores, the caecum is relatively short as their diet contains little plant materials.

44. Mention the main differences between digestive system of amphibians and mammals.
Answer:

	Amphibians	***Mammals***
(a)	Teeth are homodont, acrodont and polyphyodont.	Teeth are heterodont, thecodont and diphyodont.
(b)	Epiglottis is absent.	Epiglottis is present.
(c)	Stomach is differentiated into two parts, viz., cardiac stomach and pyloric stomach.	Stomach is differentiated into three parts, viz., cardiac, fundic and pyloric.
(d)	Ileum is short and coiled.	Ileum is very long and coiled.
(e)	Caecum is absent.	Caecum is present.
(f)	Sacculus rotundous is lacking.	Sacculus rotundous is present.
(g)	Vermiform appendix is absent.	Vermiform appendix is present.
(h)	Cloacal aperture is present.	Cloacal aperture is absent.

45. Name the gland that is both exocrine and endocrine in nature. Mention the name of animal in which this gland is not found.

    Answer: The pancreas is an exocrine as well as an endocrine gland, which is endodermal in origin. The lancelet lacks a pancreas.

46. What are the characteristics of avian lungs?

    Answer: (a) Avian lungs are small, spongy and inelastic.
    (b) Alveoli are not found in avian lungs.
    (c) Air sacs (diverticula of lung) extensively distributed through the body
    (d) Air flow through lungs is unidirectional.

47. Name the vertebrates that are without a swim bladder or lungs.

    Answer: Cyclostomes, cartilaginous fishes and some teleosts are vertebrates without a swim bladder or lungs.

48. What is a gill?

    Answer: A gill is a respiratory organ of aquatic organisms which extracts oxygen from water and excretes carbon dioxide into water. In vertebrates, gills usually develop from the wall of the pharynx.

49. Name the different modes of respiration in amphibians.

    Answer: In amphibians, the following three types of respiration occur:
    (a) Cutaneous respiration (b) Buccopharyangeal respiration
    (c) Pulmonary respiration Besides, in some forms, respiration also occurs by gills.

50. What is difference between the respiratory structures and functions of cartilaginous and bony fishes?

    Answer: (a) Cartilaginous fishes lack a countercurrent mechanism.
    (b) Cartilaginous fishes lack operculum, which is present in bony fishes.
    (c) Interbranchial septum is present between two demibranchs of the gill arch in cartilaginous fishes. In bony fishes, interbranchial septa are very short or absent.
    (d) In bony fishes, spiracle is closed or lost in adults, but is present in cartilaginous fishes.

51. Name those snakes in which the left lung is not very small or absent.

    Answer: In pythons and boas, the left lung is normal like the right lung.

52. What is the basic difference between the lungs of lower vertebrates and those of higher vertebrates?

    Answer: In lower vertebrates, the diameter of air space (alveoli) is more than the diameter of alveoli of higher vertebrates (mammals). The diameter of the alveolus of frog is 10 times that of the human alveolus. The smaller alveoli provide greater surface area for gaseous exchange.

53. Name the animals in which external gills are found.

    Answer: External gills are found in the larvae of lampreys, lung fishes and amphibians.

54. Mention the names of mammals in which lungs are simple and not lobed.

    Answer: In elephants, whales, hyrax, sirenians and many prissodactyles lungs are simple and without lobes.

55. How does the respiratory system of birds differ from that of other vertebrates?

    Answer: Birds have relatively small lungs in comparison to other vertebrates, which are non-elastic. Besides, there are found nine air sacs which play an important role in respiration. Air sacs do not participate in gaseous exchange. They permit unidirectional flow of air, i.e., moving mainly fresh air having higher oxygen contents.

56. How many air sacs are found in birds?

    Answer: Most birds have nine air sacs, viz., one interclavicular, two cervical, two anterior thoracic, two posterior thoracic and two abdominal.

57. How does ventilation occur in mammals?

    Answer: In mammals, ventilation occurs through a negative pressure pump, which is made possible by the thoracic cavity and diaphragm.

58. What is syrinx?
Answer: Syrinx is the voice box of birds.
59. How does air move inside and outside in birds?
Answer: In birds, air moves inside and outside through pressure changes in the air sacs.
60. How does the respiratory system of birds differ from that of mammals?
Answer: (a) Lungs of birds are relatively small and non-elastic while the lungs of mammals are relatively large and elastic.
(b) Birds have nine air sacs but mammals lack air sacs.
(c) Birds lack diaphragm but it is present in mammals.
(d) In birds, air moves inside and outside through pressure changes in air sacs. But in mammals, ventilation occurs through a negative pressure pump, which is made possible by the thoracic cavity and the diaphragm.
(e) In mammals, gaseous exchange occurs between alveoli but in birds, it occurs in the walls of microscopic tubules called air capillaries.
(f) The respiratory system of birds is more efficient in comparison to mammals, delivering more and oxygen at each breath in comparison to mammals.
(g) In birds, air flow is unidirectional while in mammals, it is bidirectional.
(h) In birds, respiratory cycle requires two cycles (inspiration, expiration, inspiration) while in mammals, the respiratory cycle requires one cycle.
(i) The air sacs of birds extend into bones and lack a diaphragm, so respiratory infection can easily spread into the abdomen and bones.
(j) Birds have larynx but they are not used for sound production. In birds, the sound producing organ is the syrinx.
61. What is necessary for a fish to be able to breathe?
Answer: For a fish to be able to breathe, it is necessary that water must flow over its gills, which is achieved by continuous swimming or pumping water over the gills.
62. What is the location of lungs in reptiles?
Answer: In reptiles, lungs are located in the abdomen.
63. What is the name of the cartilaginous structure that supports gill region of the *Petromyzon?*
Answer: The cartilaginous structure that supports gill region of the *Petromyzon* is called branchial basket.
64. Draw a well-labelled diagram of the respiratory organs of birds

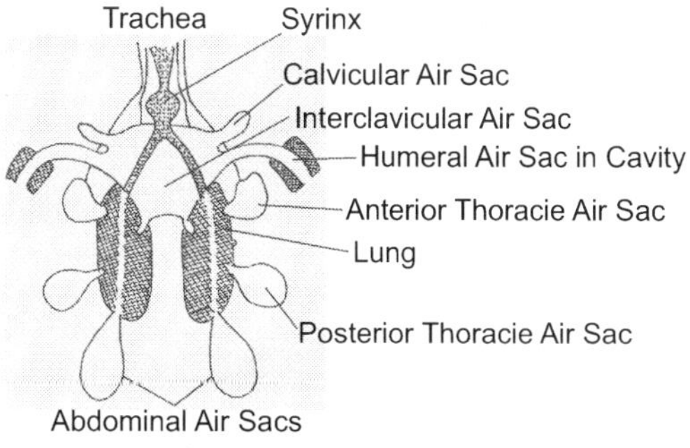

Fig.1 Respiratory organs of bird

65. Name different parts of the heart of vertebrates.
Answer: In lower vertebrates, the heart consists of auricle, ventricle, sinus venosus and conus arteriosus. In higher vertebrates, the sinus venosus and conus arteriosus disappear, while the auricle and ventricle have a tendency to divide.

66. What is sinus venosus?
Answer: Sinus venosus is a thin-walled sac-like structure, which is the first chamber to receive deoxygenated blood in fishes, amphibians and reptiles and pump this blood to the atrium. In birds and mammals, it is completely incorporated in the right auricle.

67. What is a venous heart?
Answer: A heart containing deoxygenated blood is called venous heart (e.g., heart of *Scoliodon*).

68. Name the animal having:
(a) Two-chambered heart
(b) Three-chambered heart
(c) Four-chambered heart
(d) Incompletely divided auricle
(e) Incompletely divided ventricle
Answer: (a) Two-chambered heart – *Scoliodon*
(b) Three-chambered heart – Frogs
(c) Four-chambered heart – Birds and mammals
(d) Incompletely divided auricle – Dipnoi
(e) Incompletely divided ventricle – Lizards, snakes

69. What is the basic difference between the heart of amphibians and that of reptiles?
Answer: In both amphibians and reptiles, the heart is three-chambered, but in reptiles there is a tendency of division of ventricle, resulting in less mixing of arterial and venous blood.

70. Name the reptiles in which the heart is three-chambered but functionally becomes four-chambered during contraction.
Answer: Pythons and monitor lizards have a three-chambered heart, but during contraction it becomes a four-chambered due to the muscular ridge that subdivides the ventricle into two parts during ventricular diastole and completely divides it during ventricular systole.

71. What is foramen of Panizza?
Answer: Foramen of Panizza is an aperture that connects the left and right systemic arches in the heart of members of the order of crocodilia.

72. What is conus arteriosus?
Answer: Conus arteriosus is a cone-shaped projection in the heart of the chordates derived from the right ventricle from which the pulmonary trunk arises.

73. Compare the heart of fish and amphibians.
Answer:

	Heart of fish	***Heart of amphibia***
(a)	Heart of fishes is S-shaped.	The heart is almost triangular.
(b)	Consists of an atrium, a ventricle and a conus arteriosus (in elasmobranchs) or a bulbous arteriosus (in teleosts).	Consists of two auricles, a ventricle, a sinus venosus and a truncus arteriosus.
(c)	The auricle is single-chambered which is placed on the dorsal of the ventricle. In Dipnoi, the auricle incompletely divided.	The auricle is divided by inter auricular septum into a large right auricle and a small left auricle.

Contd.

	Heart of fish	Heart of amphibia
(d)	Auriculoventricular aperture is guarded by paired membranous valves.	Auriculoventricular aperture is guarded a pair of flap-like valves.
(e)	The auricle does not receive pulmonary veins as lungs are absent.	The auricles receive pulmonary veins.
(f)	The heart receives only deoxygenated blood.	The heart receives both oxygenated and venous blood.

74. What is bulbous arteriosus?

Answer: Bulbous arteriosus is a pear-shaped structure in the heart of teleosts, which is made up of noncontractile smooth muscle and elastic fibres. It lacks cardiac muscle. Bulbous arteriosus absorbs the pressure shock of ventricular contraction.

75. Mention four differences between the reptilian heart and the mammalian heart.

Answer: (a) In reptiles, the heart is three-chambered, i.e., two auricles and one ventricle while in mammals, the heart is four-chambered, i.e., two auricles and two ventricles.

(b) In reptiles, sinus venosus is present and is the first chamber to receive blood but in mammals, it becomes incorporated in the right auricle (present during embryonic condition).

(c) In reptiles, there are two systemic arches, while in mammals there is only one systemic arch (the left one).

(d) In reptiles, there is no complete separation of oxygenated and deoxygenated blood as is the case in mammals.

76. From which germ layer does the heart develop?

Answer: The heart develops from mesoderm as a simple, straight and cylindrical cardiac tube.

77. In which vertebrates does the heart contain papillary muscles?

Answer: Papillary muscles are present in the heart of birds and mammals.

78. Name the vertebrates in which caval veins directly open into the right auricle.

Answer: In birds and mammals, caval veins directly open into right auricle as sinus venosus is absent.

79. Mention the chief differences between the amphibian heart and mammalian heart.

	Amphibian heart	Mammalian heart
(a)	The heart is located on the midline in the cavity anterior body.	The heart is located slightly left to middle in the thoracic cavity body
(b)	Heart consists of three chambers, viz., two auricles and one ventricle.	The heart consists of four chambers, viz., two auricles and two ventricles.
(c)	Sinus venosus is well developed.	In adults, sinus venosus is not present.
(d)	Both auricles open into a ventricle by a common auriculoventricular aperture.	Each auricle opens into its respective by a ventricle by a separate opening.
(e)	Auriculoventricular valve is bicuspid, while left auricuventricular is tricuspid.	The right auriculoventricular valve is tricuspid while left auricuventricular is bicuspid.
(f)	There is only one contraction centre, i.e., SA node.	There are two contraction centres SA node, i.e., and AV node.
(f)	There are two systemic arches.	There is only one systemic arch (the left one).
(h)	Coronary artery is absent.	Coronary artery is present.
(i)	There is no complete separation of and oxygenated and deoxygenated blood.	There is complete separation of oxygenated deoxygenated blood.

80. In which amniotes is sinus venosus the first chamber to receive blood?
Answer: In reptiles, sinus venosus is the first chamber to receive blood.

81. Mention the characteristics of the avian heart.
Answer: (a) The avian heart consists of four chambers, viz., two auricles and two ventricles.
(b) Sinus venosus is lacking.
(c) There is complete separation of pure and impure blood.
(d) The aorta curves down to the right.

82. What are aortic arches?
Answer: Aortic arches are the blood vessels that connect the dorsal aorta with the ventral aorta. In vertebrate embryos, six pairs of aortic arches are present, which are numbered as 1 to 6.

83. What is the function of aortic arches?
Answer: Aortic arches carry blood from the dorsal blood vessel to the ventral blood vessel.

84. How many aortic arches are present in *Scoliodon* and teleosts?
Answer: In *Scoliodon,* five pairs (second, thrid, fourth, fifth and sixth) of aortic arches are found while in teleosts, there are only four pairs of aortic arches (thrid, fourth, fifth and sixth).

85. How many aortic arches are present in lung fishes?
Answer: In lung fishes four pairs (thrid, fourth, fifth and sixth) of aortic arches are found. Besides, a set of pulmonary artery arises from the sixth aortic arch near the dorsal aorta.

86. How many afferent and efferent arteries are present in each aortic arch in each gill of elasmobranchs, teleosts?
Answer: In elasmobranchs, each aortic arch has one afferent artery and two efferent arteries in each gill while in teleosts, each aortic arch has one afferent and one efferent artery in each gill.

87. In which reptiles are ductus caroticus and ductus arteriosus present?
Answer: Ductus caroticus is present in certain snakes and lizards while ductus arteriosus is present in certain turtles and *Sphenodon.*

88. How many aortic arches are present in adult amniotes?
Answer: In adult amniotes, only three pairs (thrid, fourth and sixth) of aortic arches exist.

89. In which urodeles is the 5th aortic arch absent?
Answer: The fifth aortic arch is absent in Siren and Amphiuma.

90. What is ductus arteriosus?
Answer: Ductus arteriosus is a channel that connects the pulmonary artery to the aorta.

91. Mention the changes in aortic arches that have occurred during the course of evolution.
Answer: Changes in aortic arches are associated with changes in the heart. The main changes in aortic arches are as follows:
(a) Loss of arches first, second and third.
(b) Modification of arches third and sixth (pulmonary circulation).
(c) Retention of both sides of aortic arch fourth in reptiles.
(d) In birds, the systemic arch of the left side has been lost while in mammals, the systemic arch of the right side has vanished.

92. Why is the third aortic arch termed as the carotid arch?
Answer: Because it constitutes the commencement of the internal carotid artery.

93. Compare the aortic arches of birds and mammals.
Answer: (a) In both birds and mammals, six pairs of aortic arches are present in embryo but in adults, only three pairs (third, fourth and sixth) of aortic arches persist.
(b) A single systemic aorta on the right side in birds and on the left side in mammals is present.

(c) The third arch represents a systemic aorta.

(d) The systemic aorta unites with the radix aorta to form the dorsal aorta.

(e) The subclavian artery is present on the left side in birds and on the right side in mammals.

(f) The sixth aortic arch carries deoxygenated blood from the right ventricle to the lungs and arises from a single pulmonary aorta.

(g) In both birds and mammals, ductus caroticus and ductus arteriosus (ductus Botalli) disappear.

94. From which germ layer does the brain develop?
Answer: The brain develops from the ectoderm as an invagination.

95. Name the principal structures of the brian of adult vertebrates.
Answer: The brain of adult vertebrates consists of the following three parts:
(a) Prosencephalon (b) Mesencephalon (c) Rhombencephalon

96. What is foramen of Monro?
Answer: Lateral ventricles open into the diocoel (third ventricle) by a small aperture called foramen of Monro.

97. What is iter?
Answer: Diocoel and metacoel remain connected with each other by a narrow structure called iter.

98. What is corpus callosum?
Answer: In mammals, the two cerebral hemispheres are joined across the midline by transverse fibres called corpus callosum.

99. Distinguish between the amphibian brain and the mammalian brain.

	Amphibian brain	***Mammalian brain***
(a)	The brain is relatively simple.	The brain is relatively complex.
(b)	Cerebral hemispheres form one-third part of the brain.	Cerebral hemispheres form two-third part of the brain .
(c)	The two cerebral hemispheres are not completely separated from each other.	The two cerebral hemispheres are divided into two parts by a median fissure.
(d)	Corpus callosum is absent.	Corpus callosum is present.
(e)	The pineal stalk is small.	The pineal stalk is large.
(f)	There are two optic lobes.	There are four optic lobes.
(g)	Corpus albicans is absent.	Corpus albicans is present.
(h)	The cerebellum is poorly developed.	The cerebellum is well developed.
(i)	Arbor vitae is absent.	Arbor vitae is present.
(j)	Pons varolii is absent.	Pons varolii is present.
(k)	There are 10 pairs of cranial nerves.	There are 12 pairs of cranial nerves.

100. Mention the characteristic features of the reptilian brain.
Answer: (a) The reptilian brain is relatively small and simple.
(b) The cerebral hemispheres are small and almost smooth.
(c) A cortex showing some of the characteristics of mammalian cerebral cortex is present in all reptiles.

(d) Olfactory structures are well developed in some chelonia while they are relatively small in other reptiles. A characteristic feature of squamata is the accessory development of the olfactory bulb.
(e) The crura cerebri is poorly developed.
(f) The cerebellum is poorly developed.
(g) The pineal eye is present.

101. Distinguish between the reptilian brain and the avian brain.

	Reptilian brain	*Avian brain*
(a)	The brain is relatively simple.	The brain is relatively complex and large
(b)	The cerebral hemispheres are oval, small and do not overlap the diencephalon.	The cerebral hemispheres are large and overlap overlap the diencephalon.
(c)	The olfactory bulbs are large.	The olfactory bulbs are small.
(d)	Optic lobes are small.	Optic lobes are large.
(e)	Crura cerebri is poorly developed.	Crura cerebri is well developed.
(f)	The cerebellum is poorly developed.	The cerebellum is more developed.
(g)	The pineal eye is present.	The pineal eye is absent.

102. What is a parietal organ?
Answer: A parietal organ is a part of epithalamus, which is also known as parietal eye or third eye found in some animal species such as *Sphenodon*, lizards, frogs and tuna. It is absent in birds, mammals, etc.

103. Define crura cerebri.
Answer: Crura cerebri are a band of nerves on each side of the brain connecting the cerebrum and cerebellum.

104. Define jaw suspension.
Answer: The attachment of the upper and lower jaws with the brain case is called jaw suspension.

105. Which type of jaw suspension is found in mammals?
Answer: Craniostylic jaw suspension is found in mammals.

106. Name the jaw suspension in which the upper jaw has no direct connection with the brain case.
Answer: Hyostylic jaw suspension is found in most cartilaginous and fishes.

107. Name the jaw suspension in which the mandibular arch is not supported by hyomandibular arch.
Answer: In autostylic jaw suspension, the mandibular arch is not supported by the hyomandibular arch. Such jaw suspension is found in Dipnoi and tetrapods.

108. What is the characteristic of streptostylic jaw suspension?
Answer: In streptostylic jaw suspension, the quadrate is movable and the articular bone of lower jaw articulates with the quadrate of the upper jaw. Such type of jaw suspension is found in snakes, lizards and birds.

109. Name the embryonic kidney of vertebrates.
Answer: Pronephros is the functional kidney of the embryo of all vertebrates.

110. Name the functional kidney of fishes and amphibia.
Answer: Mesonephros is the functional kidney in fishes and amphibia.

111. Name the functional kidney of amniotes.
Answer: Metanephros is the functional kidney of amniotes.

112. What is urinary bladder? Name the vertebrates that lack a urinary bladder.
Answer: Urinary bladder is a muscular membrane sac that acts as a temporary reservoir for the storage of urine formed by the kidneys. Urinary bladder is absent in agnathans, snakes, crocodilians, some lizards and birds, except ostriches.

113. What is urethra?
Answer: The urinary bladder opening towards the outside via a short tube is called the urethra.

114. Name the mammal in which the ureters do not directly open into the urinary bladder.
Answer: Monotremes

115. What is a sexual kidney?
Answer: In some animals (males of jawed fish, amphibians and some amniotes) the mesonephric tubules carry sperms from the testis to the mesonephric ducts. This part of the kidney is called sexual kidney.

116. Define glomeruli.
Answer: Glomeruli are masses of capillaries that filter blood along with the Bowman's capsule.

117. Define mesorchium and mesovarium.
Answer: The membrane that connects testes to the body wall is called mesorchium, while the membrane that connects ovaries to the body wall is called mesovarium.

118. What are hemipenes?
Answer: Hemipenes are intromittent organ of snakes and lizards. They are a pocket-like diverticula of wall of cloaca.

119. In which animal is duplex uterus found?
Answer: Duplex uterus is found in marsupials.

120. Why are testes usually smaller than the ovaries?
Answer: Because sperms though numerous in number, are much smaller than the egg (especially egg containing yolk).

121. Name the animals in which testes are larger than the ovaries.
Answer: Testes are larger than the ovaries in mammals.

122. Which type of uterus is found in primates?
Answer: The simplex type of uterus is found in primates.

123. Draw a diagram showing the male urinogenital system of a frog.

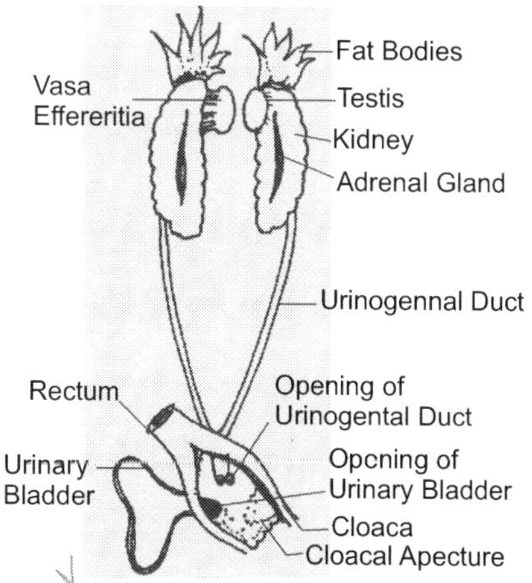

Fig. 2 Urinogenital system of a male frog

124. Why ureters are called urinogenital ducts in a male frog?
Answer: Because ureters carry both sperms as well as urine, and are therefore called urinogenital ducts.

125. Where do ureters open in reptiles and birds?
Answer: In reptiles and birds, ureters open in the dorsolateral wall of the cloaca.

126. Name the animals in which ovaries are thin-walled and hollow structures.
Answer: In some teleosts, amphibians, lizards and snakes ovaries are thin-walled and hollow sac-like structures.

127. In which animals is only the left ovary functional?
Answer: In all birds and *Orinithorhynchus* only the left ovary is functional.

128. What is the fate of the Mullerian ducts in mammals?
Answer: In mammals, Mullerian ducts give rise to oviducts, uteri and vagina.

129. Name the birds in which functional right ovaries are found.
Answer: In birds, functional right ovaries are found in accipitridae, falconidae, cathartidae and brown kiwis.

130. Draw a well-labelled diagram of TS of the mammalian ovary.

131. Name a hermaphrodite vertebrate.
Answer: Hag fishes and some teleosts are hermaphrodites.

132. Name the birds that possess a urinary bladder.
Answer: Ostriches and rheas have a urinary bladder.

133. Name the different compartments of cloaca of birds.
Answer: The cloaca of birds consists of the following three compartments:
(a) Coprodeum, which receives the faeces.
(b) Urodeum, which receives the ureters and the ducts of gonads.
(c) Proctodeum, which is the common discharge area.

134. Draw a well-labelled diagram of the urinogenital system of male birds.

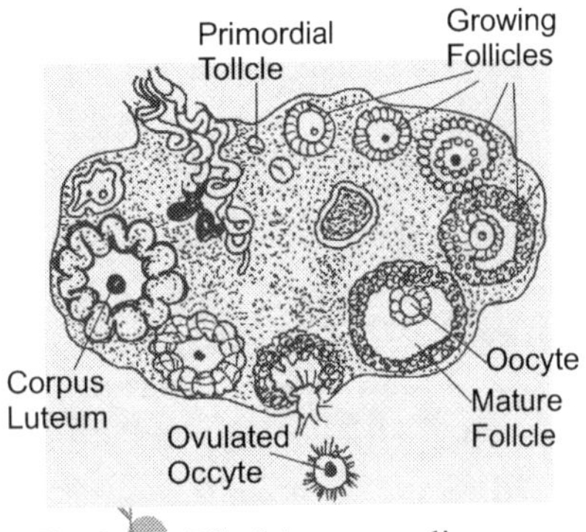

Fig. 3 TS of the mammalian ovary

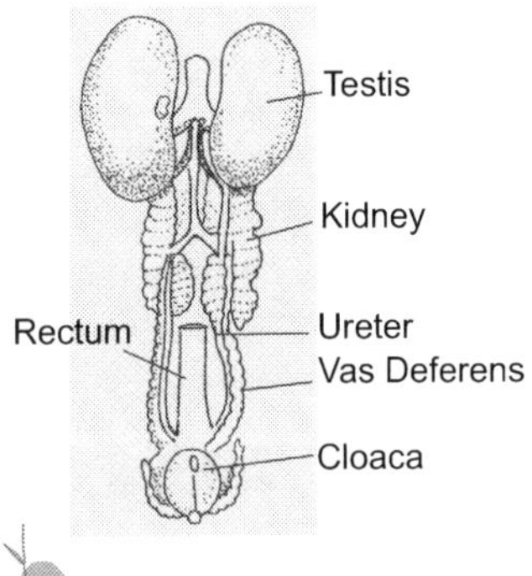

Fig.4 Urinogenital system of male birds

Long-Answer Questions

1. Define integument. Give an account of integument in vertebrates.
2. Describe epidermal structures in vertebrates.
3. Define aortic arches. Trace the evolution of aortic arches in vertebrates.
4. Describe aortic arches in tetrapods. Name the factors that led to the gradual modification of aortic arches during the course of evolution.
5. Give a comparative account of respiratory organs in vertebrates.
6. Compare the respiratory organs of reptiles, birds and mammals.
7. Describe the respiratory system of birds and compare it with that of mammals.
8. Trace the evolution of the heart in vertebrates.
9. Give a comparative account of the heart of fishes, amphibians and mammals.
10. Describe the structure of the mammalian heart. How does it differ from the amphibian heart?
11. Give a comparative account of the brain in vertebrates.
12. Describe the structure of the amphibian brain. Give a description of the advancement of the mammalian brain over the amphibian brain.
13. Compare the brains of amphibia, birds and mammals.
14. Define jaw suspension. Give an account of jaw suspension in vertebrates.
15. Give a comparative account of the kidney in vertebrates.
16. Draw a labelled diagram of VS of the skin of mammals and describe the derivatives of the integument in them.
17. Draw a labelled diagram of the brain of mammals and describe its advancement over the reptilian brain.
18. Write short notes on the following:
    (a) Scales
    (b) Poison glands of amphibians
    (c) Venous heart
    (d) Ductus Botalli
    (e) Mullerian ducts
    (f) Corpus callosum
    (g) Foramen of Panizza
    (h) Sinus venosus
    (i) Air sacs

COMPARATIVE ANATOMY

Multiple-Choice Questions

1. The aortic arches connect the:
   (a) Ventral aorta to the dorsal aorta
   (b) Ventral aorta to the radices aorta
   (c) Afferent branchial artery to the efferent branchial artery
   (d) Ductus Botalli to pulmonary artery
2. In which one of the following is the number of aortic arches 19 pairs?
   (a) *Amphioxus* (b) *Bdellostoma stouti* (c) *Petromyzon* (d) *Herdmania*
3. In teleosts ______________ aortic arches persist:
   (a) Second, third and fifth
   (b) Fourth, fifth and sixth
   (c) Third, fourth, fifth and sixth
   (d) First, third, fifth and sixth
4. Which one of the following aortic arches are redyced in anurans?
   (a) Fifth (b) Sixth (c) Fourth (d) None of these
5. Which one of the following is present in anurans?
   (a) Ductus Botalli (b) Ductus caroticus (c) Fourth aortic arch (d) Sixth aortic arch
6. Ductus caroticus and ductus Botalli persist in:
   (a) *Icthyophis* (b) *Rana* (c) *Echidna* (d) Humans
7. The basic number of aortic arches in vertebrate is ________pairs:
   (a) 2 pairs (b) 4 pairs (c) 5 pairs (d) 6 pairs
8. The ligamentum arteriosum connects the commencement of the:
   (a) Right pulmonary artery to the aortic arch
   (b) Left pulmonary artery to the aortic arch
   (c) Both (a) and (b)
   (d) Ductus Botalli to the ventral aorta
9. The 4th right aortic arch forms the right:
   (a) Subclavian artery (b) Carotid artery (c) Pulmonary artery (d) Meningeal artery
10. Persistent right aortic arch occurs in:
    (a) German shepherds
    (b) Dogs (especially Irish setters)
    (c) Both (a) and (b)
    (d) None of these
11. Which one of the following is the largest diameter branch of the aortic arch?
    (a) Subclavian artery (b) Pulmonary artery (c) Brachiocephalic artery (d) Meningeal artery
12. The fifth aortic arch is lost, except in:
    (a) Birds (b) Adult gill salamanders (c) Frog (d) *Sphenodon*
13. The end of the ___________ aortic arch gives rise to the stapedial artery:
    (a) First (b) Second (c) Third (d) Fourth
14. In lungs of fishes:
    (A) There are five pairs of aortic arches
    (B) Arches III and IV lack gills
    (C) Efferent artery of arch VI gives rise to pulmonary artery to the lung
    (D) Spiracle is present

The incorrect statements are:
(a) None of these (b) A and B (c) B and C (d) B and D

15. Ductus arteriosus is derived from the __________ aortic arch:
(a) First (b) Second (c) Third (d) Fifth

16. In adult fishes, the __________ arches persist as branchial arteries:
(a) Second (b) Second and fourth (c) Fourth and sixth (d) Sixth

17. The aortic arch develops from the __________ pharyngeal arch:
(a) First (b) Second (c) Fourth (d) Sixth

18. Birds lack:
(a) Ductus Botalli (b) Ductus caroticus (c) Left systemic arch (d) All of these

19. Ductus caroticus persists in:
(a) *Sphenodon* (b) Alligator (c) Both (a) and (b) (d) None of these

20. Which one of the following statements is correct?
(a) Mammals lack ductus arteriosus and ductus Botalli
(b) In mammals, the third, fourth and sixth aortic arches persist
(c) In mammals, only the left systemic arch is present
(d) All of these

21. In which one of the following tetrapods is the maximum number of arches found?
(a) Anura (b) Urodela (c) Apoda (d) *Phrynosoma*

22. Ductus caroticus and ductus arteriosus are present in:
(a) Snakes (b) *Uromastix* (c) *Sphenodon* (d) All of these

23. In which one of the following is the 5th aortic arch absent, reduced or incomplete:
(a) *Siren* (b) *Amphiuma* (c) *Salamandra* (d) All of these

24. Match column I with column II and select the correct answer using answer codes:

Column I	Column II
(A) Pterygopodial glands	1. Toad
(B) Cloacal scent glands	2. Alligator
(C) Femoral glands	3. Male lizards
(D) Parotid glands	4. Male *Scoliodon*

Answer codes:

	A	B	C	D
(a)	4	3	2	1
(b)	2	1	4	3
(c)	4	2	3	1
(d)	3	4	1	2

25. Which one of the following is formed by stratum corneum?
(a) Horns (b) Hooves (c) Nails and claws (d) All of these

26. Stratum corneum first appears in:
(a) Teleosts (b) Dipnoi (c) Amphibians (d) Reptiles

27. Peal organs are found in:
(a) Elasmobranches (b) Teleosts (c) Amphibians (d) Reptiles

28. Ecdysis is shown by:
(a) Amphibians (b) Snakes (c) Lizards (d) All of these

29. In which one of the following might hair be modified into spines?
(a) *Echidna* (b) Hedgehog (c) Porcupine (d) All of these
30. Luminous glands are modified:
(a) Salivary glands (b) Mucous glands (c) Poison glands (d) Dermal cells
31. Cosmoid scales are found in:
(a) Actinopterygians and crossopterygians (b) Crossopterygians and dipnoi
(c) Dipnoi (d) Teleosts
32. Cyclostomes lack:
(a) Horney ectodermal teeth (b) Bladder cells and thread cells
(c) Mucous cells (d) None of these
33. In tooth, which one of the following is epidermal in origin?
(a) Enamel (b) Dentine (c) Pulp cavity (d) All of these
34. In which one of the following fishes are scales are confined to some areas?
(a) *Chimaera* (b) *Polyodon* (c) *Acipenser* (d) All of these
35. The placoid scales appeared first in the ancestral shark of the:
(a) Upper Devonian period (b) Middle Devonian period
(c) Lower Devonian period (d) Permian period
36. Gill rackers are modified scales in:
(a) *Cetorhinus* (b) *Syngnathus* (c) *Tetradon* (d) None of these
37. Which one of the following is applicable to stratum germinativum?
(a) Undifferentiated (b) Proliferative (c) Stem cells (d) All of these
38. Which one of the following is the correct sequence from the lower to upper epidermis?
(a) Stratum granulosum → Stratum lucidum → Stratum germinativum →Stratum corneum →Stratum spinosum.
(b) Stratum germinativum → Stratum spinosum →Stratum granulosum →Stratum lucidum →Stratum corneum
(c) Stratum corneum→Stratum granulosum→Stratum lucidum→Stratum germinativum→Stratum spinosum
(d) Stratum corneum → Stratum granulosum→Stratum lucidum→Stratum spinosum→ Stratum germinativum
39. Which one of the following layers of epidermis is evident only in soles and palms?
(a) Stratum corneum (b) Stratum germinativum (c) Stratum lucidum (d) Stratum spinosum
40. Which one of the following gives colour to skin?
(a) Melanin (b) Carotene (c) Haemoglobin (d) All of these
41. Which one of the following is the largest scale in birds?
(a) Scute (b) Cancella (c) Scutella (d) Reticula
42. Nails are found only in:
(a) Primates (b) Ungulates (c) Humans (d) All of these
43. Which one of the following statements about the dermis is incorrect?
(a) Made up of fibrous connective tissue (b) Mesodermal in origin
(c) Helps in the maintenance of body heat (d) None of the above
44. Hair is derived from:
(a) Stratum granulosum (b) Stratum lucidum

(c) Stratum corneum (d) Stratum spinosum

45. In which one of the following animals are there 13 mammary glands?
(a) *Echidna* (b) Virginia opossum (c) Cheetahs (d) Armadillos

46. Bovid horns:
(a) Generally grow continuously (b) Remain unbranched
(c) Are never shed (d) All of these

47. Which one of the following statements about the horns of giraffes is correct?
(a) Found in both sexes (b) Permanently covered with skin
(c) Permanently covered with hairs (d) All of these

48. Which one of the following statements is correct?
(a) A snake is born with fixed number of scales. (b) Scales increase in number as the snake matures.
(c) Gradually, there occurs a decrease in the number of scales in snakes. (d) Scales of snakes are made up of nonkeratin materials.

49. Which one of the following statements is not applicable to reptilian integument?
(a) Epidermal scales are present. (b) Scales are homologous to fish scales.
(c) There is a reduced number of integument glands. (d) Ecdysis is common.

50. Antlers are found only in:
(a) Cervids (b) Bovids (c) Ungulates (d) Cervids and Bovids

51. Langerhans and Merkel cells are found in the deep layer of:
(a) Epidermis (b) Dermis (c) Both (a) and (b) (d) Subcutaneous layer

52. Keratin is lacking in the skin of:
(a) *Amphioxus* (b) *Balanoglossus* (c) Cyclostomes (d) All of these

53. Stratum corneum is lacking in the epidermis of:
(a) Elasmobranchs (b) Teleosts (c) Dipnoi (d) All of these

54. Hair horns are found in:
(a) Rhinoceroses (b) Giraffes
(c) In male rhinoceroses only (d) Antelopes

55. In bats, sweat glands are located in the:
(a) Sides of the head (b) Abdomen (c) Ears (d) Eyes

56. Melanocytes are found in the:
(a) Stratum corneum (b) Stratum granulosum (c) Dermis (d) Stratum lucidum

57. Dermal plates are found in:
(a) Armadillos (b) Rhinoceroses (c) Zebras (d) None of the above

58. In birds, the sheath of beak is the modification of the:
(a) Stratum corneum (b) Stratum lucidum (c) Stratum granulosum (d) Dermis

59. Ecdysis is shown by:
(a) Hedgehogs (b) Armadillos (c) Scaly anteaters (d) Flying squirrels

60. Which one of the following is a modification of stratum corneum?
(a) Rattle of rattlesnakes (b) Toothless horny beak of turtles
(c) Horns of horned toads (d) All of the above

61. The secretion of which glands contains fatty acids and lactic acids?
(a) Uropygial glands (b) Sebaceous glands (c) Mammary glands (d) Mucous glands

62. Which one of the following plays a key role in the formation of jaws in gnathostomes?
(a) Neurocranium (b) Dermatocranium (c) Splanchocranium (d) All of these

63. In which one of the following jaw suspensions is the quadrate loosely attached and movable at both ends?
(a) Craniostylic (b) Streptostylic (c) Holostylic (d) Amphistylic

64. Hyoid arch is complete, independent and not attached to the skull in_________ jaw suspension:
(a) Holostylic (b) Craniostylic (c) Monimostylic (d) Hyostylic

65. Monimostylic jaw suspension is found in many tetrapods, except:
(a) Mammals (b) Birds (c) Aves (d) Amphibians

66. Consider the following statements about jaw suspension:
(A) The upper jaw has no direct connection with the brain case (except anteriorly)
(B) Both jaws are connected to brain case entirely by the hyomandibular of the hyoid arch
(C) Provides the jaws a wider movement
(D) Helps in swallowing of larger preys
This type of jaw suspension is called:
(a) Amphistylic (b) Hyostylic (c) Holostylic (d) Streptostylic

67. Which one of the following statements about craniostylic jaw suspension is incorrect?
(a) It is found in mammals.
(b) Articular and quadrate have been transformed into malleus and incus bones, respectively.
(c) The upper jaw is not completely fused with the cranium.
(d) Dentary of lower jaw and squamosal of skull provide articulation between the jaws.

68. Double suspension is applicable to jaw suspension:
(a) Autodiastylic (b) Amphistylic (c) Autostylic (d) Craniostylic

69. In which one of the following do jaw suspensions do quadrate become firmly fixed with skull?
(a) Holostylic (b) Monimostylic (c) Streptostylic (d) Craniostylic

70. Streptostylic jaw suspension is not found in:
(a) Bony fishes (b) Lizards (c) Shakes (d) Birds

71. Crop milk is applicable to:
(a) Crocodiles (b) *Scoliodon* (c) Birds (d) *Echidna*

72. Which one of the following is lacking in *Scoliodon*?
(a) Diastema (b) Internal nares (c) Eustachian openings (d) All of these

73. Frog lacks:
(a) Salivary glands (b) Sacculus rotundus (c) Vermiform appendix (d) All of these

74. In which one of the following might the colour of liver be green, orange, brown or red?
(a) Lampreys (b) Hagfishes (c) Elasmobranches (d) Teleosts

75. Consider the following statements:
(A) Gall bladder is lacking in all perissodactyla (B) Pancreas is well developed in lung fishes
(C) A true liver is lacking in protochordates (D) In elasmobranchs, liver is three lobed
The correct statements are:
(a) All (b) (A), (B) and (C) (c) (A) and (C) (d) (B) and (D)

76. Caecum is small in:
(a) Monkeys (b) Humans (c) Apes (d) All

77. Marsupials retain their all teeth, excepting the:
(a) First molar (b) Last molar (c) First premolar (d) Last premolar

78. Rectal gland of elasmobranchs of secretes:
(a) HCl (b) Sodium chloride (c) Anti bacterial substance (d) Milky substance
79. Palate is lacking in:
(a) Lizards (b) Snakes (c) Frogs (d) All of these
80. Which one of the following is lacking in lampreys?
(a) Stomach (b) Spiral valve (c) Typhlosole (d) Conus arteriosus
81. Amphibians lack:
(a) Salivary glands (b) Rectal glands (c) Caecum (d) All of these
82. Which one of the following is absent in pigeons?
(a) Caecal gland (b) Gall bladder (c) Crypts of Lieberkuhn (d) None of these
83. The longest caecum is found in the:
(a) Koala bears (b) Flying foxes (c) Lemurs (d) Bobcats
84. The tongue is movable in all groups of mammals, except:
(a) *Echidna* (b) Armadillos (c) Whales (d) Koala bears
85. Teeth are monophyodont in:
(a) Toothed whales (b) Sloths (c) Platypuses (d) All of these
86. *Calotes* lacks:
(a) Protrusible tongue (b) Gall bladder (c) Colon (d) Rectum
87. The heart of teleosts is similar to that of cartilaginous fishes, except:
(a) A bulbus arteriosus is present (b) A conus arteriosus is present
(c) A spiral valve is present in conus arteriosus (d) Ventral aorta is short
88. Two gill hearts and one systemic heart occur in:
(a) Cephalopods (b) Gastropods (c) *Amphioxus* (d) Crustaceans
89. Which one of the following is found in the mammalian heart?
(a) Foramen ovale (b) Annulus ovalis (c) Fossa ovalis (d) All of these
90. In which of the following do four pulmonary veins open by a common aperture in the left auricle?
(a) Frogs (b) Lizards (c) Pigeons (d) All of these
91. Which one of the following is not found in the heart of *Scoliodon*?
(a) Foramen of Panizza (b) Columnae carnae
(c) Papillary muscles (d) All of these
92. Truncus arteriosus is present in:
(a) Frogs (b) Lizards (c) Pigeons (d) Rabbits
93. Conus lacks a spiral valve in:
(a) *Protopterus* (b) *Lepidosiren* (c) *Neoceratodus* (d) All of these
94. Corpus callosum is the characteristic feature of the brain of:
(a) Amphibians (b) Crocodiles (c) Marsupials (d) Eutherians
95. Restiform bodies are present in the brain of:
(a) Rabbits (b) Frogs (c) Dogfishes (d) Lizards
96. Which one of the following is present in the brain of reptiles and birds?
(a) Neuropore (b) Corpus albicans (c) Pallium (d) All of these
97. Rhinocoel is lacking in:
(a) Rabbits (b) Pigeons (c) Lizards (d) Frogs

98. Parietal organ is present in:
(a) Frogs (b) Pigeons (c) Rabbits (d) None of these

99. Which one of the following structures is not associated with telencephalon?
(a) Olfactory bulbs (b) Optic chiasma
(c) Corpus callosum (d) Neo cortex on pallium

100. Modern teleosts lack:
(a) Para pineal body (b) Vagal lobes (c) Restiform bodies (d) Optic lobes

101. Tectum includes:
(a) Restiform bodies and pyramids (b) Auditory lobes and cerebral peduncles
(c) Optic lobes and auditory lobes (d) Auditory lobes and mamillary bodies

102. Restiform bodies are associated with:
(a) Hearing (b) Maintenance of balance (c) Swimming (d) Smell

103. Which one of the following is present in *Sphenodon*?
(a) Gyri and sulci (b) Restiform bodies (c) Parietal eyes (d) Vagal lobes

104. In which one of the following is there less than six layers of neurons in the outer layer of the brain?
(a) Fishes (b) Amphibians (c) Reptiles (d) All of these

105. Consider the following statements:
(A) Three groups of animals, with a few exceptions have complex brain, which are Arthropods, Molluscs, and Craniates
(B) In human corpus callosum, there are 2,50,00,000 fibres
(C) Mesencephalon develops without further subdivision and forms the tectum
(D) In *Amphioxus*, brain and spinal cord are in a straight line.

The incorrect statements are:
(a) (A) and (B) (b) (B) and (C) (c) (A) and (D) (d) None of these

106. Olfactory lobes are very small in:
(a) Catfishes (b) Hagfishes (c) Sharks (d) None of these

107. Telencephalon of fishes is equivalent to ____________ in higher vertebrates:
(a) Cerebrum (b) Diencephalon (c) Optic lobes (d) Cerebellum

108. Hippocampus, a part of the brain, is found only in:
(a) Monotremes (b) Snakes and lizards (c) Mammals (d) All of these

109. Which one of the following statements about metanephros is incorrect?
(a) Functional kidney of adult amniotes (b) Develops from the caudal part of the mesomeres
(c) Most posterior part of the kidney and is the last to develop (d) None of these

110. Which one of the following uterus is Y-shaped externally and divided internally?
(a) Bicornate uterus (b) Simplex uterus (c) Bipartite uterus (d) Duplex uterus

111. Uterine horns are lacking and oviducts open directly into the body of uterus in:
(a) Simplex uterus (b) Bipartite uterus (c) Bicornate uterus (d) Duplex uterus

112. Consider the following statements:
(A) In reptiles, uterus opens separately in the cloaca, except turtles
(B) Kidneys of male fishes are longer than the females
(C) Some marine teleosts lack external or internal glomeruli
(D) When the metanephros kidney becomes functional, Wolffian duct persists in males as urinogenital duct, while it degenerates in females

The incorrect statements are:
(a) (A) and (B) (b) (B) and (C) (c) (A) and (D) (d) None of these

113. Follicular epithelium is lacking in the ovarian follicles of:
(a) *Amphioxus* (b) *Petromyzon* (c) *Bdellostoma* (d) *Myxine*

114. In mammals, generally, one pair Cowper's glands are found, except in some:
(a) Rodents (b) Insectivores (c) Marsupials (d) Cetaceans

115. In the human embryo, there are ___________ pronephric tubules:
(a) 3 (b) 5 (c) 7 (d) 11

116. The Mullerian duct is present in all vertebrates, except:
(a) Crossopterygii (b) Cyclostomes (c) Actinopterygii (d) Perch

117. Which one of the following statements is incorrect?
(a) In cyclostomes and perch, there is only a single median ovary.
(b) In many elasmobranchs, only the right ovary is functional.
(c) In birds and *Orinthorhynchus,* only the left ovary is functional.
(d) None of these

118. In mammals, the Mullerian duct gives rise to:
(a) Oviducts (b) Uterus (c) Vagina (d) All of these

119. Nephrostomes are lacking in:
(a) Teleosts (b) Elasmobranchs (c) Urodela (d) Amniotes

120. Cloacal bladder is found in:
(a) Dipnoi (b) Dipnoi and amphibians (c) Dipnoi and teleosts (d) *Sphenodon* and frog

121. In bats, the uterus is:
(a) Duplex (b) Bicornuate (c) Bipartite (d) Simplex

122. Gelding is applicable to castrated:
(a) Horses (b) Boars (c) Bulls (d) Humans

123. Urine is voided as viscid steaks in:
(a) *Heterocephalus* (b) *Castor* (c) *Procavia* (d) *Hydrochoerus*

124. Copulatory grapples are found in many:
(a) Aquatic carnivores (b) Terrestrial carnivores
(c) Primates (d) Prissodactyla

125. Embryotrophe is uterine milk secreted by:
(a) *Sphenodon* (b) *Myxine* (c) Monotremes (d) *Dasypus*

126. Accessory sex glands occur in:
(a) Dogfishes (b) Frogs (c) Pigeons (d) All of these

127. Sperm sacs are found in:
(a) Rabbits (b) Pigeons (c) Dogfishes (d) Lizards

128. Vasa efferentia:
(a) Are thick (b) Arise from the retia testis
(c) Are only five in number (d) All of these

129. Lungs are without lobes in:
(a) Whales (b) Elephants (c) *Hyrax* (d) All of these

130. Spiracles are present in:
(a) Teleosts (b) Elasmobranchs (c) Lung fishes (d) None of these

131. In bony fishes, operculum arises from:
(a) Hyoid arch (b) Mandibular arch (c) Fourth gill arch (d) Hyomandibular arch

132. Consider the following statements:
(A) Gills and lungs are derivative of embryonic pharynx
(B) In *Amphipnous*, the first and fourth branchial arches are without gills
(C) Hippotamous lacks vocal cords
(D) In monotremes, only the right lung is lobulated

The incorrect statements are:
(a) (A) and (B) (b) (A) and (C) (c) (B), (C) and (D) (d) None of these

133. The only known animal without pleural space:
(a) Rhinoceroses (b) Hippopotamuses (c) Monotremes (d) Elephants

134. Which one of the following is a vertebrate without swim bladder or lungs?
(a) Cyclostomes (b) Cartilaginous fishes
(c) Bottom dwellers teleosts (d) All of the above

135. In birds, the respiratory exchange system is described as:
(a) Crosscurrent exchange (b) Countercurrent exchange
(c) Diffusion exchange (d) Active exchange

136. Which one of the following is involved in the adjustment of buoyancy in many teleosts?
(a) Cerebellum (b) Pons varolli (c) Swim bladder (d) Lateral line

137. The most efficient lungs are found in:
(a) Mammals (b) Birds (c) Reptiles (d) Amphibians

138. The branchial sac is the main respiratory organ of:
(a) Tunicates (b) *Petromyzon* (c) *Myxine* (d) None of these

139. Pronephros is located _________ in the body:
(a) Anteriorly (b) Posteriorly (c) At the middle (d) For the entire length

140. Mesonephros may persist after birth in:
(a) Reptiles (b) Prototherians (c) Metatheria (d) All of these

141. Consider the following statements:
(A) Choroid plexus is well developed in the myxinoid brain
(B) In *Protopterus*, optic lobes are fused to form a single rounded lobe
(C) Cerebellum is greatly reduced in the amphibian brain
(D) Monotremes lack pallial commissure

The correct statements are:
(a) All of these (b) (A), (B) and (C) (c) (B) and (C) (d) (C) and (D)

142. The main executive organ in maintaining homeostasis of the body is the:
(a) Liver (b) Brain (c) Kidney (d) Alimentary canal

143. *Sphenodon* lacks:
(a) Neopallium (b) Egg tooth (c) Sinus venosus (d) Ductus caroticus

144. Among reptiles, the enlargement of cerebellum is maximum in:
(a) Crocodiles (b) Chelonians and lizards
(c) Snakes and lizards (d) *Sphenodon* and crocodiles

145. Consider the following statements:
(A) The sweat of *Macropus* is red (B) In Koala bear, the Mullerian duct is unpaired

(C) *Tachyglossus* lacks abdominal vein (D) In mammals, only left systemic arch is present

The correct statements are:

(a) All of these (b) (A), (C) and (D) (c) (A) and (D) (d) (C) and (D)

146. Lungs are lacking in:
(a) Desmognathus (b) *Ascaphus* (c) *Icthyophis* (d) *Siphonophs*

147. Which one of the following statements is incorrect?
(a) In amphibians, the aortic arches retain bilateral symmetry.
(b) Bidder's organ is well developed in male toads.
(c) Bidder's organ has the ability to develop into an ovary after castration in either sex.
(d) None of these

148. Which one of the following statements about the mammalian brain is incorrect?
(a) Enlarged dorsal ventricular ridge.
(b) Enlarged dorsal Pallium called cerebral cortex.
(c) Cerebral cortex receives and analyses sensory information and initiates motor activity.
(d) None of these

149. In which one of the following jaw suspensions is the mandibular arch not supported by the hyomandibular?
(a) Amphistylic (b) Holostylic (c) Autostylic (d) None of these

150. Match column I with column II and select the correct answer using answer codes:

Column I	Column II
(A) Bipartite uterus	1. Ungulates, carnivores
(B) Duplex uterus	2. Primates
(C) Bicornuate uterus	3. Monotremes, marsupials
(D) Simplex uterus	4. Some ungulates, whales

Answer codes:

	A	B	C	D
(a)	2	4	1	4
(b)	1	3	4	2
(c)	3	4	2	1
(d)	4	2	1	3

Answers to Multiple-Choice Questions

1. (b)	2. (a)	3. (c)	4. (d)	5. (b)	6. (a)	7. (d)	8. (b)
9. (a)	10. (c)	11. (c)	12. (b)	13. (b)	14. (a)	15. (d)	16. (c)
17. (c)	18. (d)	19. (c)	20. (d)	21. (b)	22. (c)	23. (d)	24. (c)
25. (d)	26. (c)	27. (b)	28. (d)	29. (d)	30. (b)	31. (b)	32. (d)
33. (a)	34. (d)	35. (a)	36. (a)	37. (d)	38. (b)	39. (c)	40. (d)
41. (a)	42. (a)	43. (d)	44. (c)	45. (b)	46. (d)	47. (d)	48. (a)
49. (b)	50. (a)	51. (a)	52. (d)	53. (d)	54. (a)	55. (a)	56. (c)
57. (a)	58. (a)	59. (c)	60. (d)	61. (b)	62. (c)	63. (b)	64. (a)
65. (a)	66. (b)	67. (c)	68. (b)	69. (b)	70. (a)	71. (c)	72. (d)
73. (d)	74. (d)	75. (c)	76. (d)	77. (d)	78. (b)	79. (d)	80. (a)
81. (d)	82. (b)	83. (a)	84. (c)	85. (d)	86. (c)	87. (a)	88. (a)
89. (d)	90. (c)	91 (d)	92. (a)	93. (c)	94. (d)	95. (c)	96. (c)

97. (b)	98. (d)	99. (b)	100. (a)	101. (c)	102. (b)	103. (c)	104. (d)
105. (d)	106. (d)	107. (a)	108. (c)	109. (d)	110. (c)	111. (a)	112. (d)
113. (a)	114. (c)	115. (c)	116. (a)	117. (d)	118. (d)	119. (d)	120. (b)
121. (c)	122. (a)	123. (c)	124. (b)	125. (c)	126. (d)	127. (c)	128. (b)
129. (d)	130. (b)	131. (a)	132. (d)	133. (d)	134. (d)	135. (a)	136. (c)
137. (b)	138. (a)	139. (a)	140. (d)	141. (c)	142. (c)	143. (a)	144. (b)
145. (c)	146. (a)	147. (d)	148. (a)	149. (c)	150. (b)		

Fill in the Blanks

1. __________ pairs of aortic arches connect ventral aorta with dorsal aortas.
2. The __________ pair of aortic arches develops first.
3. In tetrapods, the embryonic ventral aorta becomes the common ____________ arteries.
4. The __________ aortic arch of all jawed vertebrates is lost.
5. In adult tetrapods, __________ and __________ aortic arches are lost.
6. In mammalian foetus, the __________ connects the pulmonary artery with the aorta.
7. The remnants of the ductus arteriosus persist as a fibrous cord called the __________.
8. In teleosts, __________, __________, __________ and __________ aortic arches persist.
9. In birds, __________ are the only integumentary glands.
10. __________ is lacking in the epidermis of fishes.
11. Nails grow from a thin area called __________.
12. The epidermis is made up of __________ cells.
13. The mammalian epidermis consists of the __________ layers.
14. __________ is the only layer of epidermis capable of cell division.
15. In human beings, the integumentary system is about __________ of the body weight.
16. In a majority of birds, scales do not overlap significantly, except __________ and __________.
17. Mice and rats have scales on the __________.
18. Teeth are mainly composed of three parts, viz., __________, __________ and __________.
19. The chordate having the simplest possible form of skin is the __________.
20. Feathers may be intermixed with scales on some bird ____________.
21. Epithelium and __________ tissue form the skin.
22. Ontogenetically, all skin derivatives are of __________ origin.
23. A claw consists of two parts, viz., __________ and __________.
24. Hooves are characteristic of __________.
25. The arrector pili muscles are attached to __________.
26. Scent glands are derived from __________ glands.
27. The process of cornification occurs everywhere on exposed skin surfaces, except over the anterior surfaces of the __________.

28. In humans, jaw suspension is _________.
29. In bony fishes, jaw suspension is _________.
30. In _________ jaw suspension, both jaws are suspended from cranium through hyomandibular.
31. All mammals lack cloaca, except _________.
32. Pancreas is _________ in origin.
33. In rabbits, the mouth opens into a narrow space bounded by jaws and lips called _________.
34. Dentine differs from bone in that it does not contain _________.
35. Heart of *Scoliodon* is ______ -shaped.
36. In fishes, blood is pumped from the ventricle through the _________ to the gills.
37. The vertebrate heart consists of _________ muscle.
38. The human embryonic heart begins beating at around _________ days after conception.
39. Conus arteriosus sends blood to _________.
40. The heart is fed by _________ circulation.
41. The subdivision of the ventricle occurred for the first time in _________.
42. In crocodiles, the only connection between two sides of the heart is through _________.
43. Corpora quadrigemina is the characteristic of brain of _________.
44. Vertebrate nervous systems are distinguished by _________ and _________.
45. The brain innervates the head through _________.
46. The only organ in the frog contained within coelom, which has its own protective covering, is the _________.
47. In fishes, the dermis consists of _________, _________ and __________..
48. Gray matter consists of _________ of neurons.
49. The kidney originates from the lateral parts of plate _________.
50. Holonephros kidneys are found in the larvae of _________.
51. _________ is the first formed kidney of a vertebrate embryo.
52. In mammals, the genital tract is divided into three regions, viz., _________, _________ and _________.
53. Epithelial cells of the somniferous tubules that play a role in the maturation of sperms are called _________ cells.
54. _________ is the ancestral kidney that develops from all of the nephrostomes.
55. _________ is a cord of tissue in mammals that guides descent of the testes.
56. In amphibians and mammals, the urinary bladder arises as evagination of ventral wall of _________.
57. _________ canal is the passage between abdominal cavity and scrotum.
58. Mesonephros is the embryonic kidney of _________, _________ and _________.
59. In mammals, the lower ends of ureters open directly into the urinary bladder, except in _________.
60. In elasmobranchs, copulation occurs through a clasping organ called _________.
61. Vagina is the fused terminal part of the _________.
62. In amniotes, the archinephric duct is termed as _________ duct
63. The urinary bladder passes through the penis in all mammals except in _________
64. In females, the Mullerian duct forms the _________.
65. The testis is surrounded by a capsule called _________.

66. In reptiles and birds, the urinary and genital systems are separate, except in the ________ region.
67. Complete fusion of two uteri leads to the formation of ________ uterus.
68. Most tetrapods have retained cloaca, except most ________.
69. White matter consists of ___________.
70. Aortic arches are ___________ to the dorsal aorta.

Answers to Fill in the Blanks

1. Six
2. First
3. Carotid
4. First
5. First, second
6. Ductus arteriosus
7. Ligamentum arteriosum
8. Third, fourth, fifth and sixth
9. Uropygial glands
10. Stratum corneum
11. Nail matrix
12. Epithelial
13. Five
14. Stratum germinativum
15. 16 per cent
16. Kingfishers, woodpeckers
17. Tail
18. Enamel, dentine, pulp cavity
19. *Amphioxus*
20. Feet
21. Connective
22. Ectomesodermal
23. Unguis, sub unguis
24. Ungulates
25. Hair shafts
26. Sweat
27. Eyes
28. Craniostylic
29. Hyostylic
30. Hyostylic
31. Monotremes
32. Endodermal
33. Vestibule
34. Living cells
35. S
36. Conus arteriosus
37. Cardiac
38. 21
39. Ventral aorta
40. Coronary
41. Crocodilia
42. Foramen of Panizza
43. Mammals
44. Encephalization, bilateral symmetry
45. Cranial nerves
46. Heart
47. Stratum laxum, stratum compactum, subcutaneous layer
48. Cell bodies
49. Mesoderm
50. Hagfishes
51. Pronephros
52. Oviduct,vagina,uterus
53. Sertoli
54. Holonephros
55. Gubernaculum
56. Cloaca
57. Inguinal
58. Reptiles, birds, mammals
59. Monotremes
60. Clasper
61. Mullerian duct
62. Wolffian duct
63. Monotremes
64. Oviduct
65. Tunica albuginea
66. Cloacal
67. Simplex
68. Mammals
69. Axons
70. Ventral

True or False

1. In gnathostomes, the maximum number of aortic arches is six pairs.
2. All vertebrate embryos have six pairs of aortic arches.
3. The first aortic arch is called the hyoid aortic arch.
4. In an amniotes, aortic arches are symmetrical.
5. In lampreys, there are six pairs of aortic arches

6. *Sphenodon* retains the fith aortic arch.
7. Ductus Botalli persists in urodels.
8. In *Protopterus*, the third and fourth aortic arches run without break.
9. Ganoid scales lack enamel.
10. Fin rays are dermal derivative.
11. Scales of fishes are epidermal derivative.
12. Birds lack dermal elements.
13. Mammary glands are branched alveolar glands.
14. The dermis is of mesodermal origin.
15. Reptiles lack claws on all toes.
16. In snakes and lizards, there is abundance of chromatophores in the upper layer of dermis.
17. In cyclostomes, epidermis is single layered.
18. In toads, a stratum corneum is thicker.
19. Generally, epidermis does not have blood and nerve supply.
20. The epidermis is thickest in mammals.
21. Moll's glands are found in the ear.
22. Stratum germinativum is the continuous layer of cells.
23. Epidermis contains many voluntary muscle cells to control facial expressions.
24. Merkel cells are keratinzed and pigmentary.
25. The penile dermis is very sensitive.
26. Fish scales may be bony or nonbony.
27. Cycloid scales have a toothed outer edge.
28. Coelocanth has modified cosmoid scale, which lacks cosmine.
29. Cellular renewal is the characteristic of vertebrate skin.
30. Keratin is formed in the epidermis of all living vertebrates.
31. In terrestrial vertebrates, the cuticle has been replaced by keratin.
32. Teeth are party derived from the dermal bone.
33. Cyanosis is more apparent in areas of thin skin.
34. Antlers are true horns.
35. Horns and scales are the product of stratum corneum.
36. Quadrate is immovable in streptostylic jaw suspension.
37. Amphistylic jaw suspension is characteristic of primitive sharks.
38. The salivary glands of mammals are the largest oral glands.
39. The intestine of teleosts is without a spiral valve.
40. Cyclostomes lack a true stomach.
41. Lips are muscular in amphibians.
42. Monotremes lack a true stomach.
43. The valves of the heart are opened and closed by pressure changes within the heart chambers.
44. In teleosts, the conus arteriosus is very long.

45. The conus arteriosus is not present in any amniotes.
46. The development of double circulatory system is necessitated by the presence of lungs.
47. In amphibians, the sinus venosus connects only to the right atrium.
48. Ductus cuvieri is lacking in the heart of *Scoliodon*.
49. *Myxine* has a portal, a cardinal and a caudal heart.
50. Monotremes lack caecum.
51. The stomach of the camel lacks the omasum.
52. The telencephalon forms the largest section of the mammalian brain.
53. The cerebellum is a part of the telencephalon.
54. Most regions of the human cerebral cortex have six layers of neurons.
55. All vertebrates do not have a bilaterally symmetrical brain.
56. Nerve fibres of lampreys lack myelin sheath.
57. The mammalian brain is completely dominated by cerebral hemispheres.
58. The behaviour, both learnt and instinctive of mammals, is controlled by the cortex.
59. Pallia are similar in agnathans, fish and amphibians.
60. A fish brain never fills the cranium completely.
61. The testes develop earlier in embryonic life than the ovaries
62. In fishes, the urinary bladder is large.
63. In reptiles, birds and monotremes, ovaries are solid but develop irregularly.
64. Monotremes have paired uteri.
65. Mesonephric kidney is metameric.
66. Genital ducts are lacking in cyclostomes.
67. In *Echidna*, teeth are lacking in all stages of life.
68. All copulatory organs of existing vertebrates are homologous.
69. Mammalian kidneys lack afferent blood supply.
70. Genital ridges are smaller than the functional adult gonads.
71. Armadillos have a simplex uterus.

Answers to True or False

1. True	2. True	3. False	4. True	5. False	6. False	7. True	8. True
9. True	10. True	11. False	12. True	13. False	14. True	15. False	16. True
17. False	18. True	19. True	20. True	21. False	22. True	23. False	24. False
25. True	26. True	27. False	28. True	29. True	30. True	31. True	32. True
33. True	34. False	35. True	36. False	37. True	38. True	39. True	40. True
41. False	42. True	43. True	44. False	45. True	46. True	47. True	48. False
49. True	50. True	51. True	52. True	53. False	54. True	55. False	56. True
57. True	58. True	59. True	60. True	61. True	62. False	63. True	64. True
65. False	66. True	67. True	68. False	69. True	70. False	71. True	

Give Reasons

1. Aortic arch fourth is always conspicuous arch in tetrapods.
   – Because it forms the major channel for flow of blood to the body.
2. In anurans, blood entering aortic arch third (the carotid arch) goes to the head.
   – Because the dorsal aorta between arches third and fourth is lost.
3. In anurans, blood entering the sixth aortic arch goes to the skin and lungs.
   – Because a segment (ductus arteriosus) of the sixth aortic arch is lost.
4. In the mammalian foetus, ductus arteriosus allows blood to bypass the inactive lungs.
   – Because it connects the pulmonary artery with the aorta.
5. Aortic arches do not break into afferent and efferent parts in tetrapods.
   – Because in tetrapods, there are no internal gills.
6. Fish scales are also called dermal bones.
   – Because they are mainly derived from the dermis.
7. Metabolic needs in the cells of stratum germinativum are fulfilled by dermis.
   – Because the epidermis lacks its own blood supply.
8. The cells of stratum lucidum are translucent.
   – Because of large amounts of keratohyaline (a precursor of keratin).
9. Hair does not fossilise well.
   – Because it decomposes readily.
10. Sebaceous glands are absent in soles and palms.
    – Because their oily secretion will create difficulty in keeping a good grip on things.
11. A little amount of ultraviolet radiation is essential.
    – Because the skin needs ultraviolet radiation to convert a steroidal related cholesterol to vitamin D.
12. The ovarian follicles of *Amphioxus* differ from those of vertebrates.
    – Because ovarian follicles of *Amphioxus* lack a follicular epithelium.
13. The stomach of monotremes is not considered a true stomach.
    – Because the lining of epithelium of stomach in monotremes is devoid of glands.
14. Lungs are considered excretory organs.
    – Because they remove metabolic wastes, especially carbon dioxide and water.
15. The epidermis of protochordates is similar to that of invertebrates.
    – Because it is thin and made up of a single layer of cells.
16. The rectum of mammals is not homologous with that of other vertebrates.
    – Because the rectum of mammals is derived from a portion of the cloaca.
17. Protochordates and cyclostomes lack peptic digestion.
    – Because they lack a stomach.
18. When a crocodile is above water and breathing air, the semilunar valve in the right atrium remains closed.
    – Because of higher pressure in the left and right aortas

19. The midbrain is the most important region in the amphibian brain.
    – Because besides sense of sight, it deals with nerve signals from most other regions of the body as well sends signal to other parts of the brain and the spinal cord.
20. The cerebellum is enlarged in fishes.
    – Because fishes are active animals, so cerebellum is enlarged so that they can efficiently control the movement and balance of the animal.
21. Many reptiles use the colon to reabsorb of water.
    – Because they lack the loop of Henle in their nephrons.
22. It is easy to study the excretory and reproductive system together.
    – Because they originate developmentally from the adjacent tissue and utilise common ducts to allow the passage of products outside of the body.
23. Generally, testes are smaller than ovaries.
    – Because sperms are much smaller than eggs, especially eggs with yolk.
24. Mammalian kidneys lack afferent blood supply.
    – Because in mammals, there is no renal portal system.
25. The third aortic arch is called the carotid arch.
    – Because it constitutes the commencement of the internal carotid artery.
26. The digestive system of snakes is elongated.
    – Because of their body type.

BIOMETRY

MEASURES OF CENTRAL TENDENCY

- The term 'measures of central tendency' refers to the determination of mean, mode and median.
- The properties of large collected statistical data are difficult to understand without further treatment.
- The vast statistical data are condensed in such a way that the basic character of the data does not change.
- An average reduces the large number of data/observations to one figure.
- The average is a number indicating the central value of a group of observations.
- The average value of any characteristics is the one central value around which lie other observations.
- Thus, 'average' is a general term that describes the centre of observations.
- Three common types of average are mean, median and mode.

Mean

- Mean is the sum of all observations divided by the number of observations.
- It is the most common measure of the central tendency.
- It is the best known and most useful form of average.
- The method of calculation of arithmetic mean depends upon the nature of data available, which may be explained as follows:

(a) When the observations are small in size (i.e., in a series of individual observations):

$$\text{Mean or } \overline{X} = \frac{\Sigma x}{N}$$

where X = Data (values of variable) and

N = Number of observations.

(b) Calculation of arithmetic mean in a discrete series:

$$\text{Mean or } \overline{X} = \frac{\Sigma fx}{N} \text{ or } \frac{\Sigma fx}{\Sigma f}$$

where f – Frequency,

x = The concerned variable and

N or Σf = Total number of observations.

(c) Calculations of arithmetic mean in a continuous series:

$$\text{Mean or } \overline{X} = \frac{\Sigma fm}{N}$$

where

Σfm = Total of the frequency of each class multiplied with the mid value of respective class and

N = Total of the frequencies.

Median

- When all the observations of a variable are arranged in either ascending or descending order, the middle observation is known as median.
- Median is neither based on the total nor is it affected by the extreme values of variables.
- Median is a point, not a score or any particular measurement.

Calculation of Median

(i) For individual observations:

Median or M = Sum of the N + 1/2th item where N = Number of items.

(ii) For Discrete Series

For calculating median in a discrete series, frequency is made cumulative and then median is calculated on the basis of above formula.

(iii) For Continuous Series

After making the frequencies cumulative, the median item is found out as N/2th item and then the median is calculated as per the following formula:

$$M = l_1 + \frac{i}{f}(m - c)$$

or

$$M = l_1 + \frac{\frac{N}{2} - c}{f} \times i$$

where M = Median,

li = Lower limit of the median class,

i = Class interval of the median class,

f = Frequency of the median class,

c = Cumulative frequency of the class preceding the median class and

m = N/2th item.

Mode

- The mode may be defined as the observation with the highest frequency.
- This is a value that occurs most frequently in a statistical distribution.

- Normally, mode is frequently used for categorical data.
- Mode – Median = 2 (Median – Mean) or Mode = Mean = 3 (Median – Mean)

Calculation of Mode

- For individual series — After converting the data into discrete series, the modal item should be picked up as the most occurring value.
- For discrete series — Mode can be located simply by inspection of the series, i.e., the size having the highest frequency will be mode of that series.
- For continuous series — In a distribution of grouped data, the mode is estimated at the midpoint of the class interval having the greatest frequency.
- Mode can be calculated in a continuous series by the following formula:

$$MO = i_1 + \frac{f_1 - f_0}{2f_1 - f_0 - f_2} \times i$$

or

$$MO = \frac{f_1 - f_0}{(f_1 - f_0) + (f_1 - f_2)} \times i$$

where li = Lower limit of the modal class,

f_1 = Frequency of the modal class,

f_0 = Frequency of the class preceding modal class,

f_2 = Frequency of the class following the modal class and

i = Class interval of the modal class.

DISTRIBUTION

- Frequency distribution is of two types, viz., observed frequency distribution and expected frequency distribution.
- Observed frequency distribution is prepared on the basis of actual data, whereas expected frequency distribution is a theoretical one.
- Calculations of theoretical distribution are useful in many ways, such as to understand the risk and uncertainty in any event, helps in forecasting, serves as benchmarks for comparison, etc.

Types of Distribution

- There are different types of theoretical frequency distribution, but the following three are of great importance:

1. **Binomial Distribution**
    - It is also known as Bernoulli's distribution.
    - It is identified by the number of the observations, n, and the probability of occurrence which is denoted by p.
    - The essential features of this distribution are as follows.

(a) The number of trials is fixed.
(b) There are two mutually exclusive possible outcomes of each trial.
(c) The trials are independent.

- The binomial distribution is used when a researcher is interested in the occurrence of the events and not in its magnitude.
- This distribution is widely used in industries for quality control.

2. Poisson Distribution

- Poisson distribution was developed by the French mathematician, Simeon Denis Poisson (1837).
- It is a very useful probability distribution.
- Poisson distribution gives the idea of probability of rare events, i.e., the number of trials is very small and the probability of success is also very small.
- Poisson distribution is a discrete distribution with a single parameter, i.e., the mean of distribution. It is widely used in insurance, spread of diseases, physiology and genetics.

3. Normal Distribution

- The pattern of distribution of data that follows the bell-shaped curve is known as normal distribution.
- Normal distribution was used by mathematicians de Moivre and Laplace in the 1700s. German mathematician and physicist, Karl Gauss, used it to analyse astronomical data, hence it is also known as Gaussian distribution.
- All normal distribution is symmetric
- Normal distribution is the most useful theoretical distribution for continuous variables.
- The shape of normal distribution resembles the bell, so sometimes it is also referred to as the bell curve.
- It is the most frequently used of all probability distributions.
- The general equation that describes normal distribution curve is as follows:

$$Y = \frac{N}{6\sqrt{2\pi}} - e\,\frac{x^2}{26^2}$$

where N = Number of measures,
Y = Frequency,
π = 3.1416,
c = 2.7183,
σ = Standard deviation of the distribution and
x = Deviation of any unit of measurement from the mean.

- Biological distribution is generally assumed as normally distributed.

CORRELATION

- Correlation is a statistical technique showing relationship between two variables.
- It is one of the most common and most useful statistics.
- The possible correlations range from +1 to – 1.
- A zero correlation indicates that there is no relationship between variables.

- A correlation of – 1 indicates that if one variable increases the other decreases, while a correlation of +1 indicates that both variables move in the same direction.
- Further, it shows the closeness or degree of relationship between the variables.
- Correlation is also a marker of interdependence between two variables.

Types of Correlation

- On the basis of nature of relationship between the variables, correlation may be of the following types:
  (a) Positive or negative
  (b) Simple, partial or multiple
  (c) Linear or non-linear

Degree of Correlation

- On the basis of coefficient of correlation, the degree of correlation may be of the following types:
  (a) Perfect
  (b) Limited
  (c) Absent
- The degree of relationship between two variables is the coefficient of correlation represented by the symbol 'r'.
- It is called Karl Pearson's coefficient of correlation and is most widely used.

Calculation of Karl Pearson's Coefficient of Correlation

- For individual series

$$rxy = \frac{\Sigma xy}{\sqrt{\Sigma x^2 \cdot \Sigma y^2}}$$

where

r = Correlation coefficient,

x = Deviation of X variables $(X - \overline{X})$ and

y = Deviation of Y variables $(Y - \overline{Y})$.

Product-Moment Correlation

- When number of observation (N) is small, their correlation can be calculated by product-moment method, according to the formula given below:

$$\frac{\Sigma xy - \dfrac{\Sigma x \Sigma y}{N}}{\sqrt{\left[\Sigma x^2 - \dfrac{(\Sigma x)^2}{N}\right] \times \left[\Sigma y^2 - \dfrac{(\Sigma y)^2}{N}\right]}}$$

- When the values of two variables are grouped and the frequencies of different groups are given, the correlation coefficient is calculated as per the following formula:

$$\frac{\Sigma fdxdy - N\left(\frac{\Sigma fdx}{N}\right)\left(\frac{\Sigma fdy}{N}\right)}{\sqrt{\left[\frac{\Sigma fdx^2}{N} - \left(\frac{\Sigma fdx}{N}\right)^2\right] \times \left[\frac{\Sigma fdy^2}{N} - \left(\frac{\Sigma fdy}{N}\right)^2\right]}}$$

REGRESSION ANALYSIS

- Regression analysis is the technique for the prediction of the relationship of a particular variable with another, on the basis of its relationship with a third variable.
- The variable to be estimated is called the dependent variable and the variable that provides the basis for estimation is called the independent variable.
- In multiple regression, there are two or more independent variables and one dependent variable.
- In other words, from regression analysis, we can estimate the value of one variable from the given value of the other variable. For example, we can find out the expected weight of a fish from a given length.
- The relationship between the independent variable (X) and the dependent variable (Y) is expressed as regression equation. The regression equation expresses the regression lines.
- Since there are two regression lines, there are two regression equations.
- The regression equation X on Y shows the variation in the values of X for changes in Y.
- Likewise, regression equation Y on X describes the variation in the values of Y for changes in X.

  Regression equation:

  $$X = a + by \ (x \text{ on } y)$$
  $$Y = a + bx \ (y \text{ on } x)$$

  where 'a' is a constant (the point where regression line touches (Y-axis) and 'b' is also a constant called regression coefficient.

  The multiple regression equation shows the effect of a number of independent variables at the same time which may be written as follows:

  $Yc = a + b_1x_1 + b_2x_2 + b_3x_3$ ....

  Where Yc = Value of dependent variable

  x_1x_2, x_3 .... = Independent variable

  b_1, b_2, b_3 .... = Regression coefficient

HYPOTHESIS TESTING AND TEST OF SIGNIFICANCE

- The test of significance is used by the researchers to determine whether the difference between calculated value and the hypothetical parameter is significant or not
- It establishes whether there is relationship between variables or the observed values have been produced by the chance.

- The phrase test of significance was coined by R A Fischer (1925).
- Every test of significance is associated with a basic concept known as the hypothesis.
- The hypothesis is basically a statement about the population parameters.
- It can be grouped into two types, viz., null hypothesis and alternative hypothesis
- Statistical inferences are drawn on the basis of information we get from the sample.
- In other words, it is possible to make reasonable estimates from the sample data available.
- Even if we don't know about a population, we can get reliable information about it on the basis of random sample from that population.
- The estimation deals with the methods by which population parameter/characteristics are estimated from sample information, whereas hypothesis testing deals with the process involved in the acceptance or non-acceptance of the assumption or a statement about the population parameter.
- Hypothesis testing enables us to verify whether or not such statements are in agreement with the available data.

Null and Alternative Hypothesis

- The hypothesis to be tested is called 'Null Hypothesis' and is represented by Ho.
- This may be written as follows:

  Ho : $\mu - x = 0$

  [where x = Sample mean and μ = Population mean]
- From the above equation, it can be concluded that there is no difference between the population mean and sample mean.
- Null hypothesis must be tested.
- To test the null hypothesis, there is an alternative hypothesis represented as H_1.
- If this alternative hypothesis is correct, the null hypothesis is rejected.

Errors in Testing of Hypothesis

- Since the acceptance or rejection of null hypothesis (Ho) depends on sample study, there is every chance of error.
- The error may be:
  1. Type 1(α) error — to reject null hypothesis when it is true.
  2. Type II (β) error — to accept null hypothesis when it is false.

Level of Significance

- The probability of committing α-error is called level of significance.
- 5 per cent (0.05) and 1 per cent (0.01) are the most commonly used levels of significance.
- 5 per cent level of significance shows that out of 100 times, there is a probability that 5 times correct Ho will be rejected.

Test of Significance

- An assessment of significance of difference between parameters of different samples is known as the test of significance.

- Such a test gives an idea whether observed differences between two samples are significant or have occurred due to chance.

STANDARD ERROR OF MEAN (SE$\overline{X}$)

- The standard deviation of the sample means is called the standard error of mean.

$$\text{S.E. } \overline{X} = \sqrt{\frac{\Sigma(\overline{x} - \mu)^2}{N-1}}$$

where μ = Mean of the sample mean and X = Sample mean.

- When the standard deviation of a population is known,

$$\text{S.E. of mean} = \frac{SD}{\sqrt{N}}$$

- A small value of standard error of mean is a clear indication of the fact that the various values of X are close to each other and average difference between these Xs and μ is small.
- As the sample size increases, the standard error of mean becomes smaller. At the same time, on increasing the sample size, various sample means become more uniform.
- Standard error is useful in testing a given hypothesis.
- It gives an idea about unreliability of a sample.

STANDARD ERROR OF STANDARD DEVIATION

- Standard deviation of different samples of the same population varies.
- So the standard error of standard deviation can be calculated to test the significance.
- SE of standard deviation data can be calculated as follows:

$$\text{SE } \sigma = \frac{SD}{\sqrt{2N}}$$

For grouped data, it can be calculated by the following formula:

$$\text{SE } \sigma = \frac{SD}{\sqrt{2\Sigma f}}$$

STUDENT t-TEST

- Student t-test is a small sample test.
- Student t-test was developed by W S Gosset (1908).
- Gosset published his work in pseudonym 'Student' in 1908.
- It is the most common statistical technique used to test the hypothesis based on difference between sample means.
- It is also called t-ratio because it is a ratio of difference between two means.

$$t = \frac{\overline{X} - \mu}{S / \sqrt{N}}$$

where X = mean of the sample,

S = Standard deviation of the sample and

N = Number of observations in the sample.

- A conclusion based on t-test is good if the distribution is normal or near and samples are chosen randomly.
- Fisher's table gives the highest obtainable values of 't' under different probabilities, with (P) in decimal fractions corresponding to the degrees of freedom.
- Probability of occurrence of any calculated value of 't' is determined by comparing it with the value given in the table.
- If the calculated 't' value exceeds the value given in the table, it is said to be significant.

Application of the t-Test

(a) Student 't' test for single mean is used to test a hypothesis on specific value of the population mean.
(b) Student t-test is used to test the difference between the means of two samples.
(c) The paired t-test is applied when the two samples are dependent.
(d) A t-test is used to test the significance of an observed correlation coefficient.
(e) A t-test is used for testing significance of regression coefficient.

CHI-SQUARE (X^2) TEST

- Chi-square test is the most commonly used method for comparing frequencies.
- It is a statistical test that is used to measure difference between an observed data with the data we would expect according to a given hypothesis.
- Chi-square is calculated on the basis of frequencies in a sample.
- It is used as a test of significance when the data are in forms of frequencies or percentages or proportions.
- It is one of the simplest and widely used non-parametric tests in statistical analysis.
- Chi-square test compares the observed value with the expected value and find out how far the differences between the two values can be attributed to fluctuations of simple sampling.
- The Chi-square test was developed by Prof. A R Fischer (1870) and it was further developed by Karl Pearson (1906) in its present form.
- The following are the essentials to apply x^2 test:
- Random sample
- Qualitative data
- Lowest expected frequency not less than 5
- Chi-square can be calculated by the following formula :

$$x^2 = \Sigma \frac{(O - E)^2}{E}$$

where O = Observed number of trails and

E = Expected number of trails.

- In x^2 test, the number of degrees of freedom is equal to the number of classes minus one.
- The value of x^2 depends on the degrees of freedom.
- x^2 test is also applied as a test of goodness of fit as it shows the closeness of observed and expected frequency.

Characteristics of x^2 test

(a) It is based on frequencies.
(b) It is non-negative.
(c) It is highly skewed.
(d) It is based on degrees of freedom.
(e) With the change in degree of freedom, a new chi-square distribution is created.
(f) The shape of chi-square distribution does not depend on the size of sample. It may depend upon the number of categories.

Uses

(a) A chi-square test is used as a test of homogeneity. It is a test which is used to determine whether several populations are similar or equal or homogenous in some characteristics.
(b) Chi-square test is used as test of independence. With the help of chi-square test, one can be able to know whether two attributes are associated or not.
(c) Chi-square test as a test of goodness of fit is used to determine whether the sample data are in consistent with the hypothesised data.

ANALYSIS OF VARIANCE

- To test the hypothesis whether the means of several samples have significant differences or not, a method called analysis of variance is used.
- This method is based on the comparison of variances estimated from various sources.
- The analysis of variance is based on the following assumptions:
  (a) Populations are normally distributed.
  (b) Populations from which the samples have been taken have means (μ_1, μ_2, μ_3 etc.) and variances ($\sigma_1^2 = \sigma_2^2 = \sigma_3^2$ ....)
  (c) Samples have been randomly selected.
- F-test is widely used in the analysis of variance and is calculated as follows:

  F = *variance between samples / variance within samples*
- The analysis of variance is mainly of the following two types:
  (i) One-way analysis of variance
  (ii) Two-way analysis of variance

I. One-way Analysis of Variance

- Here, analysis of variance observations are grouped on the basis of single criterion, i.e., the influence of only one factor is considered.
- In this type of analysis of variance, samples have been taken from normal populations with common variance.

II. Two-way Analysis of Variance

- Here we have to take consideration of the influence of two factors.
- The data are grouped according to the two different factors.

BIOMETRY

Short-Answer Questions

1. What is biometry?

   Answer: The branch of statistics that is related with application in biological science is called biometry.

2. Who is considered the 'Founder of the Biometric School'?

   Answer: English scientist Francis Galton (1822–1911) is considered the founder of the biometric school.

3. What do you mean by the term 'measures of central tendency?'

   Answer: The term 'measures of central tendency' refers to the determination of mean, mode and median.

4. Define arithmetic mean.

   Answer: Arithmetic mean is the sum of observation divided by the number of observations. It is denoted by ${}_1\overline{X}$. ${}_1\overline{X} = {}_1\Sigma X$ N where, ${}_1\overline{X}$ = Arithmetic mean, ${}_1\Sigma X$ = Sum of all values of variables and N = Number of observations.

5. Give the types of arithmetic mean.

   Answer: Arithmetic mean is of the following two types:

   (a) Simple arithmetic mean
   (b) Weighted arithmetic mean

6. What are merits of arithmetic mean?

   Answer: (a) Arithmetic mean is easy to calculate and simple to understand.
   (b) It is an ideal measure of central tendency.
   (c) It is capable of algebraic treatment.
   (d) It can be calculated even if some of the numbers and the number of observations are known.
   (e) It is based on all observations.
   (f) It is least affected by the fluctuations of sampling.
   (g) It is widely used in statistical methods as it is being capable of treated mathematically.

7. What are the demerits of mean?

   Answer: (a) Arithmetic mean cannot be calculated for qualitative data (such as smoking, honesty).
   (b) It cannot be calculated if one of the values is missing.
   (c) It is greatly affected by extreme observation.
   (d) It cannot be determined by inspection (mode and median) or graphical location.
   (e) It is not an appropriate mean for highly skewed distribution.
   (f) The simple mean gives greater importance to larger values and little importance to smaller values.

8. What is geometric mean?

   Answer: Geometric mean is the nth root of the product of n items of the series.

9. Mention two merits and two demerits of the geometric mean.

Answer: **Merits**

(a) Geometric mean is not much affected by fluctuation of sampling.

(b) It is very useful in studying economics and social data.

Demerits

(a) It is not easy to calculate.

(b) It cannot be calculated if any of the items is zero or negative.

10. Mention the application of the simple harmonic mean.

Answer: Simple harmonic mean is very useful in computing average rates.

11. What is mode? Mention its merits and demerits.

Answer: Mode may be defined as the value of the variable that is repeated in the maximum number of times the data.

Merits

(a) It can be easily computed if some extreme values are missing.

(b) It can be represented graphically.

(c) It is the best representative of the series as it occurs more frequently in the series.

(d) Its calculation is easy.

Demerits

(a) Mode is a vague measure of the central tendency.

(b) It is not directly based on all values.

(c) It is not as stable as the arithmetic mean.

(d) If the frequency of all items is equal, it becomes difficult to identify mode.

12. Give the empirical formula for mode.

Answer: Mode = 3 Median – 2 Mean

13. What is median?

Answer: Median is the middle value in the list numbers. For even number of data, median is calculated by obtaining the average of two middle data.

14. What are the merits and demerits of median?

Answer: **Merits**

(a) It is easy to calculate and to understand.

(b) It is most suitable for calculating qualitative data such as intelligence.

(c) It can be located graphically.

(d) It can be calculated even in some incomplete series.

(e) It is not affected by the extreme values.

Demerits

(a) It is not algebraic.

(b) The value of median is affected by the sample fluctuation.

(c) It is not as stable as mean.

(d) If the data is larger it will be time consuming to place them in order of size.

(e) Median is a less representative of the average as it does not depend on all items of the series.

15. What is range?

Answer: Range is the difference between the largest and the smallest values.

16. Mention merits of range.

Answer: (a) It is simple to understand and easy to calculate.

(b) Computation of range takes minimum time.

17. What are the limitations of range?
Answer: (a) It is not based on each and every item of the distribution.
(b) It is not possible to calculate range in case of open-end distribution.
(c) It does not throw any light about the distribution of values in the series.
(d) It changes from one sample to next in a population.
(e) It is affected by sampling fluctuations.

18. Give the name of the better average when the numbers are defined in relation to some unit.
Answer: Simple harmonic mean

19. What is standard deviation?
Answer: Standard deviation is the square root of the variance which is the squared differences from the mean.

20. Write the formula of calculation of standard deviation when the sample size is less than 30 and more than 30.
Answer:

(a) $s = \sqrt{\dfrac{\Sigma(x-\bar{x})^2}{N-1}}$

where

s = the standard deviation,
x = each value in the sample,
$\bar{x}$ = the mean of the values and
N = the number of values (the sample size).

(b) $\sigma = \sqrt{\dfrac{\Sigma(x-\bar{x})^2}{N}}$

where

σ = the standard deviation

21. Who first suggested standard deviation?
Answer: Standard deviation was first suggested by Karl Pearson (1881).

22. Mention the merits and demerits of standard deviation.
Answer: **Merits**
(a) It is rigidly defined.
(b) It is based on all the representatives of the series.
(c) It is useful in calculating standard error.
(d) It is least affected by fluctuations of sampling.
(e) Further algebraic treatment is possible.

Demerits
(a) It provides weightage to only extreme values.
(b) It is more affected by extreme items.
(c) It gives weightage to only extreme values.

23. Mention the formula of calculating standard error of mean?
Answer: Standard error of mean = $\dfrac{\text{Standard deviation}}{\text{Square root of number of observations}}$

24. Why is standard deviation also known as root mean square deviation?
Answer: Because it is the square root of the mean of the squared deviation from the arithmetic mean.

25. Who first applied student t-test?
Answer: W S Gosset (1908) first applied student t-test.

26. Define chi-square test.
Answer: Chi-square test is a statistical test commonly used to test observed data with the data we would expect according to a given hypothesis. It is also referred to as a measure of goodness of fit between data. Chi-square test is the most commonly used method for comparing frequencies.

$x^2 = \sum \frac{(O - E)^2}{E}$ where x^2 = Chi-square, O = Observed value and E = Expected value.

27. Give the formula of Yate's correction. When is it used?

$$\chi^2 = \sum \frac{[|O - E| - .5]^2}{E}$$

Answer: Yate's correction is applied when any of the observed frequency is less than 10.

28. What is denoted by Σ?
Answer: Σ denotes sum.

29. Who developed the Chi-square test?
Answer: Chi-square test was developed by Prof. A R Fisher (1870) and was further developed by Karl Pearson (1906) in its present form.

30. Mention the properties of the Chi-square test.
Answer: (a) Chi-square test is nonsymmetric.
(b) It is non-negative.
(c) There are many Chi-square distributions, one for each degree of freedom.
(d) It is based on frequencies.

31. When is the chi-square test very useful?
Answer: Chi-square test is very useful when data are not quantitative.

32. When is chi-square test used?
Answer: The chi-square test is used in two similar but distinct circumstances.
(a) For calculating how closely an observed frequency matches an expected distribution.
(b) For estimating whether two random variables are independent.

33. Give the formula for calculating expected frequency.
Answer: Expected frequency (E) = $\frac{\text{Column total} \times \text{Row total}}{\text{Grand total}}$

34. How are degrees of freedom calculated?
Answer: Degrees of freedom are calculated from the number of classes. The degree of freedom is number of classes minus one (N – 1). Thus, if there are two classes, the degree freedom is 1, (2 – 1). Likewise, if there are four classes, the degree of freedom would be 2, (3 – 1). In contingency table 2 X 2, the degree of freedom is calculated by the formula: Degree of freedom (df) = (c – 1) (r – 1) where r = Number of rows in a table and c = Number of column in a table.

35. What is a null hypothesis?
Answer: Null hypothesis is a hypothesis used in statistics that shows that no statistical relation exists in a set of given observations.

36. Define correlation.
Answer: Correlation is a statistical technique showing the degree of correlation between two variables.

37. Give the important types of correlation.
Answer: The important types of correlation are as follows:

(a) Positive and negative correlation (b) Simple, partial and multiple correlation
(c) Linear correlation

38. Mention the possible range of correlations.
Answer: The possible correlations range from +1 to – 1.

39. What value of correlation indicate a positive correlation, i.e., both variables moving in the same direction together?
Answer: A correlation of +1 indicates a positive correlation indicating that both variables move in the same direction together.

40. What are the merits and demerits of correlation?
Answer: **Merits**
(a) It is easy to calculate. (b) It is easy to understand.
(c) It helps in measuring the degree of relationships between two variables.
(d) It can be used for any type of data.

Demerits
(a) It is not convenient for large data. (b) It cannot be treated further algebraically.
(c) It is of no use to categorical data such as hair colour, gender, etc.
(d) Correlation highlights whether the two variables are related, but it does not point out that one is caused by the other.

Long-Answer Questions

1. What is an average? Describe different forms of averages.
2. Describe the merits and demerits of the following:
(a) Mean (b) Mode (c) Median
3. Mention the relation between median, mode and mean.
4. Define mean deviation with examples. Mention its merits and demerits.
5. Define normal distribution. How is the normality of a distribution tested?
6. Mention the importance of a normal distribution curve.
7. What is standard deviation? How is it calculated? Mention its merits and demerits.
8. Define standard error of mean. How is it calculated? Explain with suitable examples.
9. Define the student t-test. Explain with suitable examples.
10. What is chi-square test? How is it calculated? Describe with suitable examples.
11. What is goodness of fit? Describe two methods of testing the goodness of fit.
12 Define correlation. Explain positive correlation and negative correlation with suitable examples.
13. Write short notes on the following:
(a) Mean (b) Median (c) Mode (d) Range
(e) Geometric mean (f) Harmonic mean (g) Normal distribution (h) Student t-test
(i) Chi-square test (j) Degree of freedom (k) Mean deviation (l) Correlation
(m) Regression

BIOMETRY

Multiple-Choice Questions

1. Which one of the following statements is incorrect?
   (a) The value that is more frequent is called mode. (b) It is possible to have more than one mode.
   (c) It is possible to have no mode. (d) If there is no mode, it is represented by writing 0 (zero).
2. Mean, median and mode of 4, 6 and 5 will be:
   (a) 5, 5 and 5 (b) 5, 6 and 0 (c) 5, 6 and no mode (d) 5, 6 and 3
3. The total deviation of all data toward the arithmetic mean is:
   (a) Always zero (b) 1 (c) Always more than 1 (d) 0, 1 or more than 2
4. Consider the following statements:
   (A) It is not possible to determine the value of mode graphically
   (B) Mode is affected by extremely large or small items
   (C) Mode can be used to describe qualitative phenomenon
   (D) For the determination of mode, only the values occurring with high frequencies are required to be known

   The correct statements are:
   (a) All of these (b) A, B and C (c) B and C (d) C and D
5. Which one of the following is correct?
   (A) Mode – median = 2 (median – mean) (B) Mode – median = 3(median – mean)
   (C) Both (a) and (b) (D) None
6. According to null hypothesis, the difference in means of two sets of observations theoretically should be:
   (a) Zero (b) Not zero (c) 1 (d) More than 1
7. Match column 1 with column II and select the correct answer using answer codes:

Column I	Column II
(A) Combined mean	1. $n\sqrt{X_1.X_2.X_3.X_4 \ldots\ldots\ldots\ldots X_n}$
(B) Arithmetic mean	2. $\frac{N}{\Sigma (N/X)}$
(C) Harmonic mean	3. $\frac{N_1\overline{X}_1 + N_2\overline{X}_2}{N_1 + N_2}$
(D) Geometric mean	4. $\frac{\Sigma X}{N}$

   Answer codes:

	A	B	C	D
(a)	4	3	2	1

(b) 3 1 4 2
(c) 3 4 2 1
(d) 2 3 4 1

8. Median is suitable for expressing data on:
(a) Health (b) Colour (c) Intelligence (d) All of these
9. The geometric mean of 2 and 8 is:
(a) 0 (b) 2 (c) 16 (d) None of these
10. The harmonic mean of 3, 4 and 5 is:
(a) 4 (b) 3 (c) 0.038 (d) 60
11. The median of the observations 2, 4, 4, 5, 6, 6, 6, is:
(a) 2 (b) 4 (c) 5 (d) 6
12. The median in the even integer is calculated by the formula:
(a) $^n/_2$th observation
(b) Value of $(n+1)_2$th observation in the arrangement of observations in increasing order
(c) Value of $^n/_2$th observation + value of $(^n/_2 + 1)$th observation
(d) $^n/_2$ th observation – value of $(^n/_2 + 1)$th observation
13. Consider the following statements:
(A) The mean and median are both measures of central tendency
(B) The mean and median are the same for symmetric distribution
(C) For positively skewed distributions, generally the mean will be less than the median
(D) For positively skewed distributions, the mean is higher than the medium
The correct statements are:
(a) All (b) (A), (B) and (C) (c) (A) and (B) (d) (C) and (D)
14. What is the median value of 3, 4, 2, 5, 7, 8?
(a) 2 (b) 2.5 (c) 3.5 (d) 5
15. In which one of the following may there be more than one answer?
(a) Mode (b) Median (c) Mean (d) None of these
16. Calculate the median of the data 34, 36, 38, 40, 42, 46, 48, 50:
(a) 40 (b) 41 (c) 42 (d) 41.75
17. The t-test is a basic test that is limited to:
(a) One group (b) Two groups (c) Both (a) and (b) (d) Multiple groups
18. t-rest is used when:
(a) The sample is in large size and standard deviation is unknown
(b) The sample is in small size and standard deviation is unknown
(c) The standard deviation is known and the sample is in large size
(d) The standard deviation is known and the sample is in small size
19. The t-test to know the significance of difference of means of two samples was applied by:
(a) W S Gosset (1908) (b) A R Fisher (1870)
(c) M Fisher (1936) (d) Kari Pearson (1902)
20. Paired t-test is calculated by the formula:
(a) $t = \overline{X}$. N/S.D (b) $t = \overline{X}.\sqrt{N}$/S.D (c) $t = \overline{X}$.SD/N (d) $t = \sqrt{N} \times$ S.D/ X
21. William Gosset published his t-test work in 1908 in:
(a) *Statistical Science* (b) *Biometrika*

(c) *The American Statistician* (d) *American Naturalist*

22. Student t-test is one of the most commonly used techniques for testing a hypothesis on the basis of:
    (a) A difference between sample means
    (b) A difference between sample standard deviation
    (c) A difference between observed and expected frequencies
    (d) All of these
23. Consider the following statements:
    (A) Student t-test difference of means is used for smaller samples
    (B) Student t-test of difference of means can be used to compare the sample means between two independent samples
    (C) Student t-test of difference of means can be used to compare the sample means between two dependent samples
    (D) Student t-test of difference of means is a parametric test which does not assumes a normal distribution

    The correct statements are:
    (a) All of these (b) A, B and C (c) B and C (d) A and D
24. The t-test can be done knowing just the:
    (a) Number of data points (b) Means
    (c) Standard deviation (d) All of these
25. If the p-value associated with the t-test is not small (>0.05), there is evidence that:
    (a) The means are significantly different. (b) The means are not different.
    (c) There is evidence to reject null hypothesis. (d) None of the above
26. A paired test is generally used:
    (a) To test the hypothesis that two populations have the same mean (b) When the two samples are dependent
    (c) To test the hypothesis that two populations have the same mean (d) All of these
27. Student t-test of difference of means is used when the sample size is less than:
    (a) 10 (b) 20 (c) 30 (d) 50
28. The shape of the student t-distribution is determined by the:
    (a) Degree of freedom (b) Mean
    (c) Number of observation (d) Standard error
29. In paired t-test, the degree of freedom (df) is:
    (a) N–1 (b) N–2 (c) $N_1 + N_2 + 2$ (d) $N_1 + N_2 - 2$
30. Chi-square test was developed by:
    (a) Fisher (b) Karl Pearson (c) Gosset (d) Garret
31. For a contingency table, 2 × 2 table, the degree of freedom is:
    (a) V = (C – 1) + (r – 1) (b) V = (C – 1)/(r – 1)
    (c) V = (C – 1) × (r – 1) (d) V = (C – 1)/(r – 1)
32. Which one of the following formula is used to calculate chi-square value?
    (a) $X^2 = \frac{\Sigma(0-E)^2}{E}$ (b) $X^2 = \frac{\Sigma(0+E)^2}{E}$ (c) $X^2 - \frac{\Sigma(0-E)^2}{0+E}$ (c) $X^2 = \left\{\frac{\Sigma(0-E)^2}{E}\right\}$
33. Which one of the following statements is incorrect?
    (a) The F distribution is a ratio of two chi square distributions.

(b) F distribution is right skewed distribution.
(c) F distribution is used most commonly in analysis of variance.
(d) None of the above.

34. Expected frequency is calculated by the formula:
(a) $E = \dfrac{\text{Grand Total}}{\text{Column total} \times \text{Row total}}$
(b) $E = \dfrac{\text{Column total} \times \text{Row total}}{\text{Grand total}}$
(c) $E = \dfrac{\text{Column total} + \text{Row total}}{\text{Grand}}$
(d) $E = \dfrac{\text{Column total} - \text{Row total}}{\text{Grand total}}$

35. Karl Pearson's coefficient of skewness is calculated by the formula:
(a) $SK = \dfrac{\text{Mean} + \text{Mode}}{\text{Standard deviation}}$
(b) $SK = \dfrac{\text{Mean–Mode}}{\text{Standard deviation}}$
(c) $SK = \dfrac{\text{Mean} \times \text{Mode}}{\text{Standard deviation}}$
(d) $SK = \dfrac{\text{Mean–Mode}}{\text{Standard deviation}}$

36. Skewness is used to find out:
(a) Deviation (b) Mean (c) Mode (d) Medium

37. If the value of mean is greater than the mode, the skewness will be:
(a) Positive (b) Negative
(c) May be positive or negative (d) Zero

38. Normal distribution was discovered by:
(a) De moivre (b) La Place (c) Fisher (d) Bliss

39. Lorenz curve is a graphic method of studying:
(a) Mean (b) Mode (c) Dispersion (d) Standard deviation

40. Which one of the following tests may be evaluated against the chi-square distribution?
(a) Student t-test (b) G-test (c) Wald test (d) None of these

41. Which one of the following is an incorrect match?
(a) Coefficient of variation – Karl Pearson (b) Mean deviation – Bliss
(c) Loren curve – Max O Lorenz (d) Standard deviation – Karl Pearson

42. Which one of the following is needed for the chi-square test?
(a) Data should be qualitative. (b) Sample must be random.
(c) Observed frequency should not less than 5. (d) All of these

43. For 4´3 contingent table, the degree of freedom is:
(a) 2 (b) 4 (c) 6 (d) 8

44. The chi-square test of independence is used for:
(a) One variable (b) Two variables
(c) More than two variables (d) All of these

45. The chi-square test is not suitable for:
(a) Frequency within any phenotypic class less than 5 (b) Each expected frequency is between five and 10
(c) Where there are only two classes (d) All of the above

46. In chi-square test, Yates' correction is needed particularly where there are:
(a) No phenotypic classes (b) Phenotypic classes
(c) Two phenotypic classes (d) None of these

47. Yates' correction for continuity is needed:
    (a) Where there are two phenotypic classes
    (b) In small samples, where the numbers of individuals expected in each class is between 5 and 10
    (c) Where the sample is small
    (d) All of these
48. The chi-square test can be applied to discrete distribution such as the:
    (a) Poisson (b) Binomial (c) Both (a) and (b) (d) None of these
49. Generally, a probability of ______________ is considered to be a significant difference:
    (a) 0.1 (b) 0.05 (c) 0.05 or less (d) 0.5
50. For independence assortment, attention is giving on ___________ figures on the Chi-square table:
    (a) 0.01 and 0.05 (b) 0.01 and 0.001 (c) 0.95 and 0.99 (d) 0.05
51. For goodness of fit, attention is given on __________ figure in the chi-square table:
    (a) 0.05 (b) 0.01 (c) 0.001 (d) All of these
52. Chi-square (X^2) test of goodness of fit test and test for independence of attributes depend only on the:
    (a) Set of observed frequencies (b) Set of observed and expected frequencies
    (c) Degrees of freedom (d) All of these
53. Standard deviation can be determined by taking _________ of the variance:
    (a) Dividing it by number of observations (b) Square root
    (c) By multiplying it to number of observations (d) None of the above
54. A standard deviation of data set equal to zero manifests that all values in the set are:
    (a) Same (b) Different
    (c) May be the same or different (d) Generally different but rarely the same.
55. Alve M Tuttle is related with:
    (a) Mode (b) Mean (c) Median (d) Range
56. The variability of a statistic is measured by its:
    (a) Range (b) Median (c) Mode (d) Standard deviation
57. With an increase in sample size, the standard deviation will:
    (a) Fluctuate (b) Consistently increase (c) Consistently decrease (d) Both (b) and (c)
58. The higher the probability, the more likely it is that the two sets are:
    (a) Different
    (b) Same
    (c) Same and that any differences are just due to random chance
    (d) Different and that any similarities are just due to random chance
59. Which one of the following tests is the most important and most used members of the nonparametric family:
    (a) X^2 test (b) Student's t-test (c) Z-test (d) F-test
60. The unpaired test should not be used when there is a significant difference between ____________ of two samples:
    (a) Means (b) Standard deviation (c) Variances (d) None of the above
61. Consider the following statements about Pearson's correlation coefficient.
    (A) r only measures the strength of a linear relationship
    (B) r is always between +1 and –1
    (C) r changes if the independent and dependent variables interchanged
    (D) r changes if the scale on either variable is changed

The correct statements are:
(a) All of the above (b) A, B and C (c) A and B (d) B and D

62. Which one of the following statements is incorrect?
(a) Use of chi-square is inappropriate if any expected frequency is below 1.
(b) Use of chi-square test is inappropriate if the expected frequency is less than 5 in more than 20 per cent cells.
(c) In 2 × 2 contingency table of chi-square test of independent, expected frequencies less than 5 are generally considered acceptable if the Yates'correction is used.
(d) None of the above

63. Consider the following statements about data used in a chi-square analysis:
(A) Randomly drawn from the population
(B) Measured variables must be dependent
(C) Observed frequencies should be small
(D) Values of independent and dependent variables should not be mutually exclusive.

The correct statements are:
(a) A, B and C (b) B, C and D (c) B and D (d) None of these

64. Which one of the following may be more suitable for describing exponential growth?
(a) Arithmetic mean
(b) Geometric mean
(c) Simple harmonic mean
(d) Quadratic mean

65. The term 'multiple regression' was given by:
(a) Gossett (1908) (b) Pearson (1908) (c) Mood (1950) (d) Yates (1910)

Answers to Multiple-Choice Questions

1. (d)	2. (c)	3. (a)	4. (d)	5. (c)	6. (a)	7. (c)	8. (d)
9. (d)	10. (c)	11. (c)	12. (c)	13. (c)	14. (c)	15. (a)	16. (b)
17. (b)	18. (b)	19. (a)	20. (b)	21. (b)	22. (a)	23. (b)	24. (d)
25. (d)	26. (b)	27. (d)	28. (c)	29. (a)	30. (d)	31. (c)	32. (b)
33. (d)	34. (b)	35. (b)	36. (a)	37. (a)	38. (a)	39. (c)	40. (c)
41. (b)	42. (d)	43. (c)	44. (b)	45. (d)	46. (c)	47. (d)	48. (c)
49. (c)	50. (c)	51. (d)	52. (d)	53. (b)	54. (a)	55. (d)	56. (a)
57. (a)	58. (c)	59. (a)	60. (c)	61. (c)	62. (d)	63. (b)	64. (b)
65. (b)							

Fill in the Blanks

1. Arithmetic mean is obtained by adding up the values and then ______________ by the number of values.
2. The arithmetic mean is often called the ______________.
3. An observation occurring most frequently in data is called the ______________.
4. ______________ is the difference between largest and smallest values.
5. The ______________ is the middle value.
6. The two major types of statistical distributions are ____________ random variables and ____________ random variables.

7. A distribution with three modes is called ______________.
8. A geometric mean is obtained by multiplying values in a list and then calculating the __________ of the product.
9. The __________, _________ and _________ are used to get average of a data.
10. The range of the data 2, 1, 7, 8, 9, 6, 7 is the ______________.
11. A data set has no mode when all numbers appear in the data with the ___________ frequency.
12. The median is primarily used for______________ distributions.
13. If the number of values is even, the median is the average of the __________ values.
14. Coefficient of standard deviation is the ratio __________ its arithmetic mean.
15. t = Mean / _____________
16. In symmetrical distribution ___________, _____________ and ___________ coincide.
17. Student's t-distribution was first introduced by W S Gossett (1908) under the pen name____________.
18. x^2 tests ____________ hypothesis.
19. For the test of independence, only the __________ frequencies are used to calculate the expected frequencies.
20. Yates' correction deducts ______________ from the deviation applying to each class before it is squared.
21. The main difference in goodness of fit and independence assortments is in the use of ________ table.
22. Chi-square test for testing goodness of fit is used to conclude whether there is any difference between the _________ and _________ value.
23. The term standard deviation was first used by ______________.
24. To calculate standard deviation of a population, it is first necessary to calculate ___________ of the population.
25. The average of squared differences from the mean is called ___________.
26. The standard deviation is similar to average deviation, except that the averaging is done with _____________ instead of amplitude.
27. The standard error is generally indicated by the Greek letter____________.
28. Standard error of mean is a method to estimate the _____________ of a sampling distribution.
29. Standard error of mean = Standard deviation/_____________
30. Z-test is similar to t-test, except that it is only suitable for ____________ samples.
31. P varies from 0 to ____________ .
32. SKP stands for ______________ .
33. ______________ and _____________ distribution are ______________ distribution.
34. Standard deviation for ungrouped data is computed by the formula _____________.
35. 2×2 contingency table is prepared when there are only ____________ samples and each is divided into ___________ classes.
36. The value of the coefficient of correlation should always lie between ____________ and ____________.
37. The term regression was first proposed by _______________ (1877).
38. The first step in computing chi-square test of independence is to compute the ___________ frequency.
39. As per the null hypothesis, the difference in means of two sets of observations, theoretically should be___ .

40. To perform a statistical test, we first establish our____________ hypothesis.
41. Generally, a measure of goodness of fit summarises the discrepancy between ______________ and __________ values.
42. The x^2 test is used when there are ____________ categorical variables from a single population.
43. There are two major categories of x^2 tests, viz., ________ and ____________.
44. Geometric mean of 2 and 32 is ____________ .
45. The mean, mode and median are ______________ in symmetric distributions.
46. The two basic types of regression are ______________ regression and ______________ regression.
47. Linear regression (4) = ____________.
48. Linear regression use ________________ independent variable.
49. Multiple regression uses ______________ independent variables.
50. If r = __________, there is no correlation between two variables.
51. In zero correlation, _________ relationship may be present.

Answers to Fill in the Blanks

1. Dividing
2. Average
3. Mode
4. Range
5. Median
6. Discrete, Continuous
7. Trimodal
8. Square root
9. Mode, mean, median
10. Eight
11. Same
12. Skewed
13. Two middle
14. Standard deviation
15. Standard error
16. Mean, mode, median
17. Student
18. Null
19. Observed
20. 0.5
21. Chi-square
22. Observed, expected
23. Karl Pearson (1894)
24. Variance
25. Variance
26. Power
27. Sigma (ó)
28. Standard deviation
29. Root of the simple size
30. Large
31. One
32. Karl Pearson's coefficient of skewness
33. Binomial, Poisson, discontinuous
34. $\sqrt{x^2/N}$
35. Two, four
36. +1, –1
37. Sir Francis Galton
38. Expected
39. Zero
40. Null
41. Observed, expected
42. Two
43. Goodness of fit, contingency tests
44. Eight
45. Equal
46. Linear, Multiple
47. a + bx + O
48. One
49. Two or more
50. Zero
51. Nonlinear

True or False

1. Mean, mode and median are used in the measures of central tendency.
2. Range is a single number.
3. Mean, mode and median describe same characteristic of a data set.

4. The mode may be useful for dealing with categorical data.
5. Extreme values (outliers) affect the mode.
6. There are different types of t-test.
7. The t-distribution is similar to normal distribution.
8. t-test is used to compare means of two independent groups.
9. In student's t-test of difference of means, variances must be homogenous in nature.
10. The outcome of student's t-test is the acceptance or rejection of the null hypothesis.
11. The student's t-test is not useful for testing the statistical significance of a nonzero mean.
12. Median is often used when data are skewed, i.e., distribution is even.
13. There may be more than one mode if they are listed the same number of times.
14. When no values repeat in the data set, the mode's value is useless.
15. In median, 50 per cent of the observations are below the median value and 50 per cent above the median value.
16. Geometric mean is not capable of further algebraic manipulation.
17. Mode is based on all the observations of a series.
18. The value of median is determined by each and every observation.
19. Median is equal to mean if distribution is perfectly symmetrical.
20. Mean deviation exhibits algebraic properties.
21. Standard deviation is based on all observations.
22. Range is not based on all observations.
23. Chi-square test is popularly known as goodness of fit.
24. Any statistical test that involves use of the chi-square distribution may be termed as chi-square test.
25. Goodness of fit manifests the closeness of observed frequency with that of the expected frequency.
26. In a chi-square test, the degree of freedom is equal to the number of classes minus 2.
27. The chi-square test is most widely used on parametric tests.
28. If the chi-square test value is zero, it manifests that the observed and expected frequencies are completely coincident.
29. Chi-square test requires large observations.
30. Null hypothesis states that there is no significant difference between the observed and expected results.
31. If the p value of the calculated x^2 is $p<0.05$, the hypothesis taken into consideration is accepted.
32. The chi-square test does not give the results similar to the G-test.
33. The paired t-test may be used for comparing the strengths of the right arm with those of the left arm on a set of people.
34. W S Gossett was a chemist.
35. t-test cannot be applied for small number of cases.
36. A large chi-square statistic indicates that somewhere in the table the observed frequencies are exactly similar to the expected frequencies.
37. The chi-square test of goodness of fit needs assumption regarding distribution of parent population from which the samples are taken.

38. Standard deviation is calculated as the square root of variance.
39. Standard deviation is widely used measure of the dispersion.
40. Mode is the frequency of the most numerous score.
41. The standard error is a measure of central tendency.
42. The standard error is mainly calculated from sample attributes.
43. Standard error of the mean decreases as sample size increases.
44. Standard error of mean is also called standard deviation of the mean.
45. The t-test evaluates whether means of two groups are statistically different from each other.
46. The size of the sample influences the value of 't'.
47. The data must be in ratio for computing x^2 value.
48. Correlation analysis is a measurement showing the degree of correlation between two variables.
49. Correlation is influenced by the size of extreme items.
50. In normal distribution mean, median and mode do not coincide.

Answers to True or False

1. True	2. True	3. False	4. True	5. False	6. True	7. True	8. True
9. True	10. True	11. False	12. False	13. True	14. True	15. True	16. False
17. False	18. False	19. True	20. False	21. True	22. True	23. True	24. True
25. True	26. False	27. True	28. True	29. True	30. True	31. False	32. False
33. True	34. True	35. False	36. False	37. False	38. True	39. True	40. False
41. False	42. True	43. True	44. True	45. True	46. True	47. False	48. True
49. False							

Give Reasons

1. Mean is not a good measure of central tendency for extremely skewed distributions.
   – Because the mean is more affected by extreme scores than the median.
2. Mean reflects the value of each.
   – Because the calculation of means includes every number in the set.
3. If there are extremely high or extremely low values, it is preferable to use median for the calculation of average.
   – Because median is not affected by extreme values.
4. The mode of the series 8, 7.4, 5, 7, 12, 10 would be 7.
   – Because this value occurs more frequently than any of others
5. Range is a very crude measure.
   – Because range does not take into account all the observations.
6. Student's t-test is also called t-ratio.
   – Because it is a ratio of difference between two means.

7. Gossett's work on the t-test appear under the name 'Student'.
   - Because the brewery did not allow employees to publish their results, so Gossett's work on t-test appears under the name 'Student' (referred to as student's t-test).
8. Chi-square test is an identity test.
   - Because it is concerned about whether two samples were drawn from identical distributions.
9. The chi-square test is one of the most important statistics.
   - Because it is easy to calculate and interpret.
10. For small samples, the chi-square test is not very accurate.
    - Because the sampling distribution of statistic of the test differs from the chi-square distribution.
11. Chi-square test of goodness of fit and test for independence of attributes are also termed as nonparametric tests.
    - Because they do not involve any population parameters or characteristics.
12. The units of variance are not the same as the units of the data.
    - Because the differences are squared.
13. Standard error is important.
    - Because it shows how much sampling fluctuation a statistic will indicate.
14. F-test is also called variance ratio test.
    - Because it is based on the ratio of two variances.

ECONOMIC ZOOLOGY

APICULTURE

- Apiculture is the culture of honeybees on a large scale for the production of honey.
- It is one of the oldest forms of animal husbandry.
- Humans have been collecting honey from wild beehives for more than 8,000 years.
- Apiculture can be initiated by individuals or groups. It requires minimum capital and minimum dependence on foreign technology.
- Apiculture (beekeeping) is an agro-based enterprise which farmers can take up for additional income.
- Apiculture is a flourishing industry in many advanced country like the US, Canada, Germany, Australia, etc.
- The average output of honey from *Apis indica* colony is 3 to 5 kg per colony per year, whereas it is 15 kg per colony per year in *Apis dorsata*.

Species of Honeybees

(a) *Apis indica* (The Indian bee)
(b) *Apis mellifera* (The European bee)
(c) Apis dorsata (The rock bee)
(d) *Apis florea* (The little bee)

Life Cycle

The honeybee is a social and polymorphic insect. Its colony is known as a hive. In honeybees, the following three types of individuals are found:

(a) Queens–Queens are diploid and fertile females and develop from fertilised eggs.
(b) Workers–Workers are diploid, but sterile females and develop from fertilised eggs.
(c) Drones–Drones are haploid fertile males. They develop from unfertilised eggs (parthenogenetically).

- In honeybees, the queen lays two types of eggs, viz., fertilised and unfertilised.
- The larvae that hatch from unfertilised eggs develop parthenogenetically into drones.
- The development of diploid larvae into queens or workers depends on the quality of food.
- The larvae that obtain royal jelly as food throughout the larval period develop into queens, whereas the larvae that do not obtain royal jelly as food throughout the larval period develop into workers.

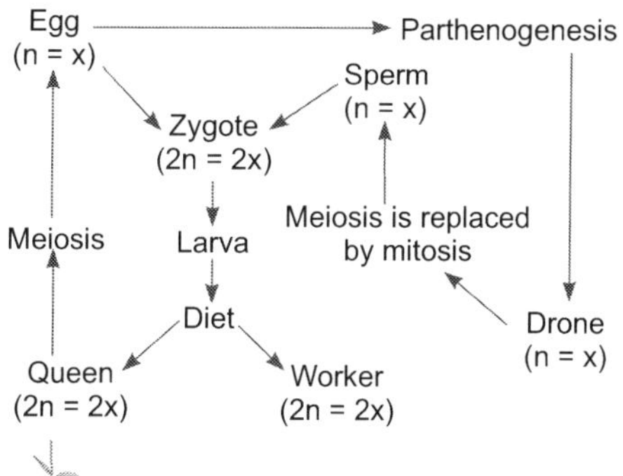

Fig. 1 Schematic representation of the formation of different castes in honeybees

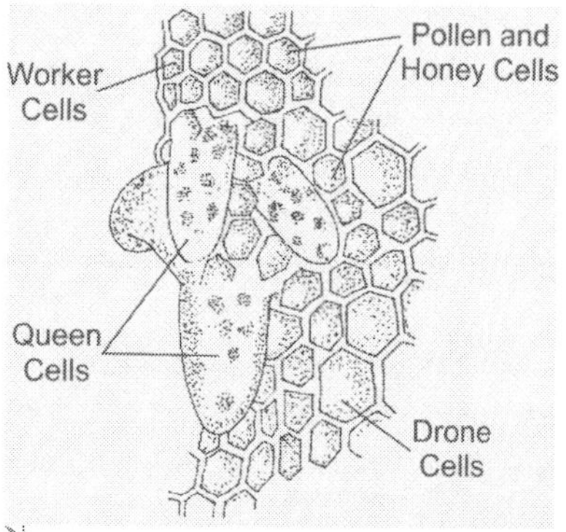

Fig. 2 Hive of *Apis indica* showing vaious cells

Table 1 Different stages of development of honeybees

S. No.	*Stage*	*Queen*	*Worker*	*Drone*
(a)	Egg	3 days	3 days	3 days
(b)	Larva	8 days	10 days	13 days
(c)	Pupa	4 days	8 days	8 days
(d)	Total	15 days	21 days	24 days

- In modern techniques of bee culture, an artificial bee-hive is used.
- The modern beehive is a 'movable frame hive'.
- The hive is made of a wooden box which may be single walled or double-walled.
- A double-walled hive gives better protection to the bees.
- Inside the hive, there is a 'brood chamber' with an opening at the bottom for the bees' entrance.
- The brood chamber has many vertical frames.
- The space between two frames is called 'bee space' and acts as a passage for the movement of bees.
- Above the 'brood chamber', there is another chamber called 'super' for storage of honey.
- The queen excluder, a frame fitted with wire net is found between the brood chamber and the super.
- A sheet of bee wax in which shape of different cells of the comb are made is called 'comb foundation'. It is cut to the size of hanging frames and is fitted inside.
- The queen is introduced into the hive through a 'queen introducing cage'.
- For extracting honey from the comb without damaging it, an instrument called 'honey extractor' is used.
- The spring season is the best season for collection of bees.
- For starting a new hive, the 'queen bee' is first introduced in the cage and then bees collected from natural hives are taken inside the hive.

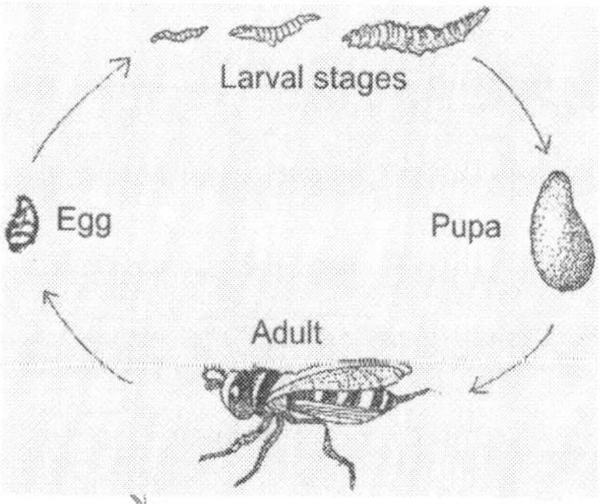

Fig. 3 Life cycle of a honeybee

- A new hive is thus established and within 24 hours, the queen and workers become familiar with the hive as well as with each other.
- A young and vigorous 'queen bee' is required for a good, functional hive.
- The queen bee feeds on royal jelly specially prepared for her.
- The worker bees feed on 'bee bread'.
- Among the workers, division of labour is seen for the purposes of fanning, cleaning, procurement of food, guarding the hive, etc.
- A hive may have 32,000 to 50,000 individuals.
- A colony may be 'weak' or 'strong' according to the number of workers.

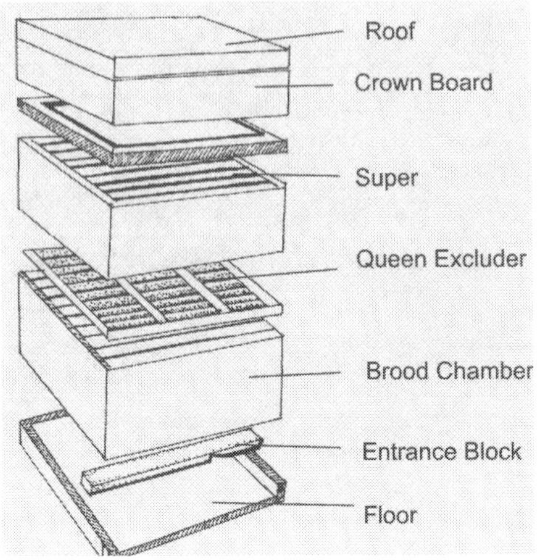

Fig. 4 Modern beehive

Diseases of Honeybees

Honeybees in an apiary suffer from the following diseases:

(a) Nosema Disease – It is caused by Protozoan *Nosema apis* which attack the epithelial layer of the midgut, leading to death.

(b) Acarine Disease – Acraine disease is caused by *Acarapis woodi* (mites) which enter the tracheal system and suck body fluids.

(c) Amoebic Disease – It is caused by *Malpghiamoeba mellificae* which attacks the malpighian tubules lining of the midgut.

(d) Brood Disease – Many bacteria and viruses attack broods of honeybees leading to the death of a large number of larvae.

Enemies and Predators

- Hawk moths (*Acherontia styx*), wax moths (*Galleria mellonella*) and wasps (*Vespa orientalis*) are predators of the honeybees' larvae.
- Ants, lizards, bears and crows, etc., are enemies of the honeybee.

Honey

- Honey is the first sweet food tasted by ancient Indians.
- It is a sweet, tasty semisolid obtained from the hives of honeybees.
- It is prepared from nectar and pollen by the honeybee in its honey sac.
- Honey contains sugar (sucrose, dextrose, levulose and dextrin), minerals (calcium, iron, manganese, etc.) and vitamin B complex.
- Honey is a good antiseptic as well as preservative.
- For collecting 500 g of honey, on an average, a honeybee makes 45,000 trips.

- The wax of bees is a byproduct of the apiculture industry and is used variously in the manufacture of cosmetics, face cream, paints, polishes, lubricants, etc.
- Apiculture has a positive ecological consequence, as bees play an important role in the pollination of many flowering plants. Thus, it increases the production of certain crops.

SERICULTURE

- Sericulture is the culture of silkworms.
- Production of silk from silkworms began in about 2700 BC.
- China was the first country to develop sericulture about 4,700 years ago and is the biggest producer of silk.
- India is the second largest country in the production of silk having an annual production of about 20,000 million tonnes.
- Karnataka is the pioneering state in the production of silk, accounting for 80 per cent of the total silk production of the country.
- There are a number of silk moths having the capacity of producing silk but only four types of silkworms have been recognised for producing silk of commercial importance.
- Mulberry silk is produced by *Bombyx mori*. It feeds on mulberry leaves. Mulberry silk is most popular variety in India, contributing about 87 per cent of the total silk production in the country.
- Tussar silk is produced by *Antharaea mylitta* and *Antharaea paphia* species. It feeds on the leaves of Arjun, sal, etc.
- Eri silk is produced by *Attacus ricini,* which feeds on the leaves of castor.
- Muga silk is produced by *Antharaea assamensis*. It feeds on champa, som, etc.

Life Cycle

- Silkworm may be univoltine (one crop in a year), bivoltine (two crops in a year) or multivoltine (more than two crops in a year).
- The *Bombyx mori* adult is about 2.5 cm and unisexual.
- Fertilisation is internal.
- Egg laying starts just after fertilisation. About 400 to 500 eggs are produced by an individual.
- After 10 days of incubation, eggs hatch into a larva called caterpillar.
- Caterpillars are voracious feeders and grow very fast.
- To attain the fully developed stage, a caterpillar takes about three weeks' time.
- After that, caterpillars stop feeding and pupation begins.
- The caterpillar starts secreting a sticky fluid from the silk gland.
- This fluid comes out through a small pore in the hypopharynx called spinneret.
- As soon as the fluid comes into contact with the air, it hardens as fine thread of silk.
- The body of the caterpillar is covered with silk threads called cocoon.
- A cocoon is made up of about 1,000 metres of silk thread.
- To obtain 500 g of silk, about 3,000 cocoons are required.
- The pupal stage lasts for about 10 days.

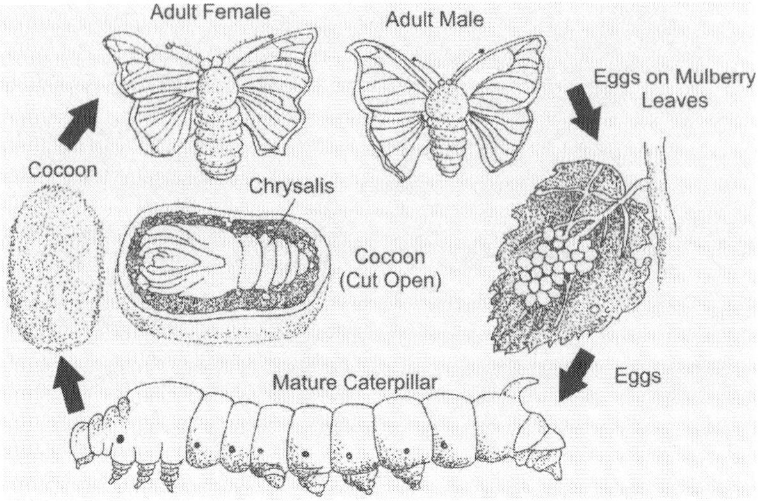

Fig. 5 Life cycle of *Bombyx mori*

- The pupa cuts the cocoon to emerge as adult moths.
- The total lifespan of *Bombyx mori* is 50 days.

Culture of Silkworm

- The eggs (commercially known as seeds) are kept in incubators (at 75° F) for hatching.
- Hatching is complete within 7 to 10 days and the caterpillars emerge.
- Continuous supply of healthy mulberry leaves is essential.
- Mud-walled thatched houses are preferred for rearing.
- Newly hatched caterpillars are kept in flat feeding trays with mulberry leaves.
- For keeping a good number of feeding trays, a machan is erected.
- Mature worms are transferred to spinning trays before the cocoon formation starts.
- Cocoons are processed to get silk.

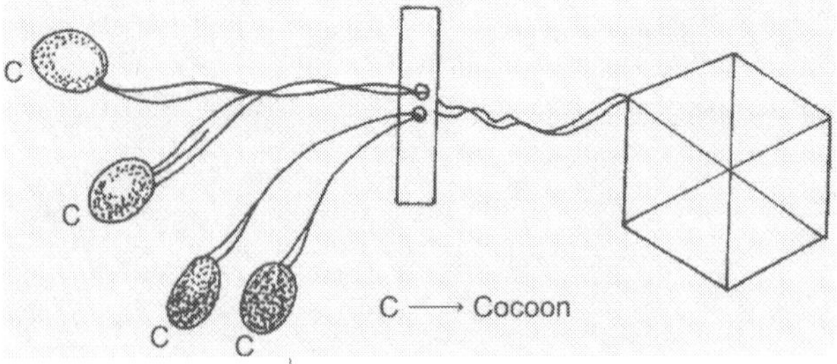

Fig. 6 Reeling of silk cocoons

- One should be very careful about the processing of cocoon, i.e., it should be done before the emergence of the silkworm by cutting the cocoons.
- For killing the pupae within the cocoons, good cocoons are dipped in hot water for about 10 minutes.
- Dipping in hot water helps in softening the adhesion of silk threads.
- Obtaining threads from cocoon is called reeling. Only about one half of the silk of each cocoon is reelable.
- The silk thus obtained is called 'raw silk' which is given various treatments for giving the final shape.

Diseases of Silkworms

- Silkworms suffer from the following diseases:
  - **(a) Pebrine Disease –** Pebrine disease is caused by the Protozoan parasite *Nosema bombycis.* The larva becomes pale and finally dies.
  - **(b) Flacherie –** Flacherie is a bacterial disease in which the body putrefies and becomes black or green.
  - **(c) Grassarie –** It is a viral disease that causes swelling of segments or rupture of skin.
  - **(d) Muscardine Disease –** It is a fungal disease and is fatal in nature.

Properties and Uses of Silk

- Silk is a natural protein having fibroin and sericin protein. Its chemical formula is $C_{12}H_{29} N_5 O$. It has the property to retain moisture.
- Silk fibres are mainly utilised in preparing silk fabrics.
- It is also used in the manufacture of fishing fibres, parachutes, insulation coils of telephones, racing car tyre, etc.
- Wastes from sericulture can be recycled as inputs in garden.
- Sericulture is an agro-based, highly remunerative cottage industry that provides employment to rural people as well as prevents their migration. Besides, it also helps in soil conservation and provides green cover.

LAC CULTURE

- The culture of lac insect using scientific techniques for the procurement of lac on large scale is known as lac culture.
- Lac is produced by lac insect (*Laccifer lacca* or *Tachardia lacca*) as a protective covering of its body.
- The lac insect was first discovered by Father Tachard.
- India is the biggest country in lac production. It produces about 85 per cent of the total production of lac in the world. India earns about Rs 120 to 130 crore of foreign exchange through export of lac every year.
- The insect is a parasite on plants. A number of plants such as Kusum, Khair, Ber, Babul, etc., are used as host by lac insect.
- The quality of lac is dependent on the quality of the host plant.
- The male lac insect is smaller than the female.
- Other Indian lac insects are *L. fici, L. indicola* etc.
- Female lac insect is responsible for large-scale production of lac.

Life Cycle

- The male enters the female cell, fertilises it and dies after copulation.
- About 200 to 500 eggs are laid by each female in incubating chambers.
- Eggs hatch into larvae after six weeks.
- Now the larvae start coming out of the female cell in search of a new area of host plant where they fix themselves.
- This mass movement of larvae is called 'swarming'.
- After getting fixed, the larvae start feeding continuously.
- Secretion of lac from larvae also starts just after the settlement of the larvae at a particular place.
- Gradually, the larvae become fully covered with the lac.
- The larvae moult in their respective cells.
- The male and female cell differs in shape, being elongated in male and oval in female.

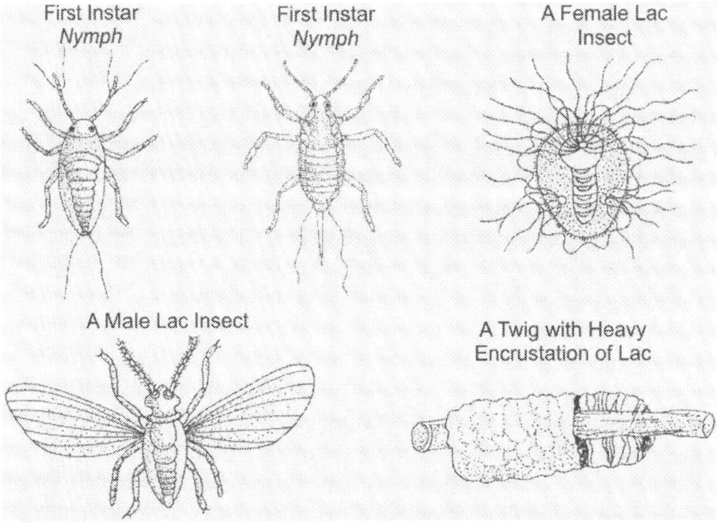

Fig. 7 Life history of lac insect

Cultivation

- The process of introducing the lac insect on the host plant is called inoculation.
- Pruning of the host plants is done before inoculation.
- The twigs having 'brood lac', i.e., lac insect larvae (just before swarming), about 20 cm in length are attached to a fresh host plant.
- The lac insect repeats its life cycle.
- There are two types of crops, viz., Rangini and Kusumi.
- The Rangini crop is also of two types, viz., Kartiki (inoculation: June to July) and Baisakhi (inoculation: October to November).

- The Kusumi crop is also of two types, viz., Agahani (inoculation: July) and Jethi (inoculation January to February).
- The collection of lac from the host plant is called harvesting.
- Harvesting may be immature (before swarming) or mature (after swarming).
- Immature harvesting produces 'Ari lac'.
- Mature harvesting produces 'mature lac'.
- Lac cut from the host plant is called 'stick lac'.
- After scraping of lac from twig, it is grinded and the unnecessary materials sorted out.
- Removal of dust and finer particles is done by cold water and the lac is called 'seed lac'.
- Seed lac is sun-dried and then melted to produce final lac shellac.
- About 30,00,000 lac insects are needed to produce 1 kg of lac.

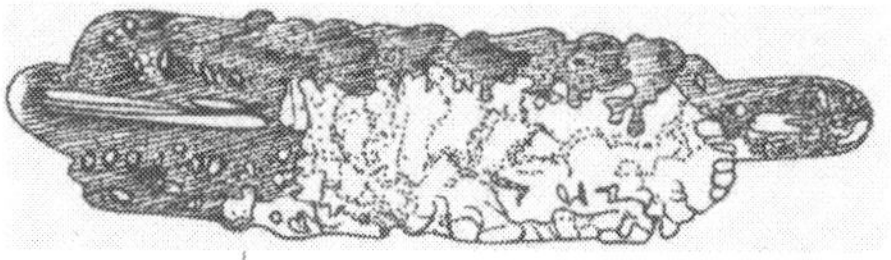

Fig. 8 A twig incrusted with lac

Natural Enemies of Lac

- *Eublemma amabilis* and *Holococera pulverea* are natural predators of lac insects. The caterpillars of these parasites feed upon lac insects.
- *Paraecthrodryinus clavicornis; Erencyrtus dewitzi; Tachardiaephagus tachardiae; Eupelmus tachardiae; Tetrasticus purpurens* insects are parasitic on lac insects. These parasites feed on lac insects showing hyperparasitism.

Properties and Uses of Lac

- Lac is a resinous complex substance having wax, dye and mineral matter.
- It is insoluble in water but soluble in alcohol.
- Lac is nontoxic and completely biodegradable.
- It is largely used as sealing wax and adhesive for optical instruments.
- It is a good insulator.
- It is used in shoe and leather polish as well as protective coating of wood.
- It is used in coating of pictures and fossils.
- It is used in painting and dyeing.
- Lac culture not only provides livelihood to millions of people but also helps in conserving stretches of forests and biodiversity associated with lac insect complex. Thus, lac culture not only prevents environmental degradation but also helps in conserving associated flora and fauna.

PEARL CULTURE

- Pearl or *moti* (as it is commonly called) is produced by bivalve Mollusca as a protection against foreign particles that have entered the body of an oyster.
- It is a white, shining globular structure.
- Although a number of bivalve Molluscs have the ability to produce pearl, pearls of high quality are pro-

duced by pearl oysters of genus Pinctada (Phylum – Mollusca; Class – Bivalvia; Family – Pteriidael).

- The important species of pearl oysters found in Indian waters are *P. vulgaris, P. atropurpurea,* etc.
- Other Molluscs such as *Heliotis, Mytilus* also produce pearl or pearl-like substances.
- Pearl culture provides the highest gross income per unit area among aquaculture systems. British biologist William Saville-Kent developed the accepted technique in Australia, which was brought to Japan by Tokichi Nishikawa and Tatsuhei Mis.

Pearl Formation

- Natural pearl begins its life cycle as a foreign object, such as parasite or sand particle.
- When a sand particle or a parasite or any foreign particle enters between the mantle and the shell, the foreign particle is enclosed in a sac made up of mantle epithelium.
- Mantle epithelium, being irritated by foreign particle starts secreting nacre (a chemical substance containing calcium carbonate and protein chonchiolin) around the foreign particle.
- The nacre (or the mother of pearl) is deposited around the foreign particle in concentric layers.
- Nacre is composed of microscopic crystals of calcium carbonate.
- The process of secreting nacre continues for several years, resulting in the formation of pearl.

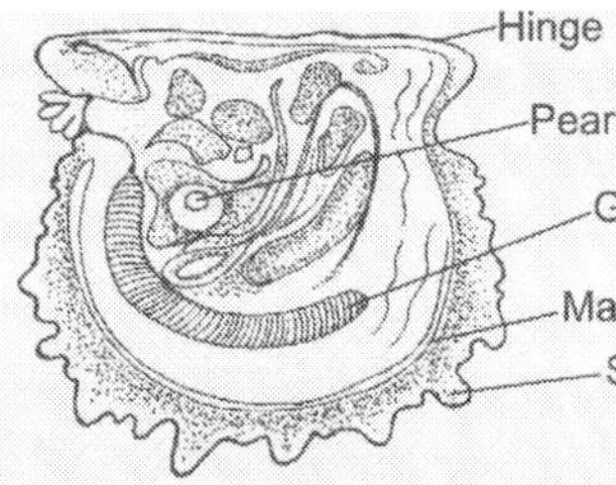

Diagrammatic sketch showing site of pearl formation in *Mytilus*

Pearl Culture

- Pearl is produced by inserting a foreign particle into the shell of oysters.
- The oysters are collected from the bottom of the sea by the divers.
- The collected oysters are reared in a special type of cage called rearing cage.
- A piece of mantle of living oyster is cut off and inserted together with a suitable nucleus (glass beads/ calcareous particle) inside the mantle of another oyster.
- The piece of mantle to be inserted in the mantle of the oyster is called 'graft tissue'.
- Calcareous nucleus gives good result.
- After insertion of nucleus, oysters are kept in seawater (enclosed in cages) at a depth of about 3 m.
- After 4 to 6 years, the pearls become ready.
- Harvesting of pearls is done between December to February.
- The best quality of pearl is called 'Lingha pearl'.

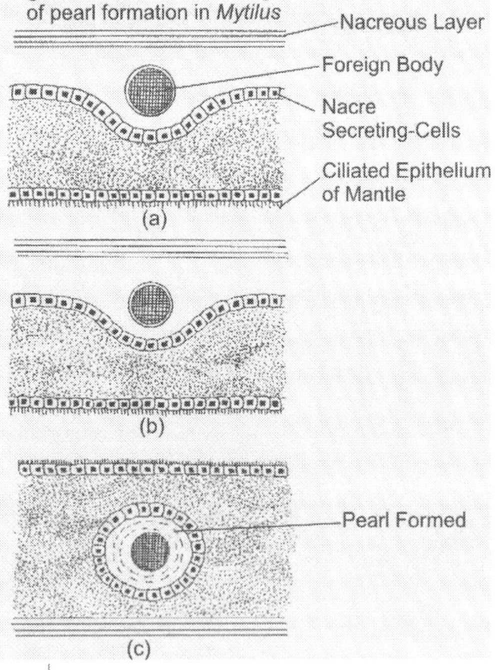

Fig. 9 Stages in pearl formation: (a) Primary stage, (b) Progressive stage and (c) Final stage

PISCICULTURE

- The culture of fish in 'inland water' is called pisciculture.
- The modern technique of pisciculture is the need of the time due to population explosion, thereby increasing the requirement of food.
- Fish is one of the most important sources of easily digestible protein, containing all the essential amino acids in desirable amounts.
- In India, fish culture was first started in West Bengal.
- Pisciculture is mainly concerned with the culture of Indian major carp, viz., Rohu *(Labe Orohita), Catla (Catla catla)* and Naini *(Cirrhina mrigala)*and exotic fishes like Silver Carp *(H. molitrix),* Grass Carp *(Ctenopharyngodon idellus),* Common Carp *(Cyprinis carpio)*, etc., introduced in Indian waters.
- Major Indian carps and the above mentioned exotic fishes do not breed in stagnant water and fish farmers have to depend on the procurement of fish seed from the rivers where they breed.
- Collection of fish seed from the natural riverine breeding sites is a difficult process and at the same time, the collection so made also contains undesirable fish seed.
- To obtain pure fish seed of the desirable fishes only, artificial methods are being adopted these days.
- For artificial breeding of major carps, perennial/seasonal/modern bundhs/ tanks are prepared.
- In bundhs, outlet and inlet of water is controlled, thus facilitating the breeding of fish.
- Apart from bundh breeding, i.e., breeding by artificially providing riverine condition, other techniques are applied for breeding of these fishes in ponds.
- One of the techniques is the artificial fertilisation in which ova and sperms from female and male fishes are mechanically squeezed out and mixed for fertilisation.
- Another technique is that of 'induced breeding' in which fishes are induced to breed in confined water by the injection of pituitary hormones.
- The process of induced breeding starts with the onset of monsoon in which 'breeders' are taken out from large ponds to smaller ponds.
- Usually one set of breeders (one female and two males) are kept in one breeding ponds after pituitary injection.
- Fish seeds (fertilised eggs, spawn, etc.) are collected from the breeding sites in the rivers or from the artificial breeding grounds in 'bundhs' or breeding 'hapa'.
- For the collection of spawn, special nets called 'Benchijal' are used.
- For further development of fish spawn, at least four types of ponds are required, viz., hatching pond or hatchery, nursery pond, rearing pond and stocking pond.
- Fertilised eggs collected from rivers or artificial breeding grounds are transferred to hatcheries for their hatching.
- Hatching may be a small sized pond (10 ft × 8 ft × 2 ft) where eggs are kept for hatching or may be 'hatching hapa' made up of cloth and bamboo poles in the river itself. Such hapas are in direct contact with the water.

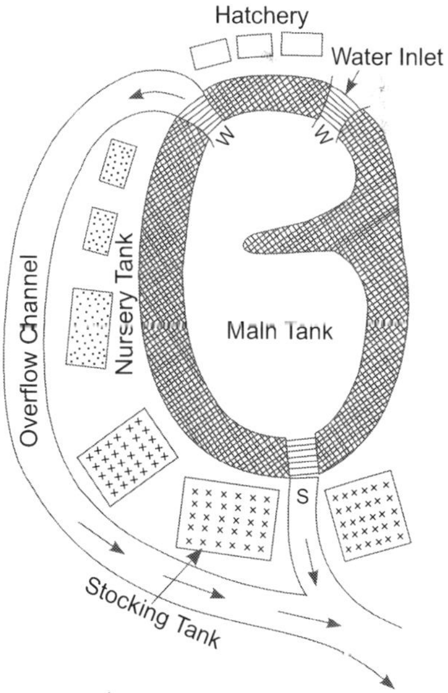

Fig. 10 Different types of ponds required for pisciculture

- Eggs hatch into larvae within 5 to 15 hours.
- After hatching, fish fry are transferred to nursery ponds.
- Nursery ponds are the ponds of about 50 ft × 30 ft × 5 ft deep size with a temporary shade above it to check the intense light.
- Nursery tanks contain clean water enriched with nutrients for proper growth of fish larvae.
- For transport of fish fry from the hatchery to the nursery pond, large sized metal containers with perforated lids or sealed metal containers with oxygen are used.
- Alkathene bags filled with oxygen are widely used for transfer of fish fry from hatchery hapas to nursery ponds.
- After attaining the size of about 2 to 3 cm., fish fry are transferred to the 'rearing ponds' which are about the size of 50 ft × 50 ft × 10 ft deep.
- In the rearing ponds, fish are kept till transfer to the stocking ponds.
- Mock netting' is also done in rearing ponds.
- After attaining the size of 10 to 15 cm., fingerlings are transferred to stocking ponds.
- Stocking ponds are manured with cow dung and inorganic fertilisers to maintain the natural food for the fish in the ponds.
- After stocking the fish, oil cakes, powdered rice, etc., are commonly used as artificial food.
- For good results, stocking of fingerlings of various major carp in the ratio of 3:3:4 (Catla, Rohu and Mrigal/Naini) is done.

PESTS OF SUGARCANE

- Sugarcane is perhaps the most important cash crop in Bihar, Uttar Pradesh, Madhya Pradesh, Maharashtra and a few other states.
- There are a number of insect pests causing damage to the production of cane crop, causing heavy losses to farmers.
- Some of the important pests of sugarcane are *Pyrilla perpusilla* (sugarcane leafhopper), *Emmalocera depresella* (sugarcane root borer), *Chilo infuscatellus* (sugarcane shoot borer), *Scirpophaga nivella* (sugarcane top borer), *Aleurolobus barodensis* (sugarcane white fly), *Bissetia steniellus* (Gurdaspur borer), etc.
- The life cycle and methods of control of some of the pests are described below.

I. *Pyrilla perpusilla*

Phylum – Arthropoda
Class – Insecta
Family – Fulgoridae
Genus – *Pyrilla*
Species – P. *perpusilla*

- It is the most destructive and foliage-sucking pest of the sugarcane.
- The sexes are separate.
- Adults are straw-coloured with head pointing forward as snout.
- It breeds in all seasons.
- Approximately 500 eggs are laid at a time.

- Eggs develop into nymphs in approximately three weeks.
- Nymphs are soft pale brown dorsally and pale orange ventrally, with two characteristic anal filaments.
- The nymphs feed on the cell sap of the leaves.
- The nymphs undergo moulting and winged adults are produced after five moulting.
- The development of nymphs into adults takes about 8 to 16 weeks.
- The lifespan of adults ranges from 5 to 8 weeks, with the lifespan of females being slightly greater.

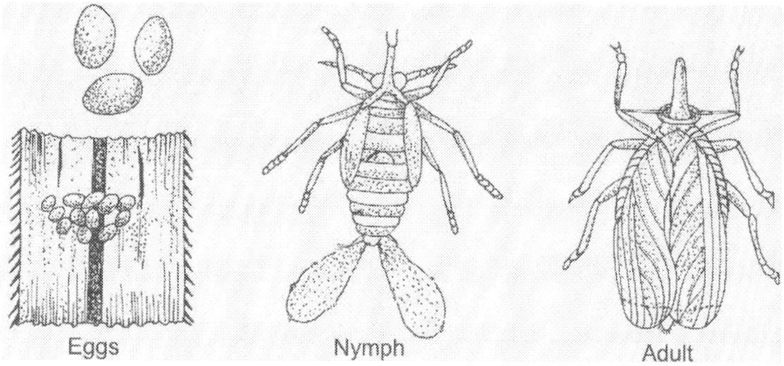

Fig. 11 Life history of Pyrilla Perpusilla

Nature of Damage

- Nymphs and adults both feed on the cell sap, thus damaging and drying up the leaves.
- Nymphs and adults pour a sticky substance called 'honeydew' on the leaves.
- Due to the presence of honeydew, a number of harmful fungi attack the leaves of sugarcane, decreasing the rate of photosynthesis.
- It causes a 30 to 35 per cent decrease in sugar production.

Control

- Excessive use of nitrogenous fertilisers should be avoided.
- Leaves with eggs should be collected and burnt.
- 0.2 per cent BHC or 0.025 per cent endosulfan or 0.01 per cent endrim should be sprayed for effective control.
- Resistant varieties of the crop should be used.

II. *Aleurolobus barodensis*

Phylum – Arthropoda
Class – Insecta
Order – Hemiptera
Family – Aleurodidae
Genus – *Aleurolobus*
Species – *A. barodensis*

- It is a polyphagous pest and is a serious pest of sugarcane.
- It is dispersed all over the country.
- Adults are pale yellow in colour, having a length of 3 mm.

- More than 50 eggs are laid by the female at a time, normally in the months of November and December.
- The eggs are conical and creamy white in colour.
- In a week or so, the eggs are transformed into nymphs.
- Nymphs start feeding on the cell sap of the leaves.
- In a span of four weeks, nymphs change into pupa.
- The pupa takes about ten days time to become an adult.
- The lifespan of adults is approximately 48 hours.

Nature of Damage

- The damage to the standing crop is by nymph. Nymphs suck sap from the leaves as a result of which leaves become yellow and pinkish, and gradually dry up.
- Heavily infested leaves are covered by a sooty mould caused by the fungus, which gradually adversely affects the photosynthesis.
- Sugar content is greatly reduced.

Control

- Infested leaves should be destroyed.
- 0.2 per cent BHC, 0.1 per cent malathion or 0.12 per cent endrin spraying controls the pest effectively.
- A resistant variety should be used.

III. *Chilo infuscatellus* (Sugarcane stem borer)

Phylum – Arthropoda
Class – Insecta
Order – Lepidoptera
Family – Crambidae
Genus – *Chilo*
Species – *C. infuscatellus*

- It is a polyphagous pest, but sugarcane is its favourite host plant.
- It is dispersed all over the country. Besides India, it is reported from Burma, Taiwan, Indonesia, Philippines and Pakistan
- Adults are 30 to 40 mm long having a wingspan of 3 mm.
- Both larvae and adults are nocturnal.
- After fertilisation, females lay egg in masses on the underside of leaves.
- Each female lays about 100 to 600 eggs.
- The eggs hatch in about a week.
- The larvae first feed on leaf epidermis and then bore into the stem.
- The larva feeds on stem tissue.
- The fully grown larva is about 3.8 cm in length and is dull white in colour, having longitudinal black stripes.
- The larval period is of about three weeks. Now the larva undergoes pupation inside the tunnel.
- The pupation period is of 6 to 10 days.
- Adults come out from the tunnel through the hole made by the larva before pupation.

- The adults survive for 8 days.
- The entire life cycle is completed within 35 to 45 days.
- There is a generation of 5 to 8 per year.

Nature of Damage

- It attacks the plant during the early period of the plant's growth.
- The first instar larva first feeds on the leaves of the plant and then bores into the shoot in the ground level and proceeds ahead in an upward direction, killing the growing points.
- It damages the sugarcane plants by forming dead hearts.
- Bore infestation during germination phase kills the mother shoot.
- It reduces both yield and quality of the sugarcane.

Control

- Affected plants should be removed and destroyed.
- Moths should be collected by light or pheromone traps.
- 10 per cent BHC in the soil should be dusted or 5 per cent endosulfan should be sprayed at three-week intervals.

PESTS OF PADDY

- Paddy is an important cereal in India and other Asian countries.
- There are a number of insect pests damaging standing crop of paddy, such as the following:
  (a) Paddy stem borer (*Tryporyza schoenobius*)
  (b) Swarming caterpillar (*Spodoptera mauritia*)
  (c) Rice bug (*Leptocorisa varicornis*)
  (d) Rice hispa (*Hispa armigera*)
  (e) Paddy grasshopper (*Hieroglyphus banian*)
  (f) Paddy gall fly (*Pachydiplosis oryzae*)
  (g) Paddy case worm (*Nymphula depunctatus*)
- The life cycle and methods of control of some pests are described below.

I. Paddy Gall Fly (*Pachydiplosis oryzae*)

Phylum – Arthropoda
Class – Insecta
Order –Diptera
Family – Cecidomyidae
Genus – *Pachydiplosis*
Species – *P. oryzae*

- This pest damages the paddy crop in India and Southeast Asia.
- The females lay their eggs on leaves and stems of the host plant.

- After hatching the, larva enters a young bud.
- Here, it damages the tissue and checks the apical growth.
- Due to the presence of maggot, an abnormal growth is produced.
- The maggot is transformed into a pupa.
- The pupa develops into an adult.
- Due to infestation of this pest, seedlings do not have ear.

Control

- Pests should be trapped and destroyed.
- 0.025 per cent endrin or parathion and diazinon spraying has effective control on the pest.

II. Rice Bug (*Leptocorisa varicornis*)

Phylum – Arthropoda
Class – Insecta
Order – Hemiptera
Family – Corcidae
Genus – *Leptocorisa*
Species – *L. varicornis*

- It is most menacing pests of paddy in India.
- They are known as gandhipoka due to their repulsive smell.
- It is about 1.8 cm long.
- The body is light green in colour and possesses one pair of fine antennae and two pairs of wings.
- The mouth parts are of the sucking type.
- The breeding is seen throughout the year.
- The female lays about 400 to 600 eggs at a time.
- Eggs hatch into nymphs within a week. Nymphs are very small but they look like miniature adult but they lack wing.
- The nymph takes 2 to 3 weeks to become an adult.
- The adult has a life span of about 5 weeks.
- The infestation of this pest is seen during the months of August to October.

Nature of Damage

- This pest causes maximum damage when the crop is in the flowering stage and the milky stage.
- The nymphs and adults both feed on the sell sap of paddy plants. They also suck the juice from the stem and leaves.
- The ears do not grow.

Control

- Pests should be trapped by net and destroyed.
- Dusting of 5 per cent BHC at 13 to 17 kg per hac is effective in controlling both nymphs and adults.
- Spraying of 0.25 per cent DDT or endrin (0.04 per cent) has also been effective.

III. *Tryporyza incertulas* (The stem borer)

Phylum – Arthropoda
Class – Insecta
Order – Lepidoptera
Family – Pyralidae
Genus – *Tryporyza*
Species – *T. incertulas*

- *Trporyza incertulas* is a monophagous pest that is endemic in paddy growing areas.
- Adults are 1.3 to 1.5 cm in length, having yellow coloured forewings and white hind wings in males, while the forewings of females have black spots.
- Females are larger than males.
- After fertilisation, the female lays eggs in masses. There are 50 to 100 eggs in each egg mass.
- Each egg mass is covered with yellow coloured hair.
- After 5 to 6 days, the larvae emerge from the eggs.
- The larvae are yellow coloured.
- The larvae begin to crawl over the plant and bore into the stem with the help of their biting and chewing type of mouth parts.
- The larvae feed on the plant tissue and bore a tunnel in the stem.
- Fully grown larvae are 20 mm long.
- The fully grown larva secretes a silken covering around itself and undergoes pupation.
- The pupal period is of 5 to 12 days.
- The adult moth emerges from the tunnel.
- The entire life cycle is completed within 40 to 60 days.

Nature of Damage

(a) The pest attacks paddy plants from seedling to the post flowering stage.
(b) The caterpillars feed on plant tissue, causing dead hearts. As a result, plants become hollow and dry.
(c) Due to the attack of this pest, the ear heads become white and empty.

Control

(a) The affected plants should be removed and destroyed.
(b) Dusting of DDT and BMC is beneficial.
(c) There should be cultivation of resistant variety of paddy.

PESTS OF WHEAT

Wheat is the main rabi crop of India. It is attacked by many pests such as the following:

(a) *Nanagria uniformis* (Wheat stem borer)
(b) *Mathymna separata* (Wheat Armyworm)

(c) *Schizaphis graminum* (Wheat Aphid)
(d) *Anapothrips flavicinctus* (Wheat Thrips)
(e) *Odontotermes obesus* (White ants)
(f) *Tanymecus indicus* (Gujhia weevil)

- The life cycle and methods of control of some pests of wheat are described below.

I. *Tanymecus indicus* (Gujhia Weevil)

Phylum –Arthropoda
Class – Insecta
Order – Curculionidae
Genus – *Tanymecus*
Species – *T. indicus*

- It is a polyphagous pest attacking wheat, barley, gram, pea, rice, etc. The wheat plant is its favourite host.
- It is distributed all over India, particularly in wheat growing areas (Punjab, Haryana, Uttar Pradesh and Bihar.
- Adults are earthen grey weevil, having a length of 4 to 8 mm and width of 2 to 3 mm.
- The female lays 6 to 80 eggs.
- After 6 to 7 weeks, the grubs emerge from the eggs.
- Grubs feed on the root and become fully grown within 8 to 10 days.
- The fully grown grubs pupate.
- The adult weevil emerges from the pupa after one week.
- The adults start feeding on the tender parts of the wheat plants.
- There is only one generation in a year.

Nature of Damage

(a) Both grubs and adults damage the wheat plants.
(b) The grubs feed on roots while adults cut the growing points at the margins of leaves, which is more severe at the seeding stage. As a result, the yield is very poor.

Control

(a) Fields should be ploughed during summer to expose and kill the pupae.
(b) 5 per cent aldrin or heptachlor should be mixed 12 to 25 cm deep with the layer of the soil.
(c) A resistant variety of wheat should be used.

II. *Mathymna separata* (ArmyWorm)

Phylum – Arthropoda
Class – Insecta
Order – Lepidoptera
Family – Noctuidae
Genus – *Mathymna*
Species – *M. separate*

- It is a polyphagous pest, occurring throughout the year in wheat growing areas.
- It attacks the crop that is sown late.
- Adults are pale brown moths.
- Females lay eggs on leaves and leaf sheaths near the ground.
- A female lays about 500 eggs in clusters.
- Larvae emerge from the eggs within 3 to 5 days.
- The larvae are very active and feed on tender shoots and leaves.
- The larvae are found in cracks in the soil. They feed at night or early in the morning. However, on damp days they may feed all day.
- The larvae become fully mature within 15 days.
- The fully mature larva is 3.5 cm long.
- The fully grown larva undergoes pupation by spinning cocoon.
- The pupal period is of about two weeks.
- The adult moth comes out from the pupa.
- During winter, the last instar larva enters diapause.

Nature of Damage

Larvae feed on tender leaves chewing from the edges to midribs causing heavy loss of the crop.

Control

(a) Soil should be treated with insecticides to kill larvae.
(b) Spraying 0.05 per cent malathion is useful.

III. *Schizaphis graminum* (Wheat Aphids)

Phylum – Arthropoda
Class – Insecta
Order – Hemiptera
Family – Aphididae
Genus – *Schizaphis*
Species – *S. graminum*

- It is a polyphagous pest and is dispersed all over India.
- It is found in colonies on plants.
- Adults are transparent and soft bodied, having a length of 1.5 to 2 mm.
- It has darker stripe running along the back. Its legs and antennae are black.
- They multiply extremely fast, forming large colonies.
- It reproduces both asexually (parthenogenetically) as well as sexually (in cold climate).
- Where winter is mild, the female produces 3 to 5 nymphs per day parthenogenetically.
- After three moults in 6 to 10 days, these nymphs mature and are able to give birth to nymphs.
- In colder regions, sexual reproduction occurs in which after mating, the female lays eggs.
- Nymphs emerge from eggs, which mature and start asexual reproduction.
- The life span is of 3 to 4 weeks.

Nature of Damage

(a) Nymphs and adults suck sap from leaves, tender shoots and immature grain.

(b) During feeding, it injects its saliva to plants, disintegrating the plant's cell wall and chlorophyll, thus causing toxic effect on plant tissues.

Control

Dusting with 5 per cent BHC or spraying 0.02 per cent Phosphamidon, 0.03 per cent Dimethoate or Diazinon is useful.

PESTS OF STORED GRAINS AND THEIR CONTROL

- Apart from pests that damage standing crop, there are pests that damage the stored grains and make them unfit for human use.
- It has been estimated that about one quarter to one third of world grain is destroyed each year during storage.
- The major factor behind this damage the attack of insects, which not only destroy stored grains, but also reduce the percentage of protein content of the food grains, thus lowering the percentage of the seeds that germinate.
- The important pests that effect stored grains are *Sitophilus oryzae, Corcyra cephalonica, Trogoderma granarium, Tribolium custaneum, Callosobronchus chinensis, Callosobronchus analis, Sitotroga cerealella,* etc.
- The life history and the methods of control of a few pests are given below.

I. Rice Weevil (*Sitophilus oryzae*)

Classification:

Phylum – Arthropoda
Class – Insecta
Order – Coleoptera
Family – Curculionidae
Genus – *Sitophilus*
Species – *S. oryzae*

- This pest has been found to be most destructive and causes damage to up to 50 per cent of stored rice.
- Adults are reddish brown, having a length of 3 mm and a forward projected slender pointed head.
- Females are larger than the males.
- Adults do not fly but try to crawl away, when disturbed.
- Adults as well as larvae feed upon entire seeds.
- The larval stage has been found to be more destructive.
- The adult is about 3 mm long; the male being smaller than the female.
- The female lays about 400 eggs during April to October.
- Eggs are whitish oval and are about 0.7 mm long.

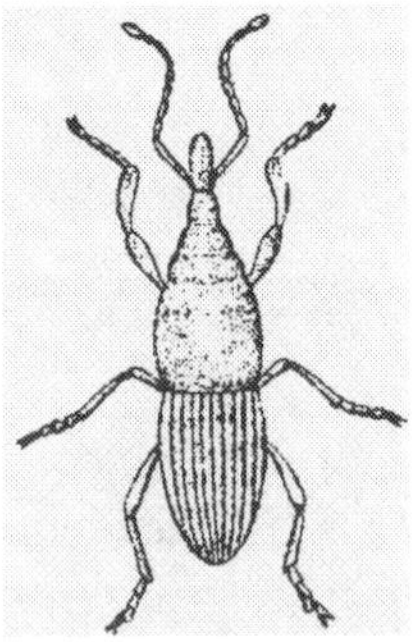

Fig. 12 Rice weevil

- Eggs hatch into larvae (called grubs) within a week.
- The grubs are tiny, fleshy, white and legless and are about 3 to 4 mm long.
- The larvae start feeding and transform into pupae.
- Pupation occurs inside the grain.
- The pupa is light yellowish, but later becomes dark brown.
- Within a week or two, the pupa develops into adult.
- The adult emerges by boring a hole in the grain.
- The adult pest has a lifespan of 4 to 5 months.
- There are three to four generations per year.

Damage

(a) Both larvae and adults feed on the grains, though larvae are more destructive.
(b) Larvae feed inside the grain, making it hollow and exit by making a circular hole on the surface.
(c) The maximum damage occurs in August to October, when moisture content is high while minimum damage occurs in summer.

Control

(a) Godowns should be clean, dry and insect proof.
(b) Godowns should be sterilised.
(c) Godowns should be heated up to 150^0 F for 10 hours.
(d) Fumigation of grain-filled granaries by 0.25 to 5 per cent HCN gives a good result.
(e) Aerosol controls surface infestation.

II. Rice Moth (*Corcyra cephalonica*)

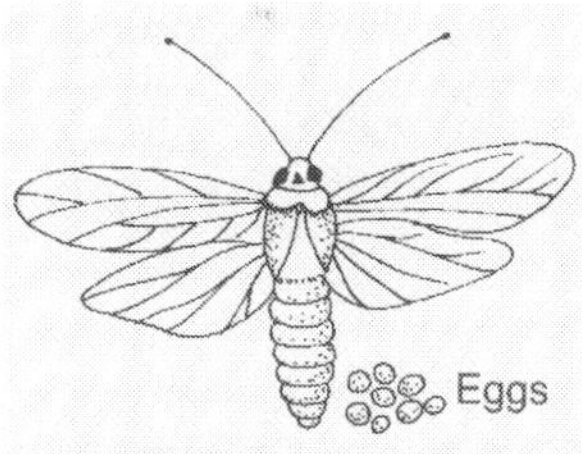

Fig. 13 Rice moth

Phylum – Arthropoda
Class – Insecta
Order – Lepidoptera
Family – Galleriidae
Genus – *Corcyra*
Species – *C. cephalonica*

- This is also a serious pest of rice.
- It damages stored wheat in India.
- Adults are greyish brown in colour, having a length of about 12 mm.
- The female lays about 300 eggs at a time.
- Eggs are whitish and oval in shape, having a length of about 0.5 mm.
- The eggs hatch into larvae within 2 weeks.
- Larvae start feeding after hatching.
- Fully grown larva is pale yellow in colour, having a length of about 15 mm.
- Development of larvae becomes complete in a month, after which pupation starts.
- Pupation occurs in a tough, opaque and whitish cocoon.
- The pupal state continues for a month.
- After about one month, the pupa develops into an adult.

- Depending upon temperature and humidity, the time required for the competition of life cycle varies from 40 to 90 days.

Damage

(a) The damage on stored grains is inflicted mainly by the larvae.
(b) In case of heavy infestation, the entire stock is converted into a mass of silken web.

Control

(a) Godowns should be sterilised.
(b) Fumigation by carbon tetrachloride and carbon bisulphide before and after the storage of grain is very effective.
(c) The infested grains should be removed.

III. Wheat Weevil (*Trogoderma granarium*)

Phylum – Arthropoda
Class – Insecta
Order – Coleoptera
Family – Dermestidae
Genus – *Trogoderma*
Species – *T. granarium*

- This is an important pest of stored wheat.
- Adults are small and sluggish, having a length of 2 to 4mm.
- The female starts laying eggs after 5 to 6 days of mating.
- On an average, the female lays 300 to 500 eggs.
- Eggs are dull whitish in colour and oval in shape, having a length of 0.7 mm.
- Within a week, the eggs hatch into larvae.
- Larvae are straw coloured, having a length of 4.5 mm.
- Full maturity of the larvae is attained within a month.
- After that, pupation starts and continues for a week.
- The pupa is oval in shape and dark brown in colour.
- The pupa develops into an adult.
- The entire life cycle is completed within 30 to 50 days.

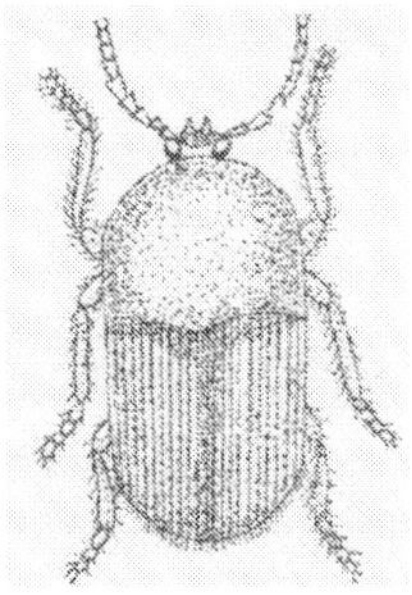

Fig. 14 Wheat weevil (*Trogodena granaium*)

Damage

(a) The damage is mainly caused by larvae.
(b) The marketability of the grain is reduced as larvae leave the cast skin in the grains.
(c) Maximum infestation is seen during the months July to October.

Control

(a) The larvae and adults should be collected and destroyed.
(b) Storage sites should be sterilised.

ECONOMIC ZOOLOGY

Short-Answer Questions

1. Define sericulture.
   Answer: The rearing of silkworms for the production of raw silk is known as sericulture.
2. What is moriculture?
   Answer: Moriculture is the cultivation of the mulberry plants.
3. When did production of silk from mulberry silk start?
   Answer: Production of silk from mulberry silk began in about 2700 BC.
4. In which country did silk production originate?
   Answer: Silk production originated in China.
5. Name two countries which are the biggest producers of silk.
   Answer: China and India are the biggest producers of silk in the world.
6. Define silk reeling.
   Answer: Extraction of silk filament from cocoons by employing a variety of processes is termed as silk reeling.
7. How is silk reeling presently done?
   Answer: Presently, silk reeling is done by three devices, viz., charka, cottage basins and multi-end basins.
8. Define raw silk.
   Answer: The silk obtained as a result of reeling is termed as raw silk.
9. Name the host plant of *Bombyx mori.*
   Answer: Mulberry
10. When is *Bombyx mori* reared?
    Answer: *Bombyx mori* is reared throughout the year.
11. What is the common name of *Antheraea assamensis*?
    Answer: Muga silk
12. Why cannot the cocoon of eri silk be reeled?
    Answer: Because it is made up of uneven fibres.
13. Name the pioneering state in India in the production of silk.
    Answer: Karnataka is the pioneering state in the production of the silk, accounting for about 80 per cent of the total silk production in the country.
14. How many types of silk are produced in India?
    Answer: In India, four types of silk are produced, viz., mulberry, tussar, eri and muga.
15. Name the different phases of activities of sericulture.
    Answer: Sericulture involves the following four different phases of activities:

(a) Cultivation of mulberry (b) Rearing of silkworms
(c) Reeling (d) Weaving and processing

16. In which state is the Central Silk Board located?
Answer: The Central Silk Board is located in Karnataka (Mysore) state.

17. Describe the chemical nature of silk.
Answer: Silk is a protein having fibroin and sericin protein. Fibroin forms the central structure while sericin is a sticky material surrounding the fibroin.

18. Name the acid that dissolves silk.
Answer: Sulphuric acid

19. What is tussah?
Answer: Tussah is a wild silk obtained from the *Antherea mylitta* caterpillar, which is found in China, Korea and India.

20. What is the chemical formula of silk?
Answer: The chemical formula of silk is $C_{12} H_{29} N_5 O_{11}$.

21. Define denier.
Answer: Denier is a unit of measure to determine the fineness and weight of silk yarn.

22. Which country is the largest consumer of silk?
Answer: India is the largest consumer of silk.

23. Name the diseases of silkworms and their causative agents.
Answer: (a) Pebrine disease – It is caused by the Protozoan parasite *Nosema bombycis*
(b) Grassarie – It is a viral disease
(c) Flacherie – It is a bacterial disease
(d) Muscardine disease – Muscardine is a fungal disease

24. Define apiculture.
Answer: The culture of honeybees to obtain honey and other allied products is known as apiculture.

25. What is an apiary?
Answer: Apiary is a place where beehives of honeybees are kept.

26. How many types of individuals are found in a honeybee colony?
Answer: A honeybee colony consists of the following three types of individuals:
(a) Queen – The queen is the diploid fertile female that develops from fertilised eggs. In a colony, there is only one queen. The queen performs only one function, i.e., laying eggs.
(b) Workers – Workers are diploid sterile females that develop from fertilised eggs. Workers perform all the functions of the honeybee colony, except laying eggs.
(c) Drones – Drones are fertile males that develop from unfertilised eggs (parthenogenetically). The drone fertilises the queen.

27. What determines the flavour, aroma and colour of honey?
Answer: The floral source determines the flavour, aroma and colour of honey.

28. How is the integrity of the honeybee colony maintained?
Answer: The integrity of the honeybee colony is maintained by s chemical secretion called pheromones.

29. Mention the ecological consequences of beekeeping.
Answer: Honeybees play a key role in pollination of many flowering plants, thus increasing the production of many crops (such as fruits, sunflower, etc.).

30. In a honeybee colony, how is the colony scent recognisable by bees of the same colony?
Answer: Because of the presence of the unique combination of nectar and pollen.

31. What is queen substance?
Answer: Queen substance is a pheromone produced by the mandibular gland of the queen that inhibits the formation of ovaries in worker bees.

32. Name the first sweet food tasted by ancient Indians.
Answer: Honey is the first sweet food tasted by the ancient Indians.

33. What is the chemical composition of honey?
Answer: Honey contains sugar (80 per cent), water (18 per cent), minerals, vitamins, pollens and amino acids (2 per cent).

34. What are pests?
Answer: Organisms that are harmful to humans or their properties are called pests.

35. What are monophagous, oligophagous and polyphagous pests? Give examples.
Answer: (a) Monophagous pest – The pests that feed on a single plant are called monophagous pests (e.g., *Tryporyza incertulas*).
(b) Oligophagous pest – They feed upon a limited variety of food (e.g., cabbage butterfly).
(c) Polyphagous pest – Polyphagous pests feed on different plants of different families (e.g., *Orseolia oryzae* (gall fly).

36. Name three pests of paddy.
Answer: (a) Swarming caterpillar, *Spodoptera mauritia*
(b) Gundhi bugs, *Leptocorisa varicornis*
(c) Tissue-borers, *Tryporyza incertulas*

37. Name a paddy pest that is monophagous.
Answer: *Tryporyza incertulas,* which feeds only on paddy plants.

38. Which stage of *Tryporyza* damages the paddy crop?
Answer: *Tryporyza* damages the paddy crop in the larval stage.

39. Why is *Tryporyza incertulas* is called a stem borer?
Answer: Because some caterpillars crawl down the plants and reach the central stem where they make small bores and migrate into the central tissue. This is why *Tryporyza* is called a stem borer.

40. Name the different stages of *Leptocorisa varicornis* and the damage caused by it.
Answer: The life cycle of *Leptocorisa varicornis* involves eggs, nymphs and adults. Both adults and nymphs suck milk from the paddy grains, as a result of which no formation of seeds occurs. They also suck juice from the stem and leaves.

41. Name three pests of sugarcane and parts of plant infested by them.
Answer:

	Pests	Plant parts infested
(a)	*Pyrilla perpusilla*	*Leaf and leaf sheath Notin*
(b)	*Chilo infuscatellus*	Young shoots and during draught the internodes
(c)	*Emmalocera depressella*	*Young plants Notin*

42. The caterpillars of which silkworm feed on oak, fig, ber and sal plants?
Answer: The tussar silkworm

43. Name the silk produced by *Antherea royeli* and *Antherea perniyi.*
Answer: Oak tussar silk

44. Name the species of honeybee found in India from which honey is collected.
Answer: Three species of honeybees are found in India from which honey is collected. These are:
(a) *Apis dorsata* (b) *Apis indica* (c) *Apis florea*

45. What is lac?
Answer: Lac is a natural resinous substance produced by female lac insect that is used in the production of varnishes, sealing wax and in the production of colouring matter.

46. Define lac culture.
Answer: The scientific management of lac insects to procure a large amount of lac is known as lac culture.

47. Who first systematically described the lac insect?
Answer: Father Tachard (1709)

48. What is the name of the Indian lac insect?
Answer: *Laccifer lacca.*

49. Name the host plants of *Laccifer lacca.*
Answer: *Zizyphus mauritiana* and *Zizyphus jujube.*

50. Name the major steps of the lac culture
Answer: Lac culture involves two major steps, viz., inoculation and cropping.

51. From which tree is the finest quality of shellac obtained?
Answer: The finest quality of shellac is obtained from the kusum tree *(Schleichera oleosa)*.

52. What is stick lac?
Answer: Lac scrapped from branches is known as crude lac or stick lac.

53. What is seed lac?
Answer: The harvested lac is ground in stone mills and the ground lac is repeatedly washed in cold water to remove insect parts and other soluble impurities. The material thus obtained is called seed lac.

54. Is lac resistant to ultraviolet rays?
Answer: Yes, lac is resistant to ultraviolet rays.

55. What is the name of the only known commercial resin of plant origin?
Answer: Shellac is the only known commercial resin of plant origin.

56. How is shellac made?
Answer: Shellac is generally made from two ingredients, viz., seed lac and ethyl alcohol.

57. Name the important trees needed for encrustation of lac.
Answer: Kusum, Ber, Palas, Babul, Sal, etc.

58. Mention the properties of lac.
Answer: (a) Lac is insoluble in water, but soluble in alcohol and weak alkalies.
(b) It has adhesive strength and hardness, and is resistant to ultraviolet rays.
(c) Lac is a bad conductor of heat.

59. What is the composition of lac?
Answer: Lac contains 68 per cent lac resin, 6 per cent lac wax, 1 to 2 per cent lac dye and 25 per cent other materials.

60. What is shellac?
Answer: Shellac is an amber-coloured resin produced by refining the secretion of the lac insect.

61. Which country is the leading producer of lac?
Answer: India is the leading producer of lac.

62. Which Indian state is the leading producer of lac?
Answer: The state of Jharkhand is the leading producer of lac.

63. Why is it safe to use lac?
Answer: Because lac is nontoxic and is completely biodegradable.

64. Give the systematic position of lac.
Answer: Phylum – Arthropoda
Class – Insecta
Order – Hemiptera
Sub-order – Homoptera
Super family – Coccoidea
Family – Lacciferidae
Genus — *Laccifer*
Species – *lacca*

65. Distinguish between male and female lac insects.
Answer: (a) Males are smaller than the females.
(b) Males lack mouth parts but females have mouth parts.
(c) Males have hyaline wings but females lack wings.
(d) Males are mobile while females are immobile.

66. How many times do the nymphs of the lac insect moult?
Answer: The nymphs of the lac insect moult three times.

67. What is pisciculture?
Answer: Pisciculture is the scientific management of fishes in their natural environment to procure good yield of fishes.

68. Mention the types of fish culture.
Answer: Fish culture may be grouped into the following three types:
Answer: (a) Inland fisheries (b) Marine fisheries (c) Estuarine fishes

69. Name the main freshwater fishes generally used for the purpose of culture.
Answer: Carps such as *Labeo*, *Catala*, *Cirrhina*, etc., are used for the purpose of culture.

70. What is the difference between fish food and other animal protein food?
Answer: The presence of omega-3 fatty acids (such as linolenic acid, decosahexaenoic acid (DHA) and eicosapentaenoic acid (EPA) differentiates fish food from other animal protein food. Omega-3 fatty acids are good for heart. DHA improves learning ability in children and memory in adults.

71. Define aquaculture.
Answer: The culture of aquatic organism such as fishes, crustaceans, molluscs and aquatic plants is known as aquaculture.

72. What is isinglass?
Answer: Isinglass is a substance obtained from the swim bladder of fishes, which is used in the clarification of wine and beer.

73. Name the factor to which spoilage of fish is directly related.
Answer: The spoilage of the fish is directly related to temperature.

74. Name three poisonous fish.

Answer: Three poisonous fish are the stonefish (*Synanceia verrucosa*) , lionfish (*Pterois miles*) and toadfish (*Porichthys*).

75. Name a poisonous amphibian and a poisonous mammal.

Answer: (a) Poisonous amphibian – *Dendrobates tinctorius* (Poison dart frog)

(b) Poisonous mammal – *Ornithorhynchus anatinus*

76. What is integrated pest management?

Answer: An effort towards solving pest problems by applying our skills to prevent them from harming crops, animals, damaging buildings and other things is called integrated pest management.

77. What is fish glue?

Answer: Fish glue is an adhesive prepared from fish bones and skin. It is used for book binding, shoe repairing, etc.

78. Mention the use of coral sand.

Answer: Coral sand is used in the manufacture of cement.

79. Name the fish that is used to control the population of mosquitoes.

Answer: Mosquito fish (*Gambusia* sp.) is used to control the population of mosquitoes. It feeds on the larvae of mosquitoes.

80. Which earthworm is commonly used for waste management?

Answer: The *Eisenia foetida*

81. Name the fish that secretes neurotoxin tetradoxin.

Answer: Puffer fish

82. Name the vitamins that are present in liver oil of codfish and sharks.

Answer: Vitamins A and D

83. What is azadirachtin? How does it help in pest control?

Answer: Azadirachtin is the neem's principal active compound which acts on insects by:

(a) Repelling them
(b) Inhibiting their feeding
(c) Disrupting their growth, metamorphosis and reproduction
(d) It changes the natural physiology of the insects by suppressing their natural hormone.

84. What is cochineal?

Answer: Cochineal is a red colouring dye made from the dried bodies of coccid, *Dactylopius coccus*

85. Why is honey a natural antiseptic?

Answer: Because it contains an antibacterial agent

86. What is nectar?

Answer: Nectar is a sugar solution produced by flowers. It contains 80 per cent water and 20 per cent sugar.

87. How many kilograms of honey are consumed by bees to produce 1 kg of wax?

Answer: Bees consume 10 kg of honey to produce 1 kg of wax.

88. Name three colonial insects.

Answer: Honeybees, ants and termites

89. Mention the importance of mosquito, *Drosophila* and honeybees.

Answer: (a) Mosquitoes –Vectors of many diseases such as malaria, filaria, etc.

(b) *Drosophila* – Model organism for the genetic experiment

(c) Honeybees – Produce honey as well as play a key role in pollination

90. Which Protozoan is used in nutritional research?
Answer: *Tetrachymena geleii* (Holotrichia) is used in nutritional research. The effects of various foods and poisons have been investigated on this insect.

91. Name two economically important lobsters.
Answer: *Panulirus polyphagus and Panulirus ornatus*

92. Name a pest of the cotton plant.
Answer: *Pectinophora gossypiella*

93. Name three pests of wheat.
Answer: (a) *Nanagria uniformis* (Wheat stem borer) (b) *Tanymecus indicus* (Gujhia weevil) (c) *Schizaphis graminum* (Wheat aphids)

94. Name three pests of vegetables.
Answer: (a) *Pieris brassicae* (b) *Crocidolomia binotalis* (c) *Thrips tabaci*

95. Name two pod borer pests.
Answer: (a) *Agromyza obtusa* belongs to the order diptera.
(b) *Helicoverpa armigera* belongs to the order lepidoptera.

96. Which vector spreads plague from rats to humans?
Answer: *Xenopsylla cheopis.*

97. What is vermiculture?
Answer: Vermiculutre is the scientific management of earthworms to decompose natural organic wastes into rich compost fertiliser.

98. Name the diseases caused by the following parasites:
(a) *Leishmania donovani* (b) *Trypanosoma gambiense*
(c) *Plasmodium vivax* (d) *Entamoeba histolytica*
Answer:
(a) *Leishmania donovani* – Kala azar
(b) *Trypanosoma gambiense* – Sleeping sickness
(c) *Plasmodium vivax* – Malaria
(d) *Entamoeba histolytica* – Amoebic dysentery

99. Name the vectors of the above mentioned parasites.

	Parasite	**Vector**
(a)	*Leishmania donovani*	*Phleobotomus argentipes*
(b)	*Trypanosoma gambiense*	*Glossina palpalis* (Tse-tse fly)
(c)	*Plasmodium vivax*	*Anopheles*
(d)	*Entamoeba histolytica*	Since it is a monogenetic parasite, there is no vector

100. Name the parasite and its vector that causes elephantiasis.
Answer: Elephantiasis is caused by the nematode parasite *Wuchereria bancrofti* whose vector is the female *Culex* mosquito.

101. Name the infective stage of the following:
(a) *Entamoeba histolytica*
(b) *Plasmodium vivax*
(c) Wuchereria bancrofti

Answer:

(a) *Entamoeba histolytica* — Tetra nucleate cyst
(b) *Plasmodium vivax* — Sporozoite
(c) *Wuchereria bancrofti* — Microfilaria

102. How does infection of *Ascaris* occur in humans?
Answer: Infection of *Ascaris* in humans occurs by swallowing of the shelled egg that contains second stage rhabditoid larvae.

103. Name two monogenetic parasites.
Answer: (a) *Entamoeba histolytica* (b) *Ascaris lumbricoides*

104. What is the fate of the zygotic stage of the *Plasmodium?*
Answer: The zygotic stage of the *Plasmodium* develops into ookenite.

105. What is malaria?
Answer: Malaria is an infectious disease caused by the Protozoan parasite *Plasmodium* transmitted from humans to humans by the bite of the infected female *Anopheles* mosquito.

106. Name the four species of *Plasmodium.*
Answer: (a) *Plasmodium vivax* (b) *Plasmodium malariae*
(c) Plasmodium ovale (d) *Plasmodium falciparum*

107. Name the species of *Plasmodium* that is confined only to warmer parts.
Answer: *Plasmodium falciparum*

108. Name the Protozoan parasites the infective stage of which is transferred from the host's blood to new hosts by biting insects.
Answer: *Plasmodium, Trypanosoma,* and *Leishmania,* etc.

109. Name the parasite that infects humans both by penetration as well as orally.
Answer: *Ancylostoma duodenale*

110. Name the parasite that shows nocturnal periodicity.
Answer: *Wuchereria bancrofti*

111. What are coral reefs?
Answer: Coral reefs are exoskeleton of polyps.

112. What is pearl fishery?
Answer: The culturing of pearl oyster in order to obtain pearls is known as pearl culturing.

113. Name the disease the germs of which are spread by the sandfly.
Answer: Kala azar

114. Name the disease the germs of which are spread by the tse-tse fly.
Answer: Sleeping sickness

Long-Answer Questions

1. Define sericulture. Give an account of sericulture in India.
2. What is lac? Describe the lac culture.

3. What is an apiary? Name the first sweet food tasted by ancient Indians. Describe apiculture and add a note on its present status.
4. Write an essay on pisciculutre.
5. What is a pest? Name three pests of sugarcane and describe the life cycle of any one of them.
6. Give names of three pests of paddy and parts damaged by them. Add a note on control of paddy pests.
7. Name three serious pests of wheat. Describe the life cycle of any two wheat pests studied by you.
8. Give an account of pests of stored grains.
9. Write short notes on the following:
   (a) Lac
   (b) Honey
   (c) Cocoon of silkworm
   (d) Biological control of pests
   (e) Induced breeding
   (f) Cage culture
   (g) Bee wax

ECONOMIC ZOOLOGY

Multiple-Choice Questions

1. Which one of the following is the oldest method of pest control?
   (a) Host plant resistance
   (b) Physical and mechanical control
   (c) Cultural control
   (d) Biological control
2. The most well-known method of eradicating screw worm is by using:
   (a) Sterile-male technique
   (b) Antifeeding compound
   (c) Repellents
   (d) Attractants
3. Which one of the following methods of pest control cannot be undertaken by an individual farmer?
   (a) Chemical control
   (b) Biological control
   (c) Host plant resistance control
   (d) Cultural control
4. Which one of the following is a male chemosterilant?
   (a) Hemel (b) Hempa (c) Tretamine (d) All of these
5. Chemosterilants act by causing:
   (a) Chromosomal break
   (b) Disruption of gametogenic cycle
   (c) Gametic disorders
   (d) All of the above.
6. Which one of the following is a genetic technique of pest control?
   (a) Cytoplasmic incompatibility
   (b) Chromosomal translocations
   (c) Hybrid sterility
   (d) All of these
7. Biological control of pests gives good results in:
   (a) Islands (b) Mountains (c) Humid climate (d) Temperate regions
8. Most parasites used in biological control are entomophagous:
   (a) Hymenoptera
   (b) Diptera
   (c) Hymenoptera and diptera
   (d) Hymenoptera and coleopteran
9. Which of the following chemosterilants prevents the development of eggs and interferes with the synthesis of folic acid?
   (a) Hemel (b) Amethopterin (c) Tretamine (d) Hempa
10. Sugarcane borer is the host pest of:
    (a) *Lixophaga diatraca*
    (b) *Prosena sibirita*
    (c) *Hamaxia incongrua*
    (d) *Amphlex compressa*
11. Which one of the following statements about *Chilo infuscatellus* is incorrect?
    (a) It belongs to order orthoptera.
    (b) It is a pest of sugarcane.
    (c) It is polyphagous.
    (d) It causes formation of dead hearts.
12. The cane white fly:
    (a) Belongs to the order hemiptera
    (b) Nymphs and adults suck sap from leaves
    (c) Disturbs photosynthesis
    (d) All of these

13. Silkworm belongs to the order:
    (a) Lepidoptera (b) Orthoptera
    (c) Hemiptera (d) Diptera
14. Match column I with column II and select the correct answer using answer codes:

Column I	Column II
(A) *Bombyx mori*	1. Muga silk
(B) *Anthrea mylitta*	2. Eri silk
(C) *Anthrea assamensis*	3. Tassar silk
(D) *Philosamia ricini*	4. Mulberry silk

Answer codes:

	A	B	C	D
(a)	2	3	4	1
(b)	3	4	1	2
(c)	4	3	1	2
(d)	4	1	2	3

15. The host plant of *Anthrea mylitta* is:
    (a) Saal (b) Oak (c) Arhar (d) Castor
16. The larva of silkworm is called:
    (a) Wriggler (b) Caterpillar (c) Grub (d) Maggot
17. Which one of the following is never found in a wild condition?
    (a) *Bombyx mori* (b) Eri silkworm (c) Tussar silkworm (d) None of the above
18. The art of sericulture was first developed in:
    (a) France (b) China (c) Lanka (d) Japan
19. Which one of the above following statements about silk glands is correct?
    (a) The anterior part is nonsecretary in function. (b) The middle part secretes sericin protein.
    (c) The posterior part secretes two proteins called fibroin and P25. (d) All of these
20. The maximum quantity of silk produced in world is the:
    (a) Eri silk (b) Mulberry silk (c) Tussar silk (d) Muga silk
21. Cocoon spinning needs _________ temperature:
    (a) 15ºC (b) 20ºC (c) 24ºC (d) 32ºC
22. In multivoltine races, spinning of cocoon occurs for:
    (a) 1 to 2 days (b) 2 to 5 days (c) 3 to 7 days (d) 5 to10 days
23. Tussar moths are:
    (a) Univoltine (b) Bivoltine (c) Multivoltine (d) All of these
24. Only about ________ per cent of silk in each cocoon is reelable:
    (a) 20 (b) 25 (c) 50 (d) 75
25. *Bombyx mori* is a moth:
    (a) Which has lost its power to fly (b) Capable only of mating
    (c) Capable of mating and producing eggs for future generation (d) All of these
26. Which one of the following is applicable to silkworms?
    (a) Fly (b) Moth (c) Termite (d) Parthenogenesis

27. The caterpillar of *Bomyx mori* lacks:
    (a) Anal horn (b) Spiracles (c) Mandibulate mouth (d) None of these
28. Spinneret helps in:
    (a) Excretion (b) Respiration
    (c) Extrusion of silk from the inner silk gland (d) None of these
29. Lyononet's glands are found in:
    (a) *Bombyx mori* (b) *Apis indica* (c) *Laccifer lacca* (d) *Sphenodon*
30. Which one of the following statements about *Bombyx mori* is correct?
    (a) Females are larger than the males. (b) Univoltines are larger than the multivoltines.
    (c) Females moths lack mouth and malpighian tubules. (d) Accessory glands are present in both sexes.
31. The molecular formula of fibroin is:
    (a) $C_{30}H_{46}N_{10}O_{12}$ (b) $C_{35}H_{55}N_{12}O_{14}$ (c) $C_{30}H_{40}N_{10}O_{16}$ (d) $C_{25}H_{30}N_{10}O_{12}$
32. Mori cultivation by pit system is done in:
    (a) Tamil Nadu (b) Karnataka (c) Jammu and Kashmir (d) West Bengal
33. Silk reeling is done by:
    (a) Charka (b) Cottage basins (c) Multi-end basins (d) All of these
34. From which one of the following reeling devices can gradable silk be obtained?
    (a) Multi-end reeling (b) Cottage basin reeling (c) Charka reeling (d) All of these
35. The most important way of silk reeling is:
    (a) Cottage basin reeling (b) Automatic silk reeling machines
    (c) Multi-end basin reeling (d) Charka reeling
36. Which one of the following silk reeling devices is most popular in Japan and China?
    (a) Multi-end basin reeling (b) Cottage basin reeling
    (c) Automatic silk reeling (d) Charka
37. The most popular variety of silk in India is the:
    (a) Mulberry silk (b) Eri silk (c) Muga silk (d) Tussar silk
38. Bivoltine silk is mostly produced in country having a:
    (a) Cold climate (b) Temperate climate
    (c) Humid and cold climate (d) All of the above
39. India ranks________ in the world in the production of silk?
    (a) First (b) Second (c) Third (d) Fourth
40. Silk noil is applicable to:
    (a) Reeling (b) Raw silk (c) Cocoon (d) High quality silk
41. Consider the following statements:
    (A) Raw silk can be used directly for weaving
    (B) Twisting is the conversion of raw silk into twisted silk
    (C) Silk weaving is done either on handlooms or power looms
    (D) China is the largest silk producing country in the world

    The correct statements are:
    (a) All of these (b) A, B and C (c) B, C and D (d) A and D.
42. In India, mulberry silk is cultivated in:
    (a) Tamil Nadu and West Bengal (b) Tamil Nadu, Karnataka and Andhra Pradesh

(c) Jammu and Kashmir (d) All of these

43. The commercial characters of cocoons for price fixation depend on:
(a) Cocoon weight (b) Cocoon colour (c) Compactness (d) All of the above

44. The colour of the cocoon depends upon the presence of pigment in the:
(a) Sericin layer (b) Fibroin layer (c) Shell (d) Type of food plants

45. Which one of the following is known as the Indian bee?
(a) *Apis dorsata* (b) *Apis indica* (c) *Apis florea* (d) *Apis mellifera*

46. The honeybee lacks:
(a) Pollen basket (b) Rake (c) Comb (d) None of these

47. Worker bees perform all functions, except:
(a) Thermoregulation (b) Guarding combs
(c) Laying of eggs (d) Collection of nectar and pollen.

48. The life span of the queen honeybee is of:
(a) 24 days (b) 21 days (c) 15 days (d) 10 days

49. Wax is secreted by the:
(a) Queen (b) Workers (c) Drones (d) All of these

50. Wax gland is located in the:
(a) Abdomen of worker bees (b) Thorax of worker bees
(c) Legs of worker bees (d) Abdomen of drones

51. The mouth parts of honeybees are of the:
(a) Sponging type (b) Siphoning type
(c) Chewing and lapping type (d) Piercing and sucking type

52. The father of modern apiculture is:
(a) Amos Root (b) Jan Dzierzon (c) Moses Quinby (d) May Berenbaum

53. Which one of the following is needed for brood development?
(a) Pollens (b) Nectar (c) Pollens and nectar (d) Wax

54. The floral source determines the:
(a) Flavour of honey (b) Colour of honey (c) Aroma of honey (d) All of these

55. Which one of the following is correct about honeybees?
(a) Social insect (b) Polymorphic (c) Division of labour (d) All of these

56. Honey is stored in the _________ part of the comb:
(a) Upper (b) Lower (c) Middle (d) All of these

57. Honey contains mainly:
(a) Citric acid (b) Gluconic acid (c) Malic acid (d) Isocitric acid

58. The abdomen has seven segments in the:
(a) Queen (b) Workers (c) Drones (d) All of these

59. The number of wax glands in worker bees is:
(a) 2 pairs (b) 3 pairs (c) 4 pairs (d) 5 pairs

60. In honeybees, the receptors for measuring distances are located in the:
(a) Eyes (b) Antennae (c) Legs (d) Abdomen

61. The wings are largest in the:
(a) Drones (b) Queen (c) Workers (d) Queen and workers

62. Which one of the following statements about *Apis dorsata* is incorrect?
    (a) Commonly called rock bee (b) Largest in size
    (c) Stings rarely (d) Not suitable for domestication
63. The movable frame hive was designed by:
    (a) Ernest Spytzner (1788) (b) Langstroth (1851)
    (c) Roberts (1897) (d) Mc Clung (1912)
64. Pine honey is rich in vitamin:
    (a) D (b) E (c) K (d) B_6
65. In honeybees, the pollen basket occurs in the:
    (a) Mesothoracic leg (b) Metathoracic leg (c) Prothoracic leg (d) Head
66. Which one of the following is not applicable to honeybees?
    (a) Caste system (b) Arrhenotoky (c) Thelytoky (d) Colonial
67. The smallest honeybee is:
    (a) *Apis indica* (b) *Apis florea* (c) *Apis mellifera* (d) *Apis dorsata*
68. Which one of the following diseases destroys the epithelial layer of the midgut of the honeybee?
    (a) Nosema disease (b) Acarine disease (c) Amoebic disease (d) Brood disease
69. The silkworm spends ________ days in spinning a cocoon:
    (a) 1 or 2 (b) 3 or 4 (c) 4 or 5 (d) 6 or 7
70. Functions change with age in the ________ of honeybees:
    (a) Workers (b) Queen (c) Drones (d) None of the above
71. Lac insect belongs to the order:
    (a) Lepidoptera (b) Hemiptera (c) Hymenoptera (d) Diptera
72. Which one of the following is not associated with lac cultivation?
    (a) Harvesting (b) Reeling (c) Inoculation (d) Swarming
73. Consider the following statements about the lac insect:
    (A) Father Tachard first described the lac insect in 1709 (B) Males lack mouth parts
    (C) The bulk of lac is secreted by females (D) Females are devoid of wings and legs
    The correct statements are:
    (a) All of these (b) A, B and C (c) B, C and D (d) C and D
74. Which one of the following is a host plant of the lac insect?
    (a) Babul (b) Ber (c) Khair (d) All of these
75. About _________ lac insects produce 1 kg of lac:
    (a) 10,000 (b) 20,000 (c) 3, 00, 000 (d) 9, 00,000
76. Chemically, lac contains:
    (a) Lac resin and wax (b) Lac resin and lactose
    (c) Albuminous matter and dyes (d) All of these
77. Female lac insects have:
    (a) Eyes (b) Wings and legs (c) Mouth parts (d) All of these
78. Which one of the following is lacking in drones of honeybees?
    (a) Sting (b) Salivary gland (c) Wax gland (d) All of these
79. Consider the following statements about an insect:
    (A) It lives on cacti (B) Its body is dark red

(C) Its dead body is used for preparing cochineal (D) It is cultured on a large scale in Mexico, Peru, Spain, etc.

This insect is the:

(a) *Laccifer lacca* (b) *Dactylopius coccus*
(c) Blister beetle (d) *Triatoma rubrofasciata*

80. Cantharidin is obtained from the dried elytra of:
(a) *Lytta* (b) *Lucilla* (c) *Phormia* (d) *Novius*

81. *Pyrilla perpusila* is a pest of:
(a) Sugarcane (b) Paddy (c) Wheat (d) Stored grains

82. Consider the following statements about a pest:
(A) Belongs to the order lepidoptera and family pyralidae
(B) Endemic in areas where repeated cropping of paddy is done
(C) Adults are with yellow coloured forewings and white hind wings and are monophagous
(D) Attacks the crop from seeding to the post flowering stage

This pest is:

(a) *Aleurolobus barodensis* (b) *Scirophaga incertulas*
(c) *Orseolia oryzae* (d) *Hieroglyphus banian*

83. Match column I with column II and select the correct answer using answer codes:

Column I	Column II
(A) *Odontoterms obesus*	1. Stored cereals
(B) *Chilo infuscatellus*	2. Rice and grasses
(C) *Nephotettix virescens*	3. Sugarcane, sorghum, maize, etc.
(D) *Sitophilus oryzae*	4. Wheat, millet and sugarcane

Answer codes:

	A	B	C	D
(a)	2	3	4	1
(b)	4	3	2	1
(c)	3	4	1	2
(d)	2	1	4	3

84. Which of the following paddy pests generally attacks the crop at its tillering stage?
(a) *Orseolia oryzae* (b) *Hieroglyphus banian* (c) *Nephotettix virescens* (d) *Spodoptera mauritia*

85. *Aleurolobus barodensis* is a serious pest of:
(a) Paddy (b) Wheat (c) Sugar cane (d) Cotton

86. Which one of the following pests of sugarcane prevents photosynthesis by discharging honeydew on leaves?
(a) *Pyrilla perpusila* (b) *Chilo infuscatellus*
(c) *Aleurolobus barodensis* (d) All of these

87. *Trogoderma granarium* is a common pest of stored:
(a) Bajara (b) Maize (c) Barley (d) Wheat

88. Which one of the following statements about *Sitophilus oryzae* is incorrect?
(a) Larvae are legless grubs. (b) Pupation occurs outside the grain.
(c) The pupation period is of 5 to 7 days. (d) Diapause may occur for 15 to 20 days.

89. Which one of the following is a common pest of stored cereals, barn and husk of seeds?
(a) *Trogoderma granarium* (b) *Tenebrio molitor*

(c) *Cadra cautella* (d) *Tribolium castaneum*

90. Short life cycle with parthenogenesis and viviparity are found in:
(a) *Apis gossypii* (b) *Empoasca devastans* (c) *Pyrilla perpusilla* (d) *Spodoptera mauritia*

91. Consider the following statements about a pest:
(A) Belongs to the order hemiptera and family cicadellidae
(B) The entire life cycle takes place on plants
(C) Hot and humid climate is favourable
(D) Nymphs and adults suck sap from green leaves
This paddy pest is:
(a) *Orseolia oryzae* (b) *Spodoptera maurtia*
(c) *Chilo polychrysa* (d) *Nephotettix virescens*

92. The first insect used as an object for the sterile male technique is the:
(a) Screwworm (b) Codling moth (c) Ballworm (d) Glowworm

93. Integrated pest management was proposed by:
(a) La Brecque et al. (1960) (b) Geler and Clark (1961)
(c) Smith and van der Bosch (1967) (d) De Bach (1978)

94. Paddy gall fly belongs to the order:
(a) Lepidoptera (b) Hemiptera (c) Homoptera (d) Diptera

95. Mature commercial lac is called:
(a) Stick lac (b) Seed lac (c) Dust lac (d) Shellac

96. The larva of the codling moth destroys:
(a) Apples (b) Mangoes (c) Guavas (d) Sugarcane

97. Allontoni is obtained from:
(a) Ants (b) Blister beetle (c) Lac insects (d) Blow flies

98. Hessian fly is the serious pest of:
(a) Sugarcane (b) Wheat (c) Paddy (d) Maize

99. Which one of the following is a monophagous pest of sugarcane?
(a) *Bissetia steniellus* (b) *Pyrilla perpusila*
(c) *Aleurolobus barodensis* (d) *Emmalocera depressella*

100. Which one of the following statements about *Laccifer lacca* is incorrect?
(a) Belongs to the order hemiptera. (b) Males lack mouth parts.
(c) One male fertilises only one female. (d) A pair of caudal setae is present in the abdomen of female.

101. The Indian Lac Research Institute is located in the state of:
(a) Uttar Pradesh (b) Jharkhand (c) Jammu and Kashmir (d) Karnataka

102. In which one of the following lac crops does the swarming period occur between October and November?
(a) Jethi (b) Baishakhi (c) Katki (d) Agahani

103. Which one of the following statements about *Antheraea assamensis* is incorrect?
(a) Belongs to the family saturniidae (b) Found all over the world
(c) Food plants are Som and Champa (d) It yields muga silk

104. The first fish hatchery was established by:
(a) Seth Green (b) Rochester (c) Zernike (d) La dua

105. Which one of the following breeds in deep water?
(a) Rohu (b) Mrigal (c) Catla (d) All of these

106. The peak period of infection of *Tryporyza* is the:
(a) July (b) August (c) September (d) November

107. *Tryporyza* damages the paddy crop in the:
(a) Egg stage (b) Larval stage (c) Nymph stage (d) Adult stage

108. The life cycle of *Hispa armigera* is completed within:
(a) 15 days (b) 30 days (c) 45 days (d) 60 days

109. Fish liver oil is a natural source of vitamins:
(a) A and E (b) A and D (c) D and E (d) B_5 and B_7

110. Aquaculture is the farming of:
(a) Freshwater organisms (b) Saltwater organisms
(c) Both (a) and (b) (d) Exotic species

111. Match column I with column II and select the correct answer using answer codes:

Column I	Column II
(A) *Thrips tabaci*	1. Cauliflower
(B) *Urentius sentis*	2. Wheat
(C) *Pieris brassicae*	3. Brinjal
(D) *Nanagria Uniformis*	4. Onion

Answer codes:

	A	B	C	D
(a)	4	3	2	1
(b)	3	1	4	2
(c)	4	1	2	3
(d)	4	3	1	2

112. Which one of the following has only a host plant?
(a) *Attacus ricinii* (b) *Antheraea assamensis* (c) *A. mylitta* (d) *Bombyx mori*

113. *Apis indica*:
(a) Is the only Indian bee, capable of domestication in artificial hives (b) Is slightly longer than *Apis florea*
(c) Is smaller than *Apis dorsata* (d) All of these

114. Which one of the following statements about the wheat thrip is incorrect?
(a) Belongs to the order hemiptera. (b) The host plants are wheat, sugarcane and other graminaceous plants.
(c) It has sexual and parthenogenetic forms. (d) Low temperature favours pest incidence at the early stages of the crop.

115. Honey is:
(a) Antioxidant (b) Anti-allergic (c) Anticancer (d) All of these

116. Unprocessed honey contains spores of:
(a) *Clostridium, tetani* (b) *Clostridium botulinum*
(c) *Diplococcus pneumoniae* (d) *E. coli*

117. In workers of honeybees, the hypopharyngeal glands:
(a) Produce royal jelly (b) Produce six enzymes
(c) Are located above the bee's eye (d) All of the above.

118. Apamin is prevalent in the honeybees':
(a) Venom (b) Honey (c) Royal jelly (d) Wax

119. Which one of the following is not a suitable characteristic of fishes for rizipisciculture?
    (a) High turbidity tolerance
    (b) High temperature tolerance
    (c) High oxygen tolerance
    (d) Ability to grow in shallow water
120. Which one of the following fishes is not cultured in rice fields in India?
    (a) *Anabas testudineus*
    (b) *Catla catla*
    (c) *Channa striatus*
    (d) *Labeo rohita*

Answers to Multiple-Choice Questions

1. (b)	2. (a)	3. (b)	4. (d)	5. (d)	6. (d)	7. (a)	8. (c)
9. (b)	10. (a)	11. (a)	12. (d)	13. (a)	14. (c)	15. (a)	16. (b)
17. (a)	18. (b)	19. (d)	20. (b)	21. (c)	22. (a)	23. (b)	24. (c)
25. (d)	26. (b)	27. (d)	28. (c)	29. (a)	30. (c)	31. (a)	32. (b)
33. (d)	34. (a)	35. (b)	36. (c)	37. (a)	38. (b)	39. (b)	40. (b)
41. (a)	42. (d)	43. (d)	44. (a)	45. (b)	46. (d)	47. (c)	48. (c)
49. (b)	50. (a)	51. (c)	52. (b)	53. (a)	54. (d)	55. (d)	56. (a)
57. (b)	58. (c)	59. (c)	60. (b)	61. (a)	62. (c)	63. (b)	64. (c)
65. (b)	66. (c)	67. (b)	68. (a)	69. (b)	70. (a)	71. (b)	72. (b)
73. (a)	74. (d)	75. (c)	76. (d)	77. (c)	78. (d)	79. (b)	80. (a)
81. (a)	82. (b)	83. (b)	84. (a)	85. (c)	86. (a)	87. (d)	88. (b)
89. (b)	90. (a)	91. (d)	92. (a)	93. (b)	94. (d)	95. (d)	96. (a)
97. (d)	98. (b)	99. (a)	100. (c)	101. (b)	102. (c)	103. (b)	104. (a)
105. (c)	106. (c)	107. (b)	108. (a)	109. (b)	110. (c)	111. (d)	112. (d)
113. (d)	114. (a)	115. (d)	116. (b)	117. (d)	118. (a)	119. (c)	120. (a)

Fill in the Blanks

1. The pupa of silk is called ________.
2. The rearing of silkworms for the production of silk is known as ________.
3. Nondiapauses eggs are laid by_______ races of silkworm.
4. Silk is made up of ________ protein.
5. The sericin is removed by placing the cocoon in ________ water.
6. ________ cements two silk filaments together.
7. Fibroin is a type of protein formed by ________.
8. The production of raw silk from silkworms began in ________ about 2500 BC.
9. The cultivation of mulberry plants is termed as ________.
10. Silkworm seed production centres are called ________.
11. ________ are the raw materials for production of silk.
12. The extraction of silk filament from cocoons is known as ________.
13. The thickness of silk filament is denoted by ________.
14. Compactness of cocoon is an important character that indicates the ______ of silk present in the cocoon.

15. ________ is the ratio of cocoon reeled without a break and the total number of cocoons utilised.
16. Humans interfere in the life cycle of the silkworm at the ________ stage to obtain silk.
17. The honeybee belongs to the family ________, subfamily ________ and genus ________.
18. pH of honey is ________.
19. Royal jelly is secreted by ________ honeybees
20. The larva of the honeybee is called ________.
21. In honeybees, ________ develop parthenogenetically.
22. Septicemia is caused by the bacteria ________.
23. Brood foul can be controlled by spraying ________ acid.
24. In worker honeybees, the ovipositor is modified into ________
25. The rearing of honeybees to obtain honey and wax is known as ________.
26. The quality and colour of lac depends on ________ and ________ present in the host plant.
27. Agahani and Jethi lac crops are called__________ lac crop.
28. The inoculation period of Jethi lac crop is in________ month.
29. ________ was a pioneer in fish farming.
30. Fish glue is prepared from fish bones and ________.
31. Isin glass is obtained from ________ of certain fishes.
32. Lac is soluble in ________ and ________.
33. In India, the most common host trees of lac are ________, ________ and ________.
34. ________ gave the first scientific description of the lac insect and its economic value.
35. The mouth parts of lac insects are of the ________ type.
36. The branch of science dealing with the life, ecology, damage and control of pests is known ________.
37. *Hispa armigera* is a pest of ________.
38. ________ a monophagous pest of paddy.
39. ________ is the commercial resin of animal origin.
40. ________ is the resin free from lac dye.
41. Lac insects are beneficial for their three valuable products, viz., ________, ________ and ________.

Answers to Fill in the Blanks

1. Chrysalis	2. Sericulture	3. Multivoltine
4. Fibroin	5. Hot	6. Sericin
7. *Bombyx mori*	8. China	9. Moriculture
10. Grainages	11. Cocoons	12. Silk reeling
13. Denier	14. Quantity	15. Reelability
16. Cocoon	17. Apidae, apinae, *Apis*	18. 3.1 to 4.6
19. Worker	20. Grub	21. Drones
22. *Bacillus apisceptious*	23. Salicylic	24. Sting
25. Apiculture	26. Gum, resin	27. Kusumi
28. January	29. Seth Green	30. Skin
31. Air bladder	32. Alcohol, weak alkalis	33. Kusum, ber and dhak
34. Kerr and Glover (1782)	35. Piercing and sucking	36. Pestology

37. Paddy
38. *Tryporyza*
39. Shellac
40. Seed lac
41. Lac dye, lac wax, lac (resin)

True or False

1. Asia is the main producer of silk in the world.
2. The silk gland is a modified salivary gland.
3. Silk is obtained from cocoons.
4. Waste from silkworm rearing can be recycled as inputs to gardens.
5. Sericulture is an agro-industry.
6. Cocoons with finer granulations are difficult to reel.
7. Cocoons with pointed ends are commercially much suitable for steady reeling.
8. The dullness of a cocoon indicates improper stifling.
9. The best-known type of silk is produced by the larvae of *Bombyx mori*.
10. Most medicines of Ayurveda are honey-based.
11. Honey is used as an astringent.
12. The conversion of nectar into honey by honeybees is a biological process.
13. All subspecies of honeybees have the same swarming characteristic.
14. Darker honey contains greater amounts of minerals than lighter honey.
15. In the queen of honeybees, the sting also acts as an ovipositor.
16. Bee wax is a waste product of honeybee.
17. The queen of honeybee lays two types of eggs, viz., fertilised and unfertilised.
18. Female lac insects are wingless and motionless.
19. Scraping of lac after swarming is known as mature harvesting.
20. The quality of lac depends on the quality of the host plant.
21. The inoculation period of the Jethi lac crop is June to July.
22. Lac insects repeat their life cycle thrice a year.
23. The lac cells produced by male and female lac insects are similar.
24. Chalcids are common parasites of lac insects.
25. For cultivation of lac, trees should be free of ants and other insects.
26. Biological control is a natural means of pest control.
27. For planning a proper strategy to control pest, knowledge of both abiotic and biotic factors affecting the population of insects is necessary.
28. A single approach to control the problem of pest control is not feasible.
29. The incrustation of the male lac insect is somewhat round.
30. Kusum trees produce the best lac.
31. The life cycle period of the lac insect depends upon ecological factors.

32. *Hispa* attacks paddy plants from the date of its germination.
33. The gall fly is a monophagous wheat pest.
34. The potato tuber moth is a monophagous pest of potato.
35. Gujhia weevil is a polyphagous pest.
36. In tussar silk, the time required for cocoon formation is shorter in comparison to mulberry silkworm.
37. The sericulture industry is well developed in America.
38. The cocoon of tussar has reelable silk.
39. *Apis florea* are gregarious.
40. Honeybees are the only pollinating insects that can be controlled by humans.
41. *Leptocorisa* eats the leaves and ears of rice.

Answers to True or False

1. True	2. True	3. True	4. True	5. True	6. False	7. False	8. True
9. True	10. True	11. True	12. True	13. False	14. True	15. True	16. False
17. True	18. True	19. True	20. True	21. False	22. False	23. False	24. True
25. True	26. True	27. True	28. True	29. False	30. True	31. True	32. True
33. False	34. False	35. True	36. False	37. False	38. True	39. False	40. True
41. True							

Give Reasons

1. It is necessary to control pests.
   – Because pests cause economic loss.
2. Silkworm rearing is considered to be an agro-based industry.
   – Because it involves the cultivation of mulberry.
3. Immature and cocoons having fungus should be avoided.
   – Because of poor reelability and wastage due to attack of fungus.
4. Malformed cocoons should be avoided.
   – Because they have less content of silk.
5. Silk production is regarded as an important tool for the economic development of a country.
   – Because it is a labour-intensive as well as a high income generating industry. It also provides employment to the rural sector and is a very good means of foreign exchange.
6. Production of honey from farmlands can be a secondary activity for farmers.
   – Because it requires less capital as well as less time in comparison to other activities. Further, it can be carried out by women and children.
7. Honeybees are being used by scientists as environmental indicators.
   – Because honey is the last product of bees to react to pollutants.
8. The queen honeybee is unable to produce honey or wax
   – Because it lacks salivary and wax glands.

9. Cocoons of silkworms are treated with hot water.
   - To remove unwanted cementing materials of the cocoon, as this method results in the easy separation of silk threads.
10. Stick lac should be immediately converted into seed lac.
    - In order to prevent the escape of parasites and predators into the field.
11. For lac culture, it is better to inoculate the tree in November for the first time.
    - Because it is possible that insects may be washed away due to heavy rain if inoculation occurs during June.
12. *Tryporyza* is called a stem borer.
    - Because caterpillars crawl down the plants and reach the central stem where they make small bores and ultimately migrate within the central tissue.
13. *Leptocorisa varicornis* is commonly called Gandhipoka.
    - Because of their repulsive smell.
14. In the queen and drones of honeybees, the mouth parts are shorter.
    - Because they do not collect nectar, unlike workers.
15. Generally, microorganisms do not grow into honey.
    - Because of less content of water in honey.
16. Honey is an ideal liver fuel.
    - Because it contains a nearly 1:1 ratio of fructose to glucose. Fructose present in honey unlocks the enzyme from the liver cell's nucleus, which is necessary for the conversion of glucose into glycogen.
17. Honey improves red blood corpuscles.
    - Because it is rich in iron, copper and magnesium.
18. Unprocessed honey causes botulism.
    - Because it contains spores of *Clostridium botulinum*.

ANIMAL DEVELOPMENT

History of Embryology

- Aristotle (384–322 BC) was the first to describe the ontogenic development of the chick.
- Volcher Coiter (1534–1576) published a detailed study of chick embryology. He is now recognised as the 'Father of Embryology'.
- William Harvey (1578–1657) coined the term 'epigenesis'.
- Leuwenhoek (1677) described the sperm of humans and other mammals.
- Casper Friedrich Wolff (1738–1794) was the first person to demonstrate morphogenesis. He saw the development of structure out of structureless yolk material.
- Friedrich Blumenbach (1742–1840) introduced the concept of 'vis vitalis', an immortal virtue of life force.
- Swammerdam (1738) observed the first cleavage of the frog.
- Carl Ernst von Baer (1827) identified the mammalian egg.
- Carl Ernst von Baer (1792–1876) was the first person to note the many similarities between the embryos of vertebrates, particularly amniotes. He is the 'Father of Modern Embryology'.
- Pander (1817) demonstrated three primary germ layers from which the chick embryo and constituent parts develop.
- Ernst Haeckel (1834–1919) developed the concept of biogenetic law or embryological parallelism, expressed as 'ontogeny recapitulates phylogeny'.
- August Weismann (1834–1914) developed the germplasm theory. In 1892, he published the idea of self-reproducing determinants as the guiding agent in morphogenesis.
- Prevost and Duman (1824) described cleavage or segmentation of the egg.
- Oscar Hertwig (1849–1922) and his brother Richard Hertwig (1850–1937) provided the first demonstration of fertilisation. They also discovered the existence of polar bodies.
- Von Bender (1833) proved that the male and female sex contributes an equal number of chromosomes in fertilised eggs.
- Hans Spemann (1869) discovered the process of induction. He was awarded the Nobel Prize in Medicine in 1935 for his discovery of the 'organiser effect'.
- Wilhelm Roux (1850–1924) established experimental embryology by performing experiments on frogs.
- Spermann and Hilde Mangold (1927) suggested that there are interactions between the parts of developing embryos.
- Vogt (1929) traced the fates of amphibian eggs with vital dyes.
- Child (1940) observed that there exist differences in the rate of general metabolism

in the ooplasm of animal and vegetal hemispheres of oocytes and cleaving eggs of sea urchins and starfishes.

- Mart Rawles (1940) transplanted neural crest cells from a pigmented strain into unpigmented strain of chicken.
- Viriginia Apgar (1953) developed the Apgar test to assess the newborn's well-being.
- Horstadius and Josefsson (1969) showed the specific animalising and vegetalising substances – one containing tryptophan and other a nucleotide, respectively – in mature unfertilised eggs.
- Ian Wilmut (1996) gave the new concept of cloning without involving a germ cell.
- Roberts G Edwards was awarded the Nobel Prize in Medicine (Physiology) in 2010 for the development of in vitro fertilisation.

PARTHENOGENESIS

- The development of an egg without fertilisation is called parthenogenesis. It is a form of reproduction.
- Individuals that have developed from parthenogenesis are called parthenotes.
- The phenomenon of parthenogenesis was discovered in the 8th century by Charles Bonnet.
- Jacques Loeb (1900) performed the first case of artificial parthenogenesis.

Types of Parthenogenesis

Parthenogenesis is of the following two types:

I. Natural parthenogenesis

II. Artificial parthenogenesis

I. Natural Parthenogenesis

- The development of eggs under natural conditions is known as natural parthenogenesis.
- Natural parthenogenesis occurs in aphids, *Daphnia*, rotifers, nematodes and many other invertebrates. Besides, natural parthenogenesis also occurs in komodo dragons, hammer-headed and black-tip sharks, many genera of fishes, amphibians and reptiles.
- Natural parthenogenesis is of the following two types:

1. Haploid Parthenogenesis

- It is also known as arrhenotoky.
- In several animals, the males develop from unfertilised eggs and are therefore haploid (n), while females develop from fertilised eggs and are diploid (2n). Such type of fertilisation occurs in many insects, rotifers, arachnids and water fleas.
- In honeybees, the unfertilised egg develops parthenogenetically into haploid male (drone), while the queen develops from the fertilised egg and is diploid.

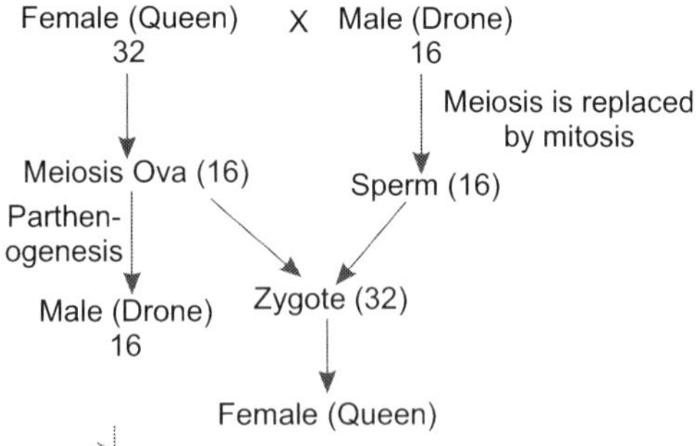

Fig. 1 Haploid parthenogenesis in honeybees

- Rotifers reproduce parthenogenetically during spring and summer.
- In bisexual forms, eggs are of two types. Eggs that have developed from diploid ovary undergo only one meiotic division, forming only one polar body. Thus, such eggs remain diploid and are called amictic and develop parthenogenetically into females.
- When a sudden change in environment occurs, diploid females give rise to mictic eggs. They undergo two meiotic divisions. The mictic eggs are haploid and are smaller.
- If there is no fertilisation, mictic eggs develop parthenogenetically into males.
- If mictic eggs are fertilised, they develop into amictic female (2n) in the next spring.

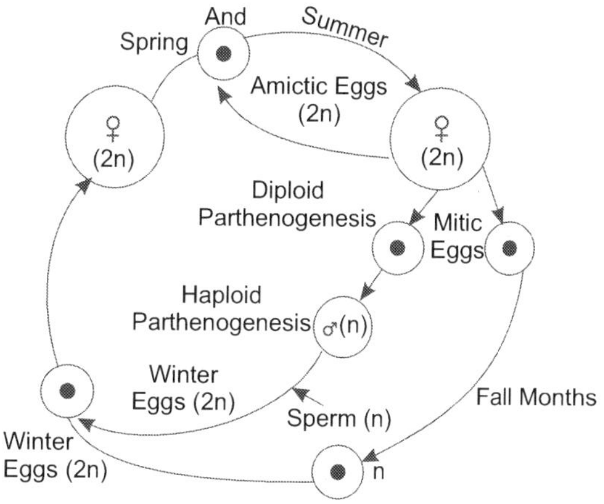

Fig. 2 Haploid and diploid parthenogenesis in rotifers

- It has been reported that parasitic bacteria (like *Wolbachia*) can induce automictic thelytoky in many insects having haplodiploidy systems. Even they result in gamete duplication in unfertilised eggs, causing them to develop into females.

2. Diploid Parthenogenesis

- It is also known as thelytoky.
- In diploid parthenogenesis, the diploid condition is retained by deleting the reduction division or by the fusion of one of the polar body with the haploid ovum.
- Diploid parthenogenesis occurs in aphids, in which young individuals develop from the unfertilised eggs.
- Diploid parthenogenesis is of the following two types:

(a) Ameiotic Thelytoky

- In some animals, the reduction division (first division) is suppressed while the second division which is mitotic in nature occurs.
- As a result of this, diploid eggs are produced, which develop into new individuals.
- Such type of parthenogenesis occurs in insects, weevils, *Daphnia pulex and Campelona rufum*.
- In *Phylloxera*, the males posses one X chromosome (2AX – X) less than female. The females have 2AX chromosomes.
- In summer, the diploid stem mother reproduces parthenogenetically to produce diploid female.
- But during spring, these females lay two types of eggs, viz., larger eggs having 2AX chromosomes that develop into females and smaller eggs having 2AX – X chromosomes that develop into males.
- During winter, these females produce eggs of only one type (AX). But males produce two types of sperms, viz., AX and A chromosomes.
- Sperms having only X chromosomes do not survive, but those having AX chromosomes fertilise the ova, resulting in the formation of zygote (2AX), which develops into the female.

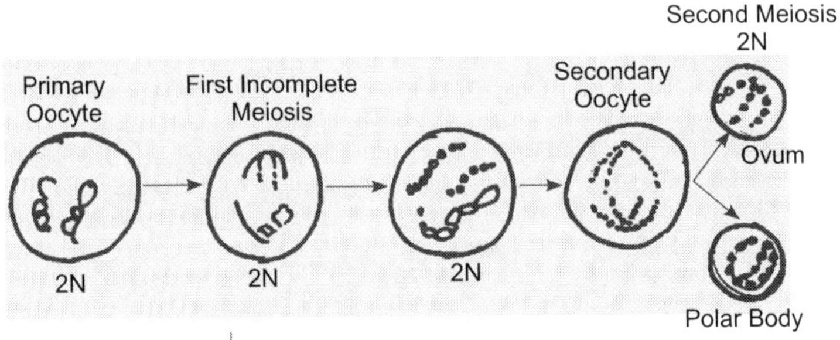

Fig. 3 Diploid parthenogenesis by restitution

(b) Meiotic Thelytoky

- In some cases, the eggs develop by normal oogenesis but the diploid number is restored due to some abnormality.
- The diploid number is restored by restitution in which karyokinesis is not followed by cytokinesis. As a result, a diploid egg is produced that develops without fertilisation (e.g., Hymenoptera).
- In some cases, the diploid number is restored due to failure of the polar body to separate from the ovum. As a result, diploid number is restored. This is known as auto fertilisation. This diploid ovum develops parthenogenetically (e.g., brachiopod crustacean).

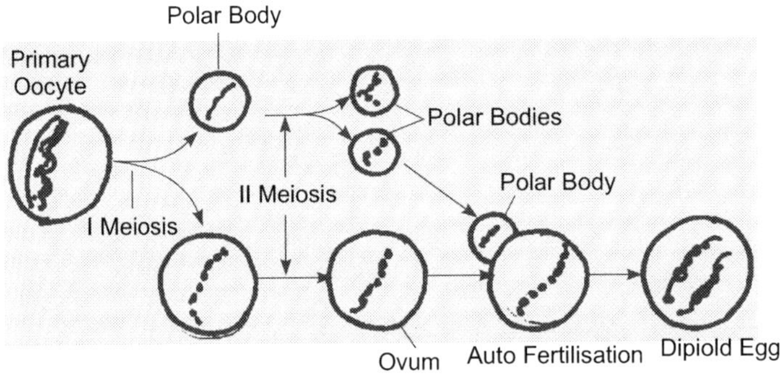

Fig. 4 Diploid parthenogenesis by autofertilisation

II. Artificial Parthenogenesis

- Parthenogenesis induced in the laboratory is known as artificial parthenogenesis.
- Artificial parthenogenesis is induced by mechanical and chemical means.

1. **Chemical Means**

   Chemically, parthenogenesis is induced by treating eggs with different chemical agents such as the following:

   - Hypotonic or hypertonic seawater
   - Blood serum
   - Urea
   - Chloroform

- Neutral salts like potassium chloride, weak acids and bases and carbon dioxide.
- Fatty acids like butyric acid, lactic acid, etc.

2. **Mechanical Means**

   Mechanically, artificial parthenogenesis is induced by the following:
   - Shaking the egg and puncturing it with a needle.
   - Electric treatment
   - High or low temperature
   - Treatment with ultraviolet light

Significance of Parthenogenesis

- Parthenogenesis is a means of reproduction and sex determination.
- It causes polyploidy in organisms.
- Parthenogenesis prevents variation in a population of offsprings.
- Parthenogenesis avoids sterility in the races.
- It lends support to the chromosomal theory of inheritance.
- Parthenogenesis prevents non-adaptive combination of genes that may be caused by mutation.
- Parthenogenesis prevents loss of energy as it does not involve mating which is an essential feature of sexual reproduction.
- It checks the distribution and transmission of harmful mutated characters.
- Recently, scientists of the International Stem Cell Corporation (ISCC) have succeeded in preparing human stem cells using parthenogenesis. These stem cells are homozygous in HLA region of DNA. They are called HLA parthenogenetic human stem cells (hpSC-Hhom). They have unique characteristic that would allow derivatives of these cells to be implanted into millions of people without immune rejection.

FERTILISATION

- Fertilisation is the fusion of gametes to produce a new organism.
- In animals, the process involves fusion of an ovum with a sperm leading to the development of an embryo.
- Depending upon the animal species the process can occur within the body of female in internal fertilisation or outside in case of external fertilisation.

Mechanism of Fertilisation

The entire process of fertilisation involves the following five steps.

I. Approximation of Gametes

- When the gametes are released, they are brought together.
- Male gametes (sperms) are always produced in large numbers so that some of them can reach near the egg.
- It is brought through dilution of decapacitation factors present over the sperms.

- The sperms become very active and swim rapidly to reach the eggs. This phenomenon is known as capacitation of sperms.
- Previously, scientists were of the opinion that the egg exerts a chemotactic effect over the sperms. But no such chemical has been found.

II. Fertilizin-antifertilizin Incompatibility

- Lillie (1909) found that surface of the egg contains a substance, which is capable of causing agglutination or clumping of sperms.
- This chemical is known as fertilizin.
- The main source of fertilizin is jelly coat or plasma membrane of the egg.
- Fertilizin is a glycoprotein, having a molecular weight of 30,00,000.
- Fertilizin functions as female attractant (gynogamone).
- It is species-specific.
- The surface of the sperm contains a male attractant (androgamone) called antifertilizin.
- Antifertilizin is an acidic protein molecule having a molecular mass of 10,000.
- Antifertilizin is smaller than fertilizin.
- Antifertilizin is also species-specific.

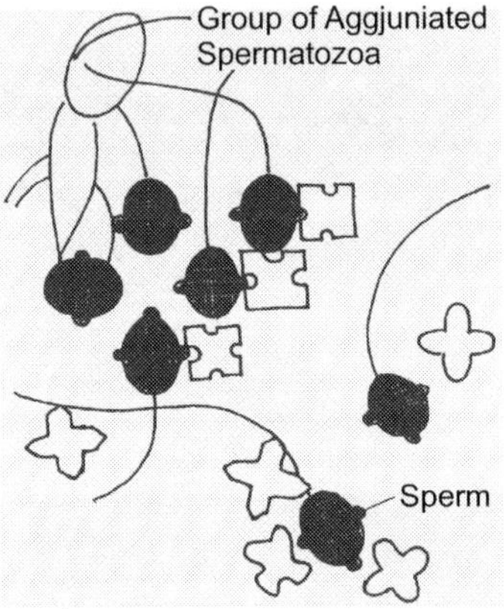

Fig. 5 Fertilizin-antifertilizin reaction

- The sperm passing near the egg is made sticky.
- The sticking of the egg to the sperm depends upon the antifertilizin present over it.
- The fertilizin and antifertilizin reaction is species-specific. This is the reason a foreign sperm does not attach to the surface of the egg.
- Fertilizin and antifertilizin reaction also prevents polyspermy, i.e., it ensures that only one sperm attaches to the surface of the egg.

III. Sperm Penetration

The following two types of reactions occur before the sperm enters the ovum. These reactions are called acrosome reaction and egg reaction.

1. **Acrosome Reaction**
   - When the sperm comes in contact with the egg surface, the membranous covering of acrosome lyses and the acrosome contents come out.
   - The acrosome content contains lytic enzymes called sperm lysins, which dissolve the egg covering.
   - The most potent sperm lysine is hyaluronidase.
2. **Egg Reaction**
   - When the sperm touches the surface of the egg, the egg cytoplasm bulges out, forming a cone-shaped structure called fertilisation cone.
   - It is located in the area of animal pole.
   - The sperm head comes in contact with the fertilisation cone.
   - The plasma membranes of sperm and fertilisation cone dissolves in the region of the contact.

- Thus, a passage is formed between sperm and egg.
- As a result, the contents of sperm head, neck and middle piece enter the ooplasm .The tail is left behind.
- After entry of the sperm, the fertilisation cone vanishes.

IV. Activation of Ovum

- As soon as the sperm enters the ovum, a wave of electrical depolarisation spreads over the egg surface.
- The cortical granules dissolve and spread over the egg surface and here it becomes transformed into a membrane called fertilisation membrane.
- Fertilisation membrane prevents entry of the second sperm into the ovum and thus prevents polyspermy.
- The fusion of a single sperm with an ovum is termed as monospermy.

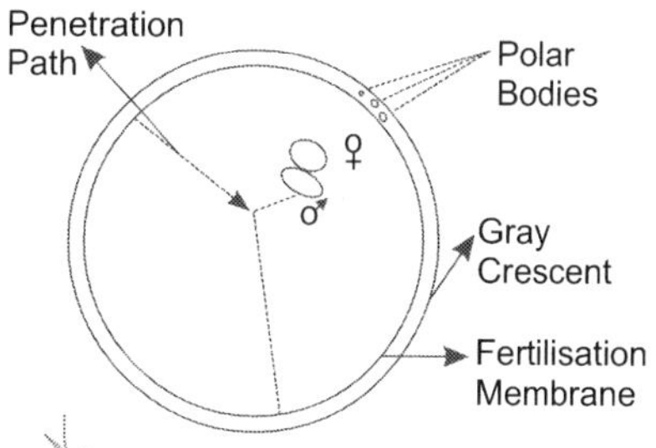

Fig. 6 Penetration and copulation paths of sperm nucleus during fertilisation

- In vertebrates (humans, rabbits, frogs, etc.), the egg completes its first reduction division in the ovary and reaches the metaphase stage of second meiotic division. The secondary oocyte is formed before fertilisation.
- The second meiotic division is completed. A second polar body is formed, which is extruded.

V. Amphimixis

- Amphimixis is the fusion of nuclei.
- During the act of amphimxis, the sperm nucleus migrates from the site of fertilisation cone to the site of amphimxis and male and female nuclei come closer are converted into pronuclei.
- As the sperm nucleus moves inward from the site of fertilisation cone, in most of the animals, it rotates at an angle of 180°, so that mitochondria and centrosome assume the leading position.

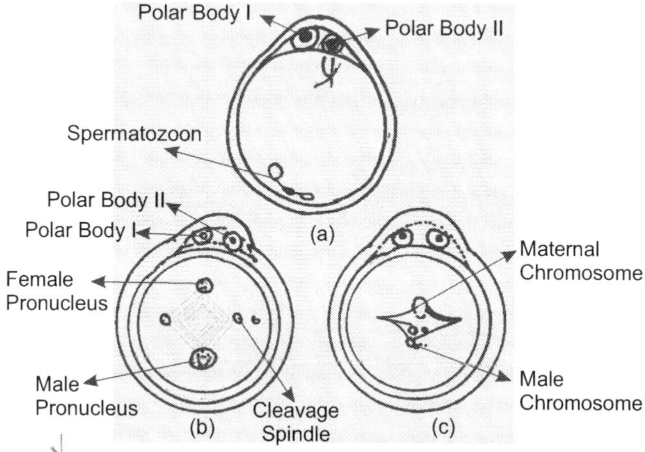

Fig. 7 Diagrammatic representation of events during fertilisation

- The proximal centriole of the sperm takes over the function of centriole of the egg.
- It divides into two centrioles that give rise to poles of the spindle.
- In the meantime, the male and female nuclei come closer.
- The envelopes of male and female nuclei disintegrate.
- Intermingling of chromosomes of male and female nuclei takes place. This completes the fusion of male and female nuclei (pronuclei).
- The actual fusion of pronuclei may differ in different animals.
- As a result of amphimxis, diploid zygote is formed.

Abnormal Fertilisation

- **Androgenesis –** Androgenesis is the nonparticipation of female pronucleus in fertilisation.
- **Gynogenesis** – Gynogenesis involves activation of the egg by sperm, but there is no fusion of female pronucleus with sperm pronucleus.
- **Polyandry** – Polyandry is the fusion of two more male pronuclei with a female pronucleus.
- **Polyspermy** – Penetration of many sperm into an ovum is called polyspermy.
- **Polygyny** – Polygyny is the fusion of two pronuclei of ovum with a male pronucleus.

Significance of Fertilisation

- Fertilisation restores the diploid number of chromosomes in the zygote.
- Fertilisation causes genetic variation in the species.
- It stimulates the egg to complete the second division.
- It causes the egg to start cleavage and further development.

TYPES OF EGGS

The animal eggs are classified on the basis of the following:

I. Amount of yolk
II. Distribution of yolk
III. Presence or absence of shell
IV. Amount of yolk

I. Amount of Yolk

- On the basis of amount of yolk, eggs are categorised into the following four types:
  1. **Alecithal** – Eggs without yolk (e. g., of eutherian mammals).
  2. **Microlecithal** – Eggs with little or negligible amount of yolk (e.g., eggs of *Amphioxus*).
  3. **Mesolecithal** – Eggs having a moderate amount of yolk (e.g., eggs of frogs).
  4. **Megalecithal** – Eggs having an enormous amount of yolk. Yolk interferes with cleavage (e.g., eggs of reptiles and birds).

II. Distribution of Yolk

- On the basis of distribution of yolk, eggs are categorised into the following three types:
  1. **Isolecithal** – The yolk is equally distributed throughout the cytoplasm (e.g., eggs of *Amphioxus*).
  2. **Telolecithal** – The yolk is concentrated at one pole. The presence of yolk at one pole of the egg imposes polarity on the egg. The pole with yolk is termed as the vegetal pole. The opposite hemisphere has the nucleus but little yolk, which is known as the animal pole (e.g., eggs of frogs).
  3. **Centrolecithal** – The yolk is concentrated at the centre of the egg, surrounded by a thin peripheral layer of cytoplasm (e.g., eggs of insects).

III. Presence or Absence of Shell

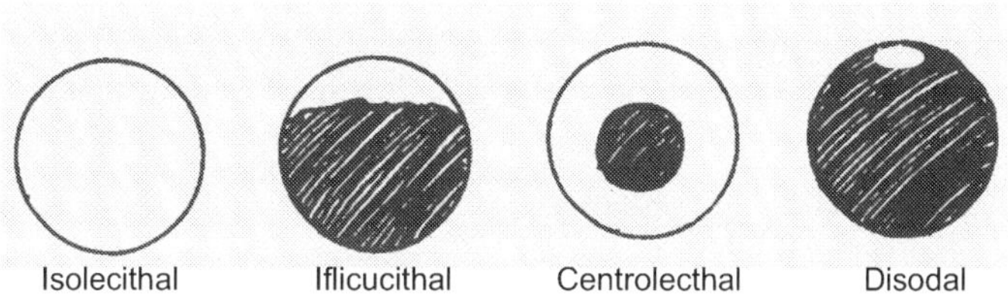

Fig. 8 Different types of eggs

- On the basis of presence or absence of shell, eggs have been categorised into the following two types:
  1. **Cleidoic –** Cleidoic eggs are fully laden with yolk and are surrounded with waterproof calcareous shell (e.g., eggs of reptiles and birds).
  2. **Noncleidoic –** Noncleidoic eggs lack a shell (e.g. eggs amphibians).

IV. Basis of Development

- On the basis of development, eggs have been categorised into the following two types:
  1. **Determinate or Mosaic Egg** – In such an egg, the definite fate of every part is predetermined. If a particular portion of the egg is removed, the developing embryo will be lacking in a particular organ (e.g., eggs of Annelids and Arthropods).
  2. **Indeterminate or regulative Egg** – In eggs of a majority of animals, there is no predetermination. If a particular portion of the egg is removed, it can develop into a normal embryo without any defect. Such a type of egg is called indeterminate or regulative egg (e.g., eggs of *Amphioxus*, reptiles, birds, etc.).

CLEAVAGE

- Cleavage is a series of rapid cell divisions without cell growth or gene expression, which occurs in early embryogenesis.
- During cleavage, the cytoplasm is divided into smaller and smaller cells called blastomeres.
- Cleavage results in a ball of cells with a central cavity called blastocoel and this stage is called blastula.

Characteristics of Cleavage

- Cleavage occurs soon after fertilisation.
- During cleavage, the total embryonic mass remains constant.
- During cleavage, there is no growth in the resulting cells (blastomeres). Cleavage results into cells of continuously diminishing size.
- In most species (mammals being the chief exception), the rate of cell division and the placement of blastomeres with respect to one another, are completely under the control of the proteins and mRNAs stored in the oocyte of the mother.
- Cleaving cells have a modified cell cycle, in which two gap phases G_1 and G_2 are completely omitted. The cell cycle rapidly occurs between M and S phases.
- There is no change in the general shape of the embryo (except for the formation of a cavity called blastocoel).
- The blastomeres remain adhered together after cleavage.
- There is rapid synthesis of nuclear matter (DNA, RNA) and increased consumption of oxygen.
- The ratio of nucleus to cytoplasm is very low in the beginning of cleavage, but this ratio becomes equal to the level found in normal somatic cells.
- There is no change in the constituent parts of ooplasm.

Patterns of Cleavages

There are several different types of cleavage patterns, which are determined by the amount and distribution of yolk protein in the cytoplasm and factors influencing the mitotic spindle.

- The cleavage occurs at a faster rate at the pole of the egg, which is yolk-free.
- The pole with the lesser yolk concentration is known as the animal pole and the pole with greater yolk concentration is known as the vegetal pole.
- The zygote nucleus is generally located in the animal pole and yolk in the vegetal pole tends to inhibit cleavage.
- Yolk is a major factor in the type of cleavage as seen in different species.

Based upon symmetry, there are five patterns of cleavage:

1. **Radial Cleavage**
    - The divisions occur in such a way that all the blastomeres are placed in a radially symmetrical fashion around the polar axis. When viewed from the poles, the blastomeres seem to be arranged in a radially symmetrical form (e.g., sea urchin and *Amphioxus*).
2. **Biradial Cleavage**
    - In biradial cleavage, four blastomeres arise from the usual two meridional cleavages, but the third cleavage plane is vertical. Such cleavage pattern is found in Ctenophores (*Beroe*).
3. **Bilateral Cleavage**
    - In bilateral cleavage, the first cleavage divides the zygote into left and right halves. The following cleavages are centred on the axis, resulting in the two halves, which are mirror images of one another.
    - In bilateral holoblastic cleavage, the divisions of the blastomeres are complete and separate. But in bilateral metroblastic cleavage, the blastomeres are partially connected (e.g., tunicate, Echinoderms, nematodes, etc.).

4. **Spiral Cleavage**
   - The divisions are unequal, but the cells are arranged in different plane within the embryo, which appears as spiral formation.
   - The spiral cleavage occurs due to oblique positions of mitotic spindles.
   - Spiral cleavage occurs in roundworms and Molluscs.
5. **Rotational Cleavage**
   - Rotational cleavage occurs in mammals.
   - In such a cleavage, the first cleavage is normal meridional. But in the second cleavage, one of the two blastomeres divides meridionally, while the other divides equatorially.

- **On the basis of amount and distribution of yolk, the following two major types of cleavages are seen:**

1. **Holoblastic Cleavage**
   - The entire egg is divided by each cleavage.
   - Holoblastic cleavage may be equal or unequal.
   - In holoblastic equal cleavage, the daughter blastomeres are almost of equal size.
   - Holoblastic equal cleavage occurs in microlecithal and isolecithal eggs (e.g., *Amphioxus* and eutherian mammals).
   - In holoblastic unequal cleavage, the daughter blastomeres are of unequal size.
   - Holoblastic unequal cleavage occurs in mesolecithal and telolecithal eggs (e.g., Dipnoi and amphibians).
2. **Meroblastic Cleavage**
   - Meroblastic cleavage occurs in zygote containing large amount of yolk. The fertilised egg undergoes partial cleavage. Meroblastic cleavage is of two types, viz., discoidal and superficial.
   - In discoidal cleavage, the cleavage furrows do not penetrate the yolk. The embryo form a disc of cells called blastodisc. The cleavage proceeds only in the cytoplasm restricted to the discoidal area. Such cleavage is found in reptiles and birds which have telolecithal eggs.
   - In superficial cleavage, cleavage occurs only in the peripheral cytoplasmic region. The nucleus present in the centre of the yolk divides and migrates to the peripheral layer of the cytoplasm, where it becomes surrounded by a small portion of the cytoplasm. In superficial cleavage, cytokinesis does not occur. It occurs in centrolecithal eggs of insects.

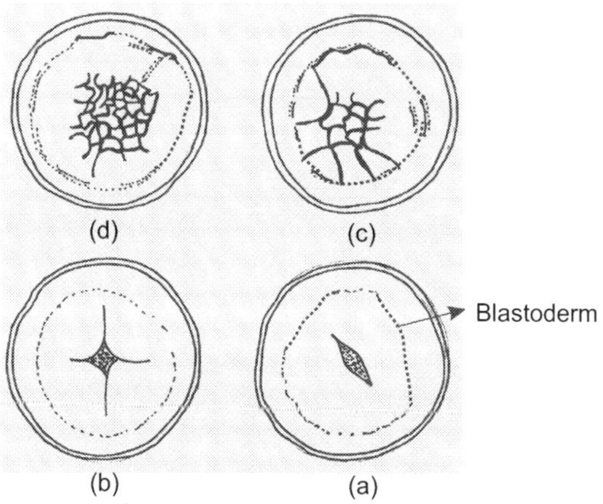

Fig. 9 Discoidal cleavage in hen's eggs

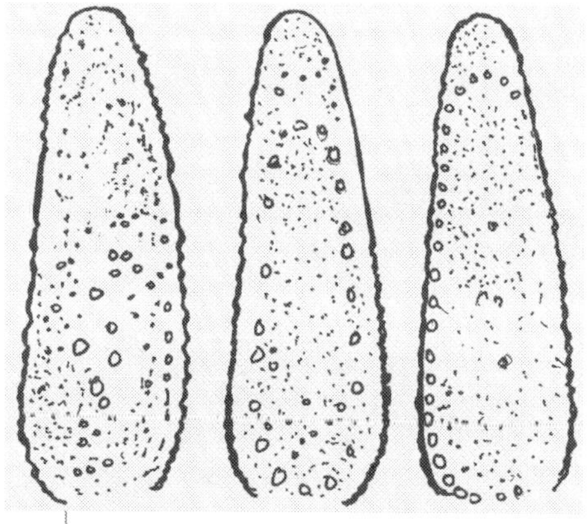

Fig. 10 Superficial meroblastic cleavage in insects

- **On the basis of amount and distribution of yolk, the following two major types of cleavages are seen:**
  1. **Determinate (Mosaic) Cleavage**
     - The pattern of cleavage in which definite blastomeres give rise to definite parts of the embryo is known as determinate cleavage. This type of cleavage is found in nematodes, Annelids and Molluscs, in which the parts of the eggs are predetermined to give rise to a particular part of the embryo.
  2. **Indeterminate (Regulative) Cleavage**
     - In such type of cleavage, the resultant blastomeres do not bear definite relation with a particular part or organ of the embryo. This type of cleavage occurs in vertebrates.

BLASTULATION

- Blastulation is the formation of blastula from the morula.
- Blastula is a hollow ball of cells (blastomeres) enclosing a fluid-filled central cavity called blastocoel.
- The blastocoel contains a jelly-like or gelatinous fluid that is secreted by the blastomeres.
- During subsequent development, each cellular areas of the wall of blastula generally produce a certain structure.
- In mammals, the blastula is called blastocyst, which consists of inner cell mass, trphoblast and blastocoel.
- The blastula stage occurs at the end of cleavage, but before gastrulation.

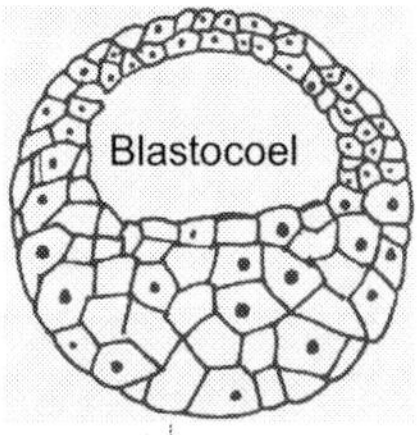

Fig. 11 Blastula

Characteristics of Blastula

- It is a ball of cells.
- The rate of cleavage slows down and a cavity is formed called blastocoel.
- Types of blastula depend upon types of eggs and types of cleavages.

Table 1 Various types of blastulae

S. No.	*Type*	*Nature of cells*	*Cleavages type*	*Size of blastocoels*	*Position of blastocoels*	*Example*
1.	Stereo blastula	Few cells but larger in size and densely packed.	Holoblastic small	Very	Central and worms	Some insets
2.	Coelo blastula	(a) One layer of many cells.	Holoblastic	Very large	Central	Starfish
		(b) Several layers of cells that are of micro and macro types.	Holoblastic	Large	On one side (i.e., towards the macromeres)	Frogs and fishes

Contd.

Table 1 *Contd.*

S. No.	*Type*	*Nature of cells*	*Cleavages type*	*Size of blastocoels*	*Position of blastocoels*	*Example*
3.	Disco blastula	In extremely telolecithal eggs	Discoidal	Small	Within the disc and separated from the yolk by a space.	Reptiles and birds
4.	Superficial blastula	Single layer of cells surrounding the central cavity.	Superficial	Absent		Some insects

GASTRULATION

- The process that converts single-layered blastula into double-layered gastrula is known as gastrulation.
- Ernst Haeckel (1872) coined the term 'gastrula' and 'gastrulation'.
- The gastrula is the double-walled stage of the embryo resulting from invagination of the blastula.
- New cavity formed during gastrulation is called archenteron.
- The archenteron is lined with endodermal cells.
- The archenteron opens to the outside through a small opening called blastopore.
- In Protosomia (Annelida, Mollusca and Arthropoda), the blastopore becomes divided into two openings, one of which develops into the mouth and other into the anus.
- In Deuterostomia (Echinodermata and Chordata), the blastopore develops into the anus.
- During gastrulation, some of the cells migrating inward contribute to the mesoderm, an additional layer between the endoderm and the ectoderm.
- Gastrulation varies in different phyla.
- Gastrulation is followed by organogenesis.
- During gastrulation:
  (a) Three primary germ layers are established.
  (b) The basic body plan is established.
  (c) Cells are brought to new positions due to movements. This allows cells to interact with cells that were not initially near to them, causing the way for inductive interactions. These inductive interactions are hallmark of neurulation and organogenesis.

Characteristics of Gastrulation

- The rate of cell divisions is decreased.
- Morphogenetic movements occur (rearrangement of blastomeres of the embryo).
- There is no growth of any type or if any growth takes place at all, it is insignificant.
- Metabolic changes and rate of oxidation become faster.
- Synthesis of new types of proteins begins.
- The influence of paternal chromosomes becomes evident.
- Nucleus becomes more active.

Significance of Gastrulation

- Gastrulation brings about the formation of three germ layers.
- Gastrulation is necessary for differentiation and morphogenesis.

Fate of Germ Layers

- During gastrulation, three primary germ layers (ectoderm, mesoderm and endoderm) are formed and organised in their proper locations. These germ layers later develop into certain body structures.

I. Ectoderm

- Ectoderm is the most exterior germ layer that forms the epidermis of skin, skin derivatives (glands, hair, nails, feathers, claws, etc.), nasal epithelium, lens of eye, cornea, conjunctiva, retina, optic nerve, inner ear (membranous labyrinth), enamel of teeth, anterior and posterior pituitary, pineal body, adrenal medulla, brain and spinal cord, sympathetic nervous system, cranial and spinal motor nerves dorsal root and sensory nerves, branchial skeleton, lining of stomodaeum and proctodaeum.

II. Mesoderm

- It is the middle germ layer that forms the dermis, muscle, skeletal system, adrenal cortex, circulatory system, kidneys, spleen gonads, reproductive ducts, coelom and coelomic epithelium, choroid and sclerotic coats of eye.

III. Endoderm

- It is the most internal layer that forms the lining of the gut (except stomodaeum and proctodaeum), gastric and intestinal glands, liver, pancreas, bile and pancreatic ducts, respiratory system, middle ear and eustachian tube, urinary bladder, thyroid, gland, parathyroid gland thymus, lining of vagina and urethra.

EXTRA EMBRYONIC MEMBRANES

- Embryos of amniotes (reptiles, birds and mammals) are provided with a series of extra embryonic membranes that protect them from desiccation, shock and assist in respiration, excretion as well as other necessary functions. These membranes are called extra embryonic membranes or foetal membranes.
- These membranes do not participate in the formation of the embryo.
- During embryonic development, four types of extra embryonic membranes are formed, viz., yolk sac, amnion, chorion and amnion.

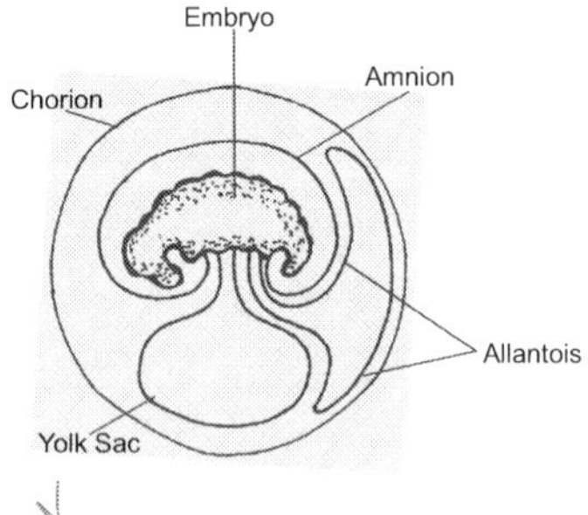

Fig. 12 Extra embryonic membranes in a generalised vertebrate

- The presence of each extra embryonic membrane varies according to the vertebrate class type.
- In fishes and amphibians, only the yolk sac is present. But in amniotes (reptiles, birds and mammals), amnion, chorion and allantois are also present in addition to the yolk sac.

I. Yolk Sac

- It is the first membrane to appear.
- It is formed from the covering of vitellus by some cells originated from the primitive gut.
- It encloses a mass of yolk.
- Yolk is the sole source of food until hatching.
- Yolk is a mixture of proteins and lipoproteins.

II. Amnion

- Amnion surrounds the embryo.
- The development of amnion is of great functional significance. Amnion provides a water pool around the embryo (amniotic fluid) which protects embryo from desiccation and mechanical shocks. Thus, it is an adaptation to terrestrial mode of life.

III. Chorion

- It is the outermost membrane.
- It covers the amnion, yolk sac and allantois.
- It is protective and respiratory in function.
- In mammals, it participates in the formation of foetal part of the placenta as well as takes additional charges of nutrition, excretion, etc.

IV. Allantois

- Allantois is mainly associated with excretion but in reptiles and birds, it also acts as a respiratory surface.
- In eutherian mammals, it joins with chorion to form the allantochorion (foetal placenta) that helps in nutrition, respiration, excretion as well as other functions of the embryo.
- The allantois is an adaptation to dry land because in embryos of oviparous terrestrial vertebrate (like reptiles and birds), the metabolic wastes cannot be immediately excreted to the aquatic surrounds (as fishes and amphibian larvae do).Thus, it was essential for these animals to develop a structure having the ability of storing excretory products till hatching.

PLACENTA

- Placenta is a temporary structure that connects the foetus with the uterine wall of the mother and serves the purpose of physiological exchange between the foetus and the mother.
- It is produced by the fusion of the extra embryonic membranes with the uterine endometrium of the mother.
- Placenta is a defining characteristic of eutherian or placental mammals.
- Placenta is classified variously.

Classification of Placenta

I. Nature of Foetal Membranes

On the basis of nature of foetal membranes, placenta is of the following three types:

1. **Chorio-vitelline Placenta**
   - It is also known as yolk sac placenta.
   - It is formed by the combination of yolk sac and chorion.
   - This type of placenta is found in marsupials (kangaroo and opossums).
2. **Chorio-allantoic Placenta**
   - It is formed by both chorion and allantois.
   - Such type of placenta is found in bandicoots and all eutherians (except humans).
3. **Chorionic Placenta**
   - It consists of only chorion.
   - Chorionic placenta is found in humans.

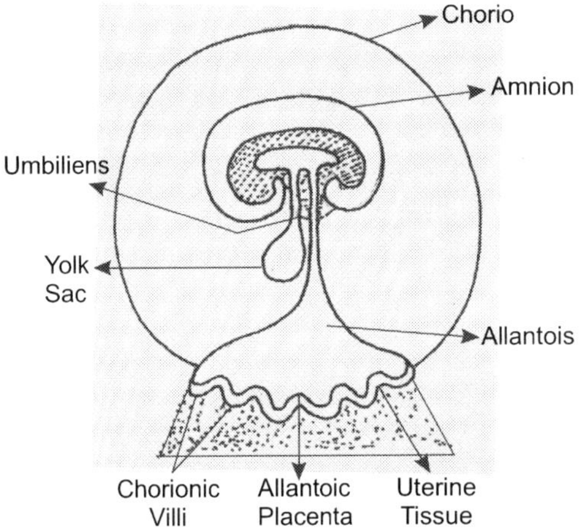

Fig. 13 Diagrammatic representation of chorio-allantoic placenta

II. Nature of Maternal and Foetal Tissues

- On the basis of nature of maternal and foetal tissues, there are five types of placenta:
  1. **Epithelio-chorial Placenta** – In such type of placenta, all the six barriers between the foetus and maternal blood are present (e.g., horses and asses).
  2. **Syndesmo-chorial Placenta** – There are only five barriers present between the foetus and maternal blood due to the breakdown of uterine epithelium (e.g., sheep and goats).
  3. **Endothelio-chorial Placenta** – There are only four barriers between the foetus and maternal blood due to breakdown of uterine connective tissue and uterine epithelium. Such type of placenta is the characteristic of the carnivores (e.g., dogs, cat lions, bears, walruses, etc.).

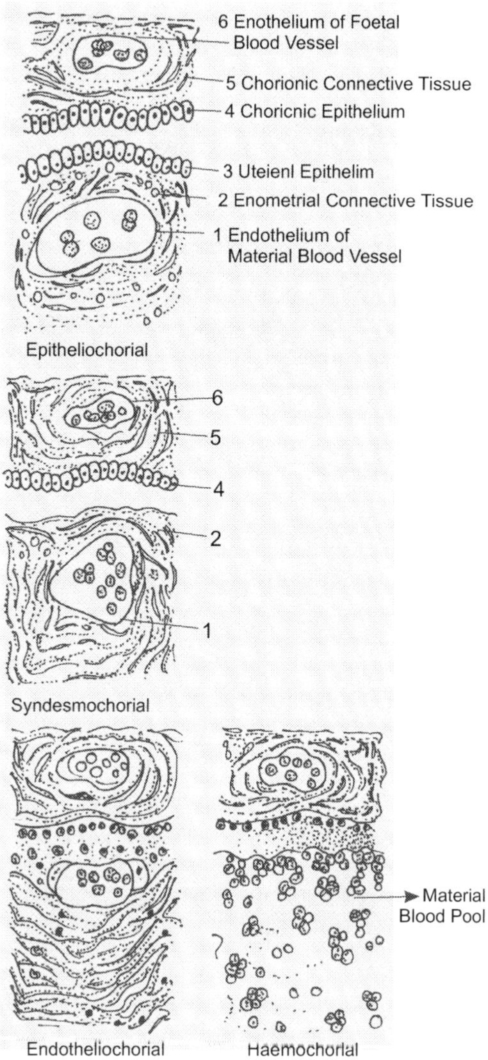

Fig. 14 Histological types of mammalian placentae

4. **Hemo-chorial Placenta** – There are only three barriers between the foetus and maternal blood as all the three maternal barriers (uterine epithelium, uterine connective tissue and endothelium of uterine capillaries) are eroded. Such type of placenta is found in primates (e.g., humans, apes, lemurs, etc.).
5. **Hemo-endothelial Placenta** – There is only one barrier between the foetus and maternal blood. In this type of placenta, three maternal and two foetal tissues (chorionic epithelium and chorionic connective tissue) are completely eroded. This type of placenta is found in rabbits, rats and guinea pigs.

III. Fate of Uterine Endometrium at Birth

Depending upon the nature of contact between maternal and foetal components, three types of placentae have been recognised.

1. **Nondeciduous Placenta** – The maternal and foetal tissues are separated without any damage to the uterine wall. Hence there is no bleeding (e.g., horses, zebras and asses).
2. **Deciduous Placenta** – At birth, a variable amount of maternal tissue (called decidua) is torn away and pulled out along with the foetus. Bleeding occurs as a result.

IV. Contra-deciduous Placenta

In such type of placenta, even the foetal part of the placenta is not passed out at birth and is retained in the uterus, where it is broken and reabsorbed (e.g., bandicoots and moles).

V. Nature and Arrangement of Villi

On the basis of nature and arrangement of villi on the surface of chorion, placentae have been grouped into six types:

1. **Diffuse Placenta** – Villi remain scattered all over the surface of chorio-allantois (e.g., pigs and mares).
2. **Cotyledonary Placenta** – Villi are present in the form of patches called cotyledons (e.g., cows, sheep and goats).
3. **Intermediate Placenta** – Villi are found both as cotyledons as well as scattered (e.g., camels and giraffes).
4. **Zonary Placenta** – Villi are found as incomplete or complete girdle around the blastocyst (e.g., dogs, cats, lions or tigers).
5. **Discoidal Placenta** – Villi are restricted to a small disc shaped area of the blastocyst (e.g., bats, rabbits, bears, porcupines).
6. **Metadiscoidal Placenta** – In the beginning, villi remain scattered but later become confined to a small disc-shaped area attached to the uterine wall. It is monodiscoidal in humans and bidiscoidal in apes and monkeys.

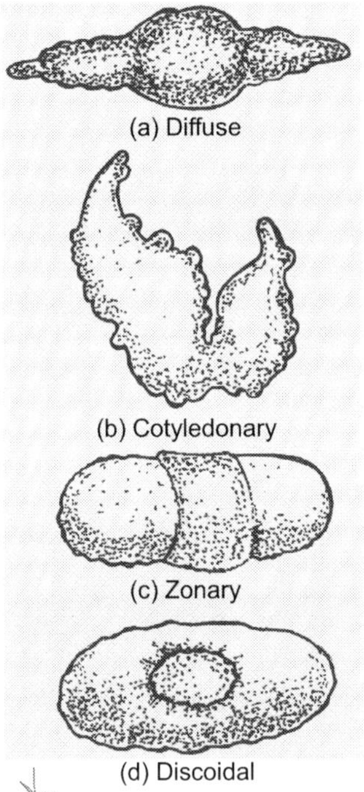

Fig. 15 Different types of placentae on the basis of arrangement of villi

Functions of Placenta

- Placenta connects the developing foetus to the uterine wall of the mother to allow nutrient uptake, gas exchange and waste elimination via the mother's blood supply.
- IgG antibodies can pass through the human placenta, thereby providing protection to the foetus in the uterus.
- Some of the drugs taken by the mother during pregnancy period may cross the placental barrier and may even cause structural deformities.
- Thalidomide, a drug taken as sedative by some women during early pregnancy causes deformities in the development of limbs and heart in children born to such women.
- Placenta stores some fat, glycogen and iron.

- Placenta also performs endocrine functions. It secretes human chorionic gonadotropin (hCG), the first placental hormone. hCG is used to test pregnancy. The human placental lactogen (hPL) secreted by placenta promotes growth of mammary glands and prepare mammary glands for lactation. Besides, placenta also secretes estrogen and progesterone.
- If the mother suffers from some diseases like syphilis, small pox, chicken pox and measles, their pathogens enter the foetus through the placenta.
- The placenta also provides a reservoir of blood for the foetus, delivering blood to it in case of hypotension and vice versa, comparable to a capacitor.
- Placenta secretes neurokinin B containing phosphocholine molecules that protect the foetus from the attack of the mother's immune system.

GESTATION PERIOD

- Gestation period is the time in which the foetus develops, starting from fertilisation and ending at birth.
- The larger the animal, the longer the gestation period.
- The gestation period varies from animal to animal.
- The gestation periods in a few animals are given below:

Animal	Days
Opossum	12 to 13
Hamster	16
Mouse	19
Skunk	31
Kangaroo	42
Domestic cat	61
Domestic dog	64
Domestic pig	114
Beaver	122
Lemur	130
Goat	145
Polar bear	240
Humans	280 to 285
Domestic cow	285
Horse	336
Zebra	365
Whale	490
Giraffe	420 to 450
Rhinoceros	540
Elephant	645
Common frilled shark*	1279.25 (3.5 years)

- Common frilled shark (*Chlamydoselachus anguineus*) has the longest gestation period.

ANIMAL DEVELOPMENT

DEVELOPMENTAL BIOLOGY

Short-Answer Questions

1. What is the name of the branch of zoology that deals with the development of animals?
   Answer: Embryology.
2. Name the first embryologist.
   Answer: Aristotle.
3. Who is the 'Father of Modern Embryology'?
   Answer: Carl Ernst von Baer
4. Who first identified the mammalian egg?
   Answer: Von Baer (1827)
5. Who first described the sperm of humans?
   Answer: Leuwenhoek (1677)
5. Who described cleavage (segmentation) of the egg?
   Answer: Prevost and Duman (1824)
6. Who first demonstrated germ layers?
   Answer: Pander (1817)
7. Name the scientist who was awarded the Nobel Prize in Medicine (Physiology) in 2010 for his contribution to in vitro fertilisation.
   Answer: Roberts Edwards
8. Who gave the concept of cloning without involving germ cells?
   Answer: Ian Wilmut (1996)
9. Who first proved that male and female contribute equal number of chromosomes in fertilised egg?
   Answer: Von Bender (1833)
10. Who discovered the existence of polar bodies?
    Answer: Oscar Hertwig and Richard Hertwig
11. Name the scientist who was awarded Nobel Prize in 1910 for the discovery of the organiser effect.
    Answer: Hans Spemann
12. Define parthenogenesis. Who performed the first case of artificial parthenogenesis?
    Answer: Development of embryos without fertilisation is known as the parthenogenesis. Jacques Loeb (1900) performed the first case of artificial parthenogenesis.
13. How is diploid condition restored in diploid parthenogenesis?
    Answer: In diploid parthenogenesis, the diploid condition is retained by deleting the reduction division or by the fusion of one of the polar bodies with the haploid ovum.

14. Give examples of haploid and diploid parthenogenesis.
Answer: (a) Haploid parthenogenesis – Honeybees (Drone)
(b) Diploid parthenogenesis – Aphids

15. What is fertilizin?
Answer: The surface of eggs contains a chemical substance capable of causing agglutination or clumping of sperms. This chemical substance is called fertilizin.

16. Define antifertilizin?
Answer: Antifertilizin is a protein molecule present on the surface of sperm which is a male attractant.

17. Name the two types of reactions that occur before the sperm penetrates the ovum.
Answer: (a) Acrosome reaction (b) Egg reaction

18. What is cortical reaction?
Answer: The formation of fertilisation cone is followed by changes in the surface of the egg cytoplasm, which is termed as cortical reaction.

19. What is the main source of fertilizin?
Answer: The main source of fertilizin is the jelly coat or plasma membrane of the egg.

20. Define amphimixis.
Answer: Amphimixis is the fusion of nuclei.

21. What term is given to the following:
(a) Nonparticipation of female gamete in fertilisation
(b) Penetration of many sperms into an ovum
(c) Fusion of two pronuclei of female with a male pronucleus
(d) Fusion of two or more male pro nuclei with a female pronucleus
Answer: (a) Androgenesis (b) Polyspermy (c) Polygyny (d) Polyandry

22. What is acrosin?
Answer: Acrosin is an enzyme produced by the acrosome that dissolves zona pellucida of the ovum. As a result, the sperm penetrates the ovum.

23. How does the sperm make contact with the egg membrane?
Answer: The sperm makes contact with the egg membrane by its acrosome.

24. Mention two significances of fertilisation.
Answer: (a) Fertilisation restores the diploid number of chromosomes in the zygote.
(b) It produces genetic variation.

25. What is an egg?
Answer: An egg is a female gamete that supplies the haploid set of chromosomes and cytoplasm to the developing zygote.

26. Define mesolecithal and megalecithal eggs.
Answer: Eggs having a moderate amount of yolk are called mesolecithal eggs (e.g., eggs of frogs), while eggs having a huge amount of yolk are called megalecithal eggs (e.g., eggs of reptiles and birds).

27. Distinguish between mosaic and regulative eggs.
Answer: In a mosaic egg, the fate of every part is predetermined (e.g., eggs of Annelids and Arthropods), while in a regulative egg, the fate of a particular portion is not predetermined (e.g., eggs of *Amphioxus*, reptiles, birds, etc.).

28. What is the function of the following:
(a) Eggshell (b) Albumen (c) Yolk (d) Chalazae

Answer: (a) Eggshell – Protection and exchange of gases
(b) Albumen – Albumen is a glycoprotein which serves as food and also provides wet environment to the embryo
(c) Yolk – Serves as food
(d) Chalazae – Keeps the yolk in position, preventing germinal disc from damaging

29. What is cleavage?
Answer: Cleavage is a series of rapid cell divisions without cell growth or gene expression, which occurs in early embryogenesis.

30. Mention the characteristics of cleavage.
Answer: (a) Cleavage occurs soon after fertilisation.
(b) During cleavage the total embryonic mass remains constant.
(c) The general shape of the embryo does not change.
(d) No growth occurs.
(e) Cleaving cells have modified cell cycle.
(f) Chemical conversion of reserve food materials (yolk, glycogen, etc.) into active cytoplasm having pool of amino acids, monosaccharides, lipid molecules and several enzymes take place during cleavage.

31. What are vegetal pole and animal pole? At which pole is the cleavage rate faster?
Answer: The pole of egg having the highest amount of yolk is termed as vegetal pole, while its opposite is termed as animal pole. The cleavage occurs at a faster rate at the pole that is yolk-free (animal pole).

32. In which type of egg is superficial meroblastic cleavage found?
Answer: Superficial meroblastic cleavage is found in the centrolecithal egg.

33. Name the pattern of cleavage in which daughter cells are placed in such a manner that the embryo has right and left halves only along one plane of division.
Answer: Bilateral cleavage

34. What name is given to different cells derived from cleavage?
Answer: The cells formed as a result of cleavage are called blastomeres.

35. What determines the pattern of cleavage?
Answer: The pattern of cleavage is determined by the amount and distribution of the yolk in the cytoplasm as well as factors influencing the mitotic spindle.

36. Which type of cleavage occurs in frogs?
Answer: In frogs, cleavage is holoblastic and unequal.

37. Define meroblastic cleavage.
Answer: The cleavage in which only a part of egg is divided into blastomeres is called meroblastic cleavage. Such type of cleavage occurs in megalecithal eggs.

38. Name the animal group that exhibits rotational cleavage.
Answer: Mammals display rotational cleavage.

39. Give examples of animals in which the following cleavage patterns are found:
(a) Bilateral (b) Discoidal (c) Radial (d) Spiral
(e) Superficial
Answer: (a) Bilateral – Tunicts, amphibians (b) Discoidal – Reptiles, birds
(c) Radial – *Amphioxus*, sea urchins (d) Spiral – Annelids, Molluscs
(e) Superficial – Insects

40. What controls the rhythm of cell division?

Answer: The rhythm of cell division is under the control of production and degradation of cyclin.

41. Name the ion that plays a key role in the degradation of cyclin.

Answer: Calcium ion

42. Define morula.

Answer: The solid mass of cells formed as a result of cleavage is known as morula.

43. What is blastula?

Answer: Blastula is a hollow ball of cells (blastomeres) enclosing a fluid-filled central cavity called blastocoel, which contains a jelly-like or gelatinous substance secreted by the blastomeres.

44. Name the animals in which following types of blastulae are found:

(a) Stereo blastula (b) Coelo blastula (c) Disco blastula

Answer: (a) Stereo blastula – Insects, worms (b) Coelo blastula – Starfishes

(c) Disco blastula – Reptiles and birds

45. What is blastocyst?

Answer: Mammalian blastula has the blastoderm, which withdraws nutrition for the embryo and is termed as the blastocyst.

46. When does the blastula stage occur?

Answer: Blastula occurs at the end of cleavage and before gastrulation.

47. Define gastrulation.

Answer: The process that converts the single-layered blastula into a double-layered gastrula is known as gastrulation.

48. What is gastrula?

Answer: The gastrula is the double-walled stage of the embryo resulting from invagination of the blastula, having a cavity called archenteron.

49. What is a blastopore?

Answer: A blastopore is the opening through which the archenteron opens to the outside.

50. When does the grey crescent appear?

Answer: The grey crescent appears when the egg is fertilised.

51. What is a fate map?

Answer: The graphic representation of the cells of the embryo that develop into specific organs in the adult is known as the fate map

52. Name the scientist who developed fate mapping.

Answer: Fate mapping was developed by Walter Vogt.

53. What is the fate of the blastopore in deuterostomia?

Answer: In deuterostomia, the blastopore develops into the anus.

54. Why is the dorsal lip of the blastopore known as the organiser?

Answer: In all chordates, the dorsal lip of the blastopore forms the notochord. But when it is transplanted to another embryo, it leads to the formation of another embryo. Therefore, the dorsal lip of the blastopore is known as the organiser.

55. Where does the formation of the grey crescent begin?

Answer: The formation of the grey crescent begins at a point opposite to the entry of the sperm in the egg.

56. Who demonstrated the importance of the grey crescent?

Answer: The importance of the grey crescent was demonstrated by Spemann.

57. Name the developmental stage during which three primary germ layers and basic body plan are established.
Answer: Three primary germ layers and basic body plan are established during the gastrula stage (gastrulation).

58. Distinguish between epiboly and emboly.
Answer: The spreading of micromeres over macromeres is called epiboly, while the inward movement of cells is called emboly.

59. What is delamination?
Answer: The separation of cells by splitting is called delamination.

60. Define area pellucida and area opaca.
Answer: Due to the presence of subgerminal cavity, the central part of the blastodisc of the chick embryo is semitransparent in appearance, which is called area pellucida and this later forms the body of the embryo proper. The peripheral region of the blastodisc surrounding the area pellucida is in direct contact with the yolk and is called area opaca, which later forms the accessory of extra embryonic structures.

61. What is the primitive streak?
Answer: The primitive streak is an elongated band of cells which is formed in the embryos of reptiles, birds and mammals during gastrulation.

62. What is Hensen node?
Answer: Hensen node is the knot at the tip of the primitive streak.

63. What marks the end of gastrulation in the chick?
Answer: Regression of primitive streak marks the end of gastrulation in the chick.

64. How can one compute age of the chick embryo?
Answer: One can compute the age of the chick embryo by counting the somites.

65. When is the first somite laid?
Answer: The first somite is laid after 21 hours of incubation.

66. How do genes control development?
Answer: Genes control development by controlling where and when proteins are synthesised.

67. Name the four stages of the embryonic development of animals.
Answer: (a) Cleavage (b) Patterning (c) Differentiation (d) Growth

68. What is germinal disc?
Answer: The spot of the yolk from which blastoderm develops after fertilisation is called germinal disc.

69. Distinguish between cleavage and mitosis
Answer:

	Cleavage	Mitosis
(a)	Cleavage occurs in the zygote.	Mitosis takes place in the somatic cells.
(b)	Total embryonic mass remains constant.	Total embryonic mass does not remain constant.
(c)	The cells formed are of decreasing size.	The cells formed remain of the same size.
(d)	The nucleo-cytoplasmic ratio gradually increase.	The nucleo-cytoplasmic ratio remains the same.
(e)	Oxygen consumption is very high.	Oxygen consumption is relatively low.
(f)	Cleaving cells have modified cell cycle in which two gap phases G_1 and G_2 are completely omitted and cell cycle occurs in S phase and M phase.	Interphase is of longer duration in which both DNA and cytoplasm are synthesised.

70. What is archenteron?

Answer: Archenteron is a cavity formed by endodermal cells during gastrulation.

71. What is the importance of gastrulation?

Answer: (a) Gastrulation leads to formation of three germ layers.
(b) Gastrulation is essential for differentiation and morphogenesis.

72. Which law states that:

(a) The cleavage is determined by the yolk in the egg.
(b) Successive cleavage occurs at right angles to each other and blastomeres tend to be of equal size unless obstructed by the yolk.

Answer: (a) Balfour's law (b) Sach's law

73. What is morphogenetic movement?

Answer: During gastrulation, there occurs a mass movement of cells from the surface of the blastula to their definite position in the embryo leading to the formation of three primary germ layers (ectoderm, mesoderm and endoderm). This mass movement of cells is called morphogenetic movement.

74. What are extra embryonic membranes?

Answer: Embryos of amniotes (reptiles, birds and mammals) are provided with a series of membranes that protect the embryo as well as assist in nutrition, respiration and excretion. These membranes are called extra embryonic membranes. These membranes are as follows:

(a) Yolk sac (b) Amnion (c) Chorion (d) Allantois

75. Name the following:

(a) First extra embryonic membrane to appear
(b) Outermost extra embryonic membrane
(c) Extra embryonic membrane that assists in excretion

Answer: (a) Yolk sac (b) Chorion (c) Allantois

76. Which extra membrane is well developed in reptiles, birds and prototherians?

Answer: The yolk sac is well developed in reptiles, birds and mammals.

77. Name a major characteristic of birds' gastrulation.

Answer: The primitive streak is the major characteristic of birds' gastrulation.

78. Define compaction.

Answer: In mammals, at each cell stage, the blastomeres develop a differential adhesion property as a result of which the outer surface becomes convex and the inner surface becomes concave. This reorganisational property is known as compaction.

79. What is placenta? How is it formed?

Answer: Placenta is a temporary structure between the foetus and the uterine wall of the mother that serves the purpose of physiological exchange between the foetus and the mother. It is formed by the fusion of the extra embryonic membranes with the uterine endometrium of the mother.

80. What is the name of the structure through which the placenta is connected to the foetus?

Answer: The placenta is connected to the foetus via the umbilical cord.

81. Name the animal in which the choriovitelline placenta is found.

Answer: The choriovitelline placenta is found in kangaroos and opossums.

82. Give examples of animals in which the following types of placentae are found:

(a) Epithelio-chorial (b) Endothelio-chorial (c) Hemo-chorial

Answer: (a) Epithelio-chorial – Horses and swines (b) Endothelio-chorial – Cats and dogs
(c) Hemo-chorial – Rodents and humans,

83. Name the placenta in which the foetal chorionic epithelium is bathed in maternal blood.
Answer: The hemochorial placenta

84. In which type of placenta are all the six barriers present between the foetus and the mother?
Answer: In the epitheliochorial placenta, all the six barriers are present between the foetus and the mother (e.g., horses and asses).

85. How is the placenta capable of transporting substances from the mother to the embryo and vice versa?
Answer: Because of its large surface area, the placenta is capable of transferring substances from the mother to the embryo and vice versa.

86. Name the hormone, the presence of which in urine is used for pregnancy test. What is its source?
Answer: The presence of human chorionic gonadotropin (HCG) in urine is used to confirm pregnancy. It is secreted by the placenta.

87. Name the major source of placental hormones.
Answer: The villus syncytiotrophoblast is the major source of placental hormones.

88. In mammals, why do foetal membranes start forming before gastrulation?
Answer: Because of immediate requirement of food by the growing embryo.

Long-Answer Questions

1. Give an account of parthenogenesis in animals.
2. Describe the process of fertilisation in animals. What is the significance of fertilisation?
3. Give an account of different types of eggs and patterns of cleavages in animals.
4. What are extra embryonic membranes? Describe extra embryonic membranes of chicks.
5. What is placenta? Describe different types of placentas in mammals.
6. Give an account of development and function of placenta in mammals
7. Write short notes on the following:
   (a) Fertilizin-antifertilizin reaction
   (b) Types of eggs
   (c) Types of cleavages
   (d) Blastula
   (e) Gastrula
   (f) Primitive streak
   (g) Fate maps
   (h) Yolk sac
   (i) Thelytoky
   (j) Primitive streak
8. Distinguish between the following:
   (a) Cleidoic egg and noncleidoic egg
   (b) Morula and blastula
   (c) Blastulation and gastrulation
   (d) Radial cleavage and bilateral cleavage
   (e) Zonary placenta and discoidal placenta
   (f) Grey crescent and yolk plug
   (g) Area opaca and area pellucida

ANIMAL DEVELOPMENT

Multiple-Choice Questions

1 The amount of yolk determines the:
(a) Type of eggs (b) Blastulation (c) Gastrulation (d) All of these

2. Which one of the following statements about von Baer is incorrect?
(a) First man to study cleavage (b) Identified the mammalian egg
(c) Gave Baer's law (d) Father of modern embryology

3. Which one of the following statements about the egg is correct?
(a) A female gamete (b) Provides haploid set of chromosomes to the zygote
(c) Provides nourishment to the developing zygote (d) All of these

4. The concept of induction was given by:
(a) von Baer (b) Horstadius (c) Spemann and Mangold (d) Harvey and Child

5. Eggs of insects are:
(a) Isolecithal (b) Centrolecithal (c) Telolecithal (d) Alecithal

6. Eggs are not megalecithal in:
(a) Reptiles (b) Birds (c) Opossum (d) Monotremes

7. The chordate having centrolecithal eggs is:
(a) *Plethodon cinereus* (b) *Hippocampus* (c) *Polyodon* (d) None of these

8. In which of the following cleavages are the furrows at right angles to each other?
(a) Radial (b) Spiral (c) Determinate (d) Indeterminate

9. In *Balanoglossus*, the cleavage is:
(a) Radial (b) Spiral (c) Superficial (d) Discoidal

10. In *Herdmania*, eggs are:
(a) Microlecithal (b) Mesolecithal (c) Megalecithal (d) Centrolecithal

11. In meroblastic cleavage, division is:
(a) Complete (b) Partial (c) Spiral (d) Horizontal

12. Amnion is found the embryos of:
(a) Frogs (b) Snakes (c) *Amphioxus* (d) Rohu

13. Meroblastic divisions are applicable to:
(a) Disco blastula (b) Coeloblastula (c) Superficial blastula (d) Stereo blastula

14. Match column I with column II and select the correct answer using answer codes:

Column I	Column II
A. Coeloblastula	1. Hens
B. Superficial blastula	2. Insects
C. Disco blastula	3. *Neries*

D. Stereo blastula	4. Frogs

Answer codes:

	A	B	C	D
(a)	4	2	1	3
(b)	3	4	2	1
(c)	4	3	2	1
(d)	2	4	1	3

15. In which one of the following is a single layer of blastomeres present around the central yolk?
(a) Disco blastula (b) Stereo blastula (c) Coeloblastula (d) Superficial blastula

16. During cleavage:
(a) Metabolism becomes fast (b) Consumption of oxygen increases
(c) There is rapid replication of DNA (d) All of these

17. Which one of the following is a solid blastula?
(a) Stereo blastula (b) Coeloblastula (c) Disco blastula (d) Superficial blastula

18. Holoblastic cleavage occurs in:
(a) Alecithal eggs (b) Homolecithal eggs (c) Mesolecithal (d) All

19. Superficial cleavage is the characteristic of:
(a) Insects (b) Birds (c) Amniotes (d) Protochordates

20. Which one of the following is not applicable to *Balanoglossus*?
(a) Blastocoel (b) Protocoel (c) Schizocoelic (d) Blastopore

21. *Amphioxus* lacks:
(a) Kolliker's pit (b) Hensen's node (c) Holoblastic cleavage (d) Emboly

22. Hensen's node is formed by the cells of presumptive:
(a) Brain (b) Notochord (c) Heart (d) Kidney

23. The initiation of primitive streak occurs after:
(a) 6 hours (b) 12 hours (c) 18 hours (d) 24 hours

24. Which one of the following acts as anembryonic urinary bladder?
(a) Yolk sac (b) Amnion (c) Chorion (d) Allantois

25. In frogs, the regularity of cleavage pattern is lost after about the ________ division:
(a) Third (b) Fourth (c) Fifth (d) Seventh

26. The early activities of cleavage are controlled by the:
(a) Mother's mRNAs (b) Mother's mRNAs and proteins
(c) Father's genome (d) Both the genome of father and mother

27. Morphogenetic movement takes place during:
(a) Gastrula (b) Blastula (c) Morula (d) All of these

28. The primitive streak is lacking in:
(a) Pisces (b) Reptiles (c) Birds (d) Mammals

29. Which one of the following statements about cleavage is correct?
(a) There is no separation of plasma membrane of daughter cells. (b) There is no formation of spindle fibres.
(c) There is no growth of cells. (d) All of these

30. Biradial cleavage occurs in:
(a) Ascidians (b) Ctenophores (c) Molluscs (d) Turbellarians

31. The pace of the cleavage is determined by the:
    (a) Nucleus (b) Cytoplasm (c) Yolk (d) Ions like Ca^{++}, Mg^{++}
32. The rate of cleavage in any region of the egg is inversely proportional to the amount of yolk present in it. This is known as:
    (a) Hertwig's law (b) Sach's law (c) Balfour's law (d) Pfluger's law
33. Morula is lacking in the:
    (a) Centrolecithal eggs of insects
    (b) Moderately telolecithal eggs of a dipnoi
    (c) Highly telolecithal eggs of reptiles
    (d) Isolecithal eggs of Echinoderms
34. Emboly includes:
    (a) Invagination
    (b) Involution
    (c) Ingression and delamination
    (d) All of these
35. In chick, retinoic acid is involved in the formation of:
    (a) Brain (b) Limbs (c) Heart (d) Eyes
36. In protostomia, blastopore becomes mouth opening, except:
    (a) Echinoderms
    (b) Hemichordates
    (c) Echinoderms and Hemichordates
    (d) Molluscs and Echinoderms
37. Archenteron is lined with:
    (a) Mesoderm
    (b) Endoderm
    (c) Ectoderm
    (d) Both mesoderm and endoderm
38. Which one of the following is applicable to the egg of a chick?
    (a) Megalecithal (b) Nucleus of Pander (c) Meroblastic cleavage (d) All of these
39. Match column I with column II and select the correct answer using answer codes:

Column I	Column II
(A) Organiser	1. Haeckel
(B) Performation theory	2. Child
(C) Metabolic axial gradient	3. Malpighi
(D) Biogenetic law	4. Spemann

Answer codes:

	A	B	C	D
(a)	4	1	2	3
(b)	4	3	2	1
(c)	2	4	3	1
(d)	3	4	2	1

40. Meroblastic cleavage occurs in:
    (a) Snakes (b) Ostriches (c) Lizards (d) All of these
41. Dorsal lip is associated with:
    (a) Morula (b) Blastula (c) Gastrula (d) Neurula
42. The first extra embryonic membrane to develop is the:
    (a) Yolk sac (b) Amnion (c) Chorion (d) Allantois
43. Torsion in the chick embryo starts at:
    (a) 24 hours (b) 48 hours (c) 72 hours (d) 96 hours
44. Yolk plug is formed during:
    (a) Blastulation (b) Gastrulation (c) Morulation (d) Neurulation

45. The waste material from embryonic metabolism is collected in:
(a) Yolk sac (b) Amnion (c) Chorion (d) Allantois

46. Which one of the following is the major characteristic of reptilian, avian and mammalian gastrulation?
(a) Delamination (b) Hensen's node (c) Primitive streak (d) Koller's sickle

47. The brain of a chick develops from:
(a) Mesoderm (b) Ectoderm
(c) Endoderm (d) From both ectoderm and endoderm

48. Allantois is vestigial sac in:
(a) Chicks (b) Rabbits (c) Pigs (d) Humans

49. Incomplete zonary placenta is found in:
(a) Raccoons (b) Lemurs (c) Cats (d) Moles

50. Eggs are oily and waterproof in:
(a) Ducks (b) Tinamou (c) Crocodiles (d) Ostriches

51. Which one of the following about an egg is incorrect?
(a) Female gamete (b) Nonmotile
(c) Usually contains abundant cytoplasm and yolk (d) None of these

52. Blastulation results in the formation of blastocyst in:
(a) Mammals (b) Birds (c) Reptiles (d) Fishes

53. During gastrulation in birds:
(a) Epiblast cells converge at the midline
(b) Epiblast cells converge at the midline and ingress at the primitive streak
(c) As gastrulation proceeds, the primitive groove extents posteriorly
(d) None of the above

54. In bilateral holoblastic cleavage, the divisions of blastomeres are:
(a) Complete (b) Complete and separate
(c) Incomplete and not separate (d) Complete and partial connected

55. In mammals, zygotic transcription starts at the __________ cell stage:
(a) 2 (b) 4 (c) 8 (d) 2, 4 or 8

56. Match column I with column II and select the correct answer using answer codes:

Column I (Cleavage)	Column II (Animal)
(A) Spiral	1. Sea urchins
(B) Bilateral	2. Molluscs
(C) Rotational	3. Amphibians
(D) Radial	4. Mammals

Answer codes:

	A	B	C	D
(a)	4	3	2	1
(b)	3	4	1	2
(c)	2	4	1	3
(d)	2	3	4	1

57. During cleavage:
(a) The number of cells increases
(b) The number of cells increases but the size of the embryo remains unchanged

(c) There is an increase both in the number of cells as well as size of the embryo
(d) Neither is there an increase in the number of cells nor in the size of embryo

58. Spemann's organiser is applicable to:
(a) Closing of blastopore (b) Replacement of archenteron by the gut
(c) Establishment of three germ layers (d) Formation of extra embryonic membranes

59. Release of _________ at the time of fertilisation causes an increase in the metabolic activity of the egg:
(a) Mg^{++} (b) Ca^{++} (c) Zn^{++} (d) Na^{+}

60. In cleaving cells:
(a) G_1 and G_2 are completely omitted (b) The cell cycle is rapid between M and S phases
(c) Both (a) and (b) (d) G2 and S phases are completely omitted

61. Monodiscoidal placenta is found in:
(a) Apes (b) Monkeys (c) Humans (d) Giraffes

62. Calcite is found in the:
(a) Eggshells (b) Shell membranes (c) Latebra (d) Chalazae

63. Archenteron is lacking in the gastrula of:
(a) *Amphioxus* (b) *Balanoglossus* (c) Chick (d) *Herdmania*

64. In *Amphioxus*, the cleavage is:
(a) Holoblastic and equal (b) Holoblastic and unequal
(c) Meroblastic (d) Biradial

65. Which one of the following dimensions of the embryo changes during cleavage?
(a) Volume (b) Appearance of blastocoel
(c) Shape (d) Size

66. Chemo differentiation is started during:
(a) Gastrulation (b) Neurulation (c) Blastulation (d) Morulation

67. Gastrula is two-layered in:
(a) Frogs (b) Cobras (c) *Hummingbirds* (d) *Amphioxus*

68. Gastrular morphogenetic Hensen's node is rich in:
(a) Linolenic acid (b) Archidonic acid (c) Retinoic acid (d) Sialic acid

69. In which one of the following placenta are five barriers present between the foetal and uterine blood streams?
(a) Haemo-endothelial placenta (b) Haemo-chorial placenta
(c) Endothelio-chorial placenta (d) Syndesmo-chorial placenta

70. In which one of the following placentae do the villi remain scattered all over the surface of the chorion?
(a) Zonary (b) Diffuse (c) Discoidal (d) Cotyledonary

71. Which one of the following is the outermost extra embryonic membrane?
(a) Yolk sac (b) Amnion (c) Chorion (d) Allantois

72. In which one of the following placentae are uterine epithelium and uterine connective tissue lacking?
(a) Epithelio-chorial (b) Endothelio-chorial (c) Hemo-endothelial (d) Hemichorial

73. Allantoic placenta is found in:
(a) Rabbits (b) Opossums (c) Pigs (d) Apes

74. Haemochorial placenta lacks:
(a) Uterine epithelium (b) Uterine connective tissue
(c) Endothelium of material blood vessels (d) All of these

75. The placenta having only one barrier is:
(a) Haemochorial (b) Endothelio chorial (c) Haemo-endothelial (d) Syndesmo-chorial

76. The placenta having six barriers is:
(a) Syndesmo-chorial (b) Epithelio-chorial (c) Endothelio-chorial (d) Hemochorial

77. Chalaza is the characteristic of the eggs of:
(a) *Sphenodon* (b) Kangaroos (c) Hens (d) *Torpedo*

78. Which one of the following provides protection against mechanical shock?
(a) Allantois (b) Yolk sac (c) Chorion (d) Amnion

79. In which one of the following is only one placental barrier found?
(a) Dogs (b) Sheep (c) Humans (d) Rabbits

80. Which one of the following does not develop from the ectoderm?
(a) Eye (b) Internal ear (c) Spleen (d) Pineal gland

81. Which one of the following set of organs is derived from the mesoderm?
(a) Gonads, kidneys and gills
(b) Connective tissue, muscular tissue and gonads
(c) Gut, endoskeleton and nasal epithelium
(d) Nasal epithelium, genital ducts and thymus

82. The middle ear develops from:
(a) Ectoderm
(b) Mesoderm
(c) Endoderm
(d) Both from mesoderm and endoderm

83. Amnion and allantois are found in:
(a) Dipnoi (b) Sharks (c) Amphibians (d) None of the above

84. Thyroid and parathyroid develop from:
(a) Ectoderm (b) Mesoderm (c) Endoderm (d) Ecto-mesoderm

85. Which one of the following is a correct match?
(a) Blastula – Archenteron
(b) Cleavage – Meroblastic
(c) Gastrula – Triploblastic
(d) Morula – Blastocoel

86. Primitive allantoic placenta is found only in:
(a) *Perameles* (b) *Didelphis* (c) *Macropus* (d) All of these

87. Which one of the following covers the embryo?
(a) Amnion (b) Chorion (c) Yolk sac (d) Allantois

88. Match column I with column II and select the correct answer using answer codes:

Column I	Column II
(A) Amnion	1. Nutrition
(B) Yolk sac	2. Excretion
(C) Chorion	3. Protection
(D) Allantois	4. Respiration

Answer codes:

	A	B	C	D
(a)	4	1	3	2
(b)	3	4	2	1
(c)	2	4	1	3
(d)	3	1	4	2

89. Amniotic eggs are characteristic of:
(a) Reptiles (b) Birds (c) Monotremes (d) All of these

90. Which one of the following statements about the allantois is incorrect?
    (a) A part of endoderm and part of lateral plate mesoderm
    (b) Known as the 'bag of waters'
    (c) Stores urine
    (d) Its blood vessels are the main vessels of the placenta
91. In eutherian mammals, the first cell types that are specified during embryogenesis are:
    (a) Embryonic structures (b) Placenta
    (c) Foetal membranes (d) Placenta and foetal membranes
92. Extra embryonic membranes of_______ have been frequently used to array potential angiogenic agent:
    (a) Snakes (b) Chicks (c) Spiny anteaters (d) *Sphenodon*
93. Which one of the following statements about the blastocyst is incorrect?
    (a) Blastocyst develops within the uterus.
    (b) Its outer cell layer forms an extra embryonic tissue.
    (c) An inner cell mass is located under the trophoblast at the end of the blastocyst, which is called embryoblast.
    (d) The blastocyst has a relatively very small fluid-filled cavity called blastocoel.
94. Allantois is prominent in:
    (a) Carnivores (b) Ungulates
    (c) Both carnivores and ungulates (d) None of these
95. The umbilical cord is first formed during the_______ embryonic week:
    (a) Third (b) Fifth (c) Seventh (d) Nineth
96. In *C. elegans*, gastrulation begins at_______ cell stage:
    (a) 8 (b) 16 (c) 26 (d) 36
97. Which one of the following provides water to the embryo?
    (a) Egg white (b) Egg yolk (c) Air sac (d) Shell lining
98. The main body axis is established by:
    (a) Morulation (b) Gastrulation
    (c) Blastulation and gastrulation (d) Morulation and blastulation
99. The process in which one cell or tissue type affects the developmental fate of another cell or tissue is called:
    (a) Ingression (b) Induction (c) Involution (d) Organiser
100. In chicks, as gastrulation proceeds, the primitive streak:
    (a) Regress posteriorly (b) Regress anteriorly
    (c) Regress posteriorly with pharyngeal endoderm (d) Regress anteriorly with pharyngeal endoderm
101. Which one of the following develops from the trophoblast of mammalian blastocyst?
    (a) Ectoderm (b) Endoderm (c) Mesoderm (d) None of these
102. Interstial implantation is found in:
    (a) Humans (b) Rabbits (c) Carnivores (d) Ungulates
103. Cells of Rauber are found in:
    (a) *Petromyzon* (b) *Amphioxus* (c) *Xenopus* (d) Rabbit
104. In humans, which one of the following acts as the site of blood cell formation until the liver takes over this function?
    (a) Placenta (b) Yolk sac (c) Chorion (d) Umbilical cord

105. Extra embryonic membranes in mammalian embryo are formed from:
(a) Trophoblast (b) Inner cell mass (c) Primitive streak (d) Both (a) and (b)

106. Coelom is formed by the:
(a) Ectoderm (b) Endoderm
(c) Mesoderm (d) Both ectoderm and mesoderm

107. Consider the following statements:
(A) Yolk sac is formed from the covering of the vitellus
(B) Extra embryonic membranes are formed from the embryo
(C) Extra embryonic membranes do not form a part of the individual organisms after its birth
(D) Extra embryonic membranes are not same in all classes of the vertebrates

The correct statements are:
(a) All of these (b) A, B and C (c) B and C (d) C and D

108. Which one of the following crosses the placenta?
(a) Nicotine (b) Cocaine (c) Heroin (d) All of these

109. In the mammalian egg, compaction occurs at the:
(a) 4 cell stage (b) 8 cell stage (c) 16 cell stage (d) 18 cell stage

110. Which one of the following prevents implantation of the blastocyst at an abnormal site?
(a) Zona pellucida (b) Corona radiatae (c) Trophoblast (d) Estrogen

111. Which one is of the following is a correct match?
(a) Haemochorial placenta – All six layers of placenta remain contact
(b) Syndesmochorial placenta – Uterine epithelium is eroded
(c) Haemoendothelial placenta – Uterine epithelium and connective tissue are eroded
(d) Contra deciduate placenta – Humans

112. In meroblastic cleavage, division is:
(a) Complete (b) Incomplete (c) Horizontal (d) Spiral

113. The spreading of micromeres over macromeres is called:
(a) Epiboly (b) Emboly (c) Involution (d) Delamination

114. Gastrulation occurs by invagination in:
(a) *Amphioxus* (b) Frogs (c) Both (a) and (b) (d) None of the above

115. The cavity of optic cup marks the future of:
(a) Lens of the eye (b) Pupil of the eye
(c) Anterior chamber of the eye (d) Posterior chamber of the eye

116. Human placenta is:
(a) Metadiscoidal /deciduate (b) Deciduate /Cotyledonary
(c) Nondesiduate /metadiscoidal (d) Cotyledonary/nondesiduate

117. Which one of the following is not applicable to *Herdmania*?
(a) Holoblastic cleavage (b) Determinate cleavage
(c) Retrogressive metamorphosis (d) Gastrulation without invagination

118. The only chordate showing retrogressive metamorphosis in their embryonic development:
(a) *Oikopleura* (b) *Pyrosoma* (c) *Herdmania* (d) *Myxine*

119. All the three uterine tissues, foetal trophoblast and connective tissue are eroded in_______ placenta:
(a) Syndesmo-chorial (b) Epithelio-chorial
(c) Endothelio-chorial (d) Haemo-endothelial

120. In which one of the following is no part of placenta shaded after birth?
(a) Horses (b) Moles (c) Humans (d) None of these

121. Which one of the following is applicable to cleavage?
(a) Meiosis (b) Amitosis (c) Mitosis (d) Fragmentation

122. In the egg, polarity is indicated by the:
(a) Amount of yolk (b) Amount of cytoplasm
(c) Position of yolk (d) Position of yolk nucleus

123. Discoidal cleavage is not found in:
(a) Monotremes (b) Insects (c) Reptiles (d) Birds

124. How many barriers are lost in human placenta?
(a) 3 (b) 4 (c) 5 (d) 6

125. The development of *Amphioxus* lacks:
(a) Extra embryonic membranes (b) Primitive streak
(c) Placenta (d) All of these

126. Which one of the following is the first organiser specific gene in *Xenopus*?
(a) *noggin*$^+$ (b) *goosecoid*$^+$ (c) Xif_3^+ (d) *Krox*$^{\pm}$20

127. In birds, which one of the following acts as a primary organiser?
(a) Anterior half of the primitive streak (b) Posterior half of the primitive streak
(c) Area pellucida (d) Area opaca

128. Spemann's organiser is centred at the:
(a) Dorsal lip of the blastopore (b) Ventral lip of the blastopore
(c) Anterior half of the blastopore (d) Posterior half of the blastopore

129. In which one of the following inductions do the responding cells adopt different fates in the presence or absence of an inducer?
(a) Appositional induction (b) Homo genetic induction
(c) Instructive induction (d) Permissive induction

130. In different animals, which one of the following has been found to cause induction?
(a) Blastomeres (b) Undifferentiated germinal layers
(c) Mesenchyme (d) All of these

131. Nieuwkoop centre is associated with:
(a) Morulation (b) Organiser
(c) Gastrulation (d) Extra embryonic membranes

132. Noggin protein induces synthesis of two proteins, which are characteristic of the:
(a) Neural tissue (b) Reproductive tissue (c) Digestive tissue (d) Eye tissue

133. In Molluscs, egg is isolecithal, except in:
(a) Gastropods (b) Bivalves (c) Cephalopods (d) Chiton

134. Match column I with column II and select the correct answer using answer codes:

Column I (Cleavage)	Column II (Animal)
(A) Rotational	1. Reptiles
(B) Bilateral	2. Eutherian mammals
(C) Spiral	3. Cephalopods
(D) Discoidal	4. Annelids

Answer codes:

	A	B	C	D
(a)	4	3	2	1
(b)	3	4	1	2
(c)	2	4	3	1
(d)	2	3	4	1

135. During gastrulation in amphibians, mesoderm and endoderm are interlised by involution over _________ lips of the blastopore:
(a) Dorsal (b) Ventral (c) Lateral (d) All of these

136. The primary mesodermal cells of the sea urchin become internal by the process of:
(a) Convergence (b) Delamination (c) Ingression (d) Divergence

137. In birds, the formation of primitive streak occurs in:
(a) Late blastula (b) Early gastrula (c) Mid gastrula (d) Late gastrula

138. Uterine milk is not rich in:
(a) Proteins (b) Glycogen
(c) Lipids (d) Mucopolysaccharides

139. Optic vesicle of *Rana esculenta* induces lens, if grafted in ______________ of *Rana* sylvatica:
(a) Ectoderm of head (b) Ectoderm of ear
(c) Endoderm of head (d) Mesoderm of notochord

140. Which one of the following is not a characteristic of morphogenesis in vertebrates?
(a) Cell death (b) Cell growth
(c) No change in the chemical composition of cell membrane (d) Cell migration

141. Which one of the following about gastrulation is incorrect?
(a) High rate of oxidation (b) Insignificant growth
(c) Increased rate of cell divisions (d) Synthesis of new types of proteins

142. In the amphibian embryo, the migration of the presumptive chorda-mesodermal cells located on the surface of the blastula towards the blastopore represents:
(a) Convergence (b) Delamination (c) Invagination (d) Involution

143. Which one of the following statements about discoidal meroblastic cleavage is correct?
(a) Occurs in insects (b) Entire yolk mass remain undivided
(c) The cleavage is restricted to vegetal pole (d) Germinal disc divides completely

144. In which one of the following does the primitive streak not act as a passage for the movement of presumptive notochordal, mesodermal as well as endodermal cells?
(a) Mammals (b) Aves
(c) Reptiles (d) Egg laying mammals

145. In the mammalian embryo, the source of hypoblast is:
(a) Primitive streak (b) Blastodisc (c) Epiblast (d) Hensen's node

146. The blastula lacks cavity in:
(a) *Xenopus* (b) Insects (c) Monotremes (d) Sea urchins

147. Consider the following statements:
(A) In mammals, all the blastomeres are of more or less equal size
(B) Cells of area opaca form extra embryonic membranes
(C) Trophoblast participates in the formation of placenta

(D) In chicks, gastrular movement does not include involution

The correct statements are:

(a) A, B and C (b) B and C (c) C and D (d) None of the above.

148. Which one of the following provides embryological basis that birds and mammals may have evolved from reptilian ancestors?
   (a) Presence of single aortic arch (b) Similar pattern of early embryonic development
   (c) Lack of archenteron (d) Lack of blastopore
149. In the primitive stage of chick embryo, the eye field is located in:
   (a) Anterior region of the open neural plate (b) Posterior region of the open neural plate
   (c) Median anterior region of the epiblast (d) Posterior region of the hypoblast
150. Massive cell death occurs in the development of:
   (a) Brain (b) Brain and eyes (c) Limbs (d) Eyes and limbs
151. In which one of the following is the placenta found?
   (a) *Salpa* (b) *Peripatus* (c) *Mustelus* (d) All of these
152. In which one of the following is the placenta only uterine epithelium eroded?
   (a) Haemo-endothelial (b) Haemo-chorial
   (c) Syndesmo-chorial (d) Endothelio-chorial
153. The embryonal disc of mammalian blastula may be considered equivalents__________ of chick:
   (a) Area opaca (b) Area pellucida
   (c) Hensen's node (d) Extra embryonic ectoderm
154. Coloured cytoplasmic regions are found in the eggs of:
   (a) Tunicats (b) Hemichordates (c) *Amphioxus* (d) *Petromyzon*
155. Which one of the following hormones participates in the development of the cornea?
   (a) Thyroxin (b) Cortisol (c) Epinephrine (d) Insulin
156. The differentiation of lens and cornea are under the control of:
   (a) Thyroxin (b) Pharyngeal endoderm (c) Neural retina (d) Optic vesicle
157. The inner cell mass becomes:
   (a) Neural crest (b) Yolk sac (c) Allantois (d) Embryo
158. Which one of the following answers is incorrect?
   (a) White egg beneath the blastodisc – Latebra
   (b) Formation of embryo proper – Area pellucida
   (c) Four pairs of somites are formed – After 24 hours of incubation
   (d) Embryonic urinary bladder – Allantois
159. Which one of the following is the first development event unique to vertebrates?
   (a) Formation of brain (b) Formation of neural crest
   (c) Formation of heart (d) Formation of notochord

Answers to Multiple-Choice Questions

1. (d)	2. (a)	3. (d)	4. (c)	5. (b)	6. (c)	7. (d)	8. (b)
9. (a)	10. (a)	11. (b)	12. (b)	13. (b)	14. (a)	15. (d)	16. (d)
17. (a)	18. (d)	19. (b)	20. (c)	21. (b)	22. (b)	23. (c)	24. (d)
25. (c)	26. (b)	27. (a)	28. (a)	29. (c)	30. (b)	31. (c)	32. (c)
33. (a)	34. (d)	35. (b)	36. (c)	37. (b)	38. (d)	39. (b)	40. (d)

41. (c)	42. (a)	43. (b)	44. (b)	45. (d)	46. (c)	47. (b)	48. (d)
49. (a)	50. (a)	51. (d)	52. (a)	53. (b)	54. (b)	55. (d)	56. (d)
57. (b)	58. (a)	59. (b)	60. (c)	61. (c)	62. (a)	63. (c)	64. (a)
65. (b)	66. (a)	67. (d)	68. (c)	69. (d)	70. (b)	71. (c)	72. (b)
73. (a)	74. (d)	75. (c)	76. (b)	77. (c)	78. (d)	79. (d)	80. (c)
81. (b)	82. (c)	83. (d)	84. (c)	85. (b)	86. (a)	87. (a)	88. (d)
89. (d)	90. (b)	91. (d)	92. (b)	93. (d)	94. (c)	95. (b)	96. (c)
97. (a)	98. (c)	99. (b)	100. (c)	101. (d)	102. (a)	103. (d)	104. (b)
105. (a)	106. (c)	107. (a)	108. (d)	109. (c)	110. (a)	111. (b)	112. (b)
113. (a)	114. (c)	115. (c)	116. (a)	117. (d)	118. (c)	119. (d)	120. (a)
121. (c)	122. (d)	123. (b)	124. (a)	125. (d)	126. (b)	127. (a)	128. (a)
129. (c)	130. (d)	131. (b)	132. (a)	133. (c)	134. (d)	135. (d)	136. (c)
137. (d)	138. (a)	139. (a)	140. (c)	141. (c)	142. (a)	143. (b)	144. (a)
145. (b)	146. (b)	147. (c)	148. (b)	149. (c)	150. (d)	151. (d)	152. (c)
153. (b)	154. (a)	155. (a)	156. (c)	157. (d)	158. (a)	159. (b)	

Fill in the Blanks

1. The portion of the telolecithal egg with little yolk is called ___________.
2. The four stages of embryonic development are the ___________ stage, ___________ stage, ___________ stage and ___________ stage.
3. The phase of development of embryos during which the morphology of the embryo is recognised to form three germ layers is known as ___________.
4. Gastrulation is followed by ___________.
5. The region of the blastoderm above the submarginal space is called the ___________.
6. The region of the blastoderm above the yolk is called the ___________.
7. The region where the area pellucida and area opaca meet is called the ___________.
8. In the human embryo, gastrula stage occurs during the ___________ week.
9. The gastrula consists of a disk that is made up of ___________, ___________ and ___________.
10. The cell division in the first stage of embryonic division is called ___________ or ___________.
11. The cells that result from the cleavage are called ___________.
12. Gastrulation is the process through which a portion of the blastula wall undergoes invagination inside the ___________.
13. The tube-like structure formed as a result of gastrulation is called ___________.
14. The cells of the inner side of the archenteron form the ___________.
15. The blastopore forms the ___________ in protostomes and ___________ in deuterostomes.
16. The mesoderm develops from differentiation of the endodermal cells which cover the ___________ region of the archenteron.
17. The opposite of ovaparity is the ___________.
18. Viviparity is applicable to all mammals excepts ___________ and ___________.

19. ___________, birds produce coloured eggs.
20. The study of eggs is called ___________.
21. Cleavage ends with the formation of ___________ and development of fertilised egg begins with ___________.
22. In amphibians, cleavage is ___________ and ___________.
23. Formation of nervous system is called___________.
24. During epiboly, micromeres entirely enclose megameres, except the region of the ___________.
25. In chick embryo, instead of blastula, a thin layer of cells formed on the surface of the yolk called ___________.
26. In the developmental sequence blastula follows the ___________ and precedes the ___________.
27. The pole of the egg having highest concentration yolk is termed as ___________.
28. Radial cleavage is the characteristic of ___________.
29. In mammals, cleavage starts at about ___________ hours of fertilisation.
30. The attachment of blastocyst on the endometrium of uterine wall is called ___________.
31. The study of fate of each blastomere after first and second division is known as ___________.
32. Yolk sac is connected with the gut of embryo by a slender ___________.
33. Yolk plug of gastrula is formed by the protrusion of ___________ cells.
34. Villi are restricted to a circular disc in ___________ placenta.
35. A temporary connection between the embryo and uterine wall of the mother is known as ___________.
36. ___________ is the region of area pellucida without mesoderm.
37. Hensen's node is formed by cells of presumptive ___________ and floor of ___________.
38. In the chick, all the presumptive areas lie within ___________.
39. In the chick, before gastrulation, the multilayered blastoderm of area pellucida becomes rearranged to form a single layer of cubical epithelial cells called ___________.
40. ___________ is the rival or bacterial infection of placenta.
41. The somatic layer of mesoderm and ectoderm are collectively called ___________.
42. The set of blood vessels that connects, the foetus with the placenta is called ___________.
43. The umbilical cord contains the right and left ___________ arteries and ___________ vein.
44. Extra embryonic membranes in humans are formed outside the embryo form the ___________.
45. The vertebrates that lack amnion in the development are called ___________.
46. Among the anamniotes, yolk sac consists of cells derived from the ___________ layers.
47. In amniotes, yolk sac is derived from the extra embryonic ___________ alone.
48. There is drastic reduction or loss of larval stages in all the major classes of vertebrates except for the ___________.
49. Yolk sac surrounds the yolk, except in ___________.
50. The amniotic egg possesses a unique set of membranes called ___________, ___________ and ___________
51. The primitive streak is analogous to amphibian ___________.
52. In humans, the allantois is vestigial, except that its blood vessels give rise to ___________.
53. Blastopore is the opening of______________.

54. The development of extra embryonic membranes begins at the moment of the differentiation of the blastocyst cells into an ___________ and a ___________.
55. ___________ antibodies can pass through the human placenta.
56. The condition in which the placenta is low-lying and may partly or fully cover the cervix is termed as ___________.
57. Prototherian and metatherian mammals have a ___________ placenta.
58. In mice, gastrulation occurs after ___________.
59. The region of the intermediate pigmentation in the marginal zone of the amphibian egg is known as ___________.
60. Walter Vogt is famous for his use of vital dyes to construct fate maps of ___________.
61. Chorionic placentae are found in__________ and ___________.

Answers to Fill in the Blanks

1. Animal pole
2. Morula, blastula, gastrula and neurula
3. Gastrulation
4. Organogenesis
5. Area pellucida
6. Area opaca
7. Marginal zone
8. Third
9. Ectoderm, mesoderm and endoderm
10. Cleavage or segmentation
11. Blastomeres
12. Blastocoel
13. Archenteron
14. Endoderm
15. Mouth, anus
16. Dorsal
17. Viviparity
18. Monotremes, marsupials
19. Passerines
20. Oology
21. Blastula, cleavage
22. Holoblastic, unequal
23. Neurulation
24. Yolk plug
25. Blastoderm
26. Morula, gastrula
27. Vegetal pole
28. Deuterostomes
29. 14 to 15
30. Implantation
31. Cell lineage
32. Yolk stalk
33. Endodermal
34. Monodiscoidal
35. Placenta
36. Pro amnion
37. Notochord, neural tube
38. Area pellucida
39. Epiblast
40. Placentitis
41. Somatopleurae
42. Umbilical cord
43. Umbilical, left umbilical
44. Trophoblast
45. Anamniota
46. Three germ
47. Splanchopleurae
48. Cyclostomes
49. Mammals
50. Amnion, chorion, allantois
51. Blastopore
52. Umbilical cord
53. Archenteron
54. Embryoblast, trophoblast
55. IgG
56. Placenta previa
57. Choriovitelline
58. Implantation
59. Gray crescent
60. Amphibian embryo
61. Men, apes

True or False

1. Telolecithal egg is either moderately telolecithal or highly telolecithal.
2. Eggs are of different sizes, colours and grades.
3. The initial step in egg production occurs prior to birth.

4. Human blastocyst is comparable to amphibian blastula.
5. The blastocyst of humans contains an inner cell mass which comparable to the amphibian blastula.
6. Blastocyst contains an embryoblast.
7. In the zygotes of mammals, cellular divisions are synchronous.
8. In superficial cleavage, mitosis occurs but there is no cytokinesis
9. Superficial cleavage occurs in centrolecithal eggs.
10. In frogs, the first cleavage is asymmetrical.
11. Gastrulation occurs by involution.
12. In discoidal cleavage, the cleavage furrows penetrate the yolk.
13. Mammalian zygotes have slow rate of divisions.
14. Cleavage is a series of divisions without gene expression.
15. In spiral cleavage, the division planes are at 90º angle.
16. In frogs, the first cleavage furrow occurs about 3 hours of fertilisation.
17. In deuterostomes, blastopore forms the anus.
18. Epiboly is a kind of cell movement.
19. In vertebrates, the plane of cleavage a more rigid.
20. During cleavage, the size of the nucleus diminishes.
21. Midblastula transition is characterised by the initiation of transcription.
22. The first cleavage in humans is meridional.
23. In frogs and *Amphioxus*, gastrulation occurs by invagination.
24. The vegetal hemisphere of frogs remains directed upwards.
25. Mother's blood circulates in the foetus and vice versa.
26. The frequency of cleavage depends on the viscosity of the cytoplasm
27. The amount of yolk present in egg affects the way of blastula formation.
28. Blastocyst is found in mammals.
29. The movements of cells during gastrulation are of reversible nature.
30. The rate of cleavage is very slow in mammalian eggs.
31. In cleavage, the interphase is long.
32. In cleavage, the nuclear-cytoplasmic ratio increases with each division.
33. Extra embryonic membranes of mammals are derived from the trophoblast.
34. Zona pellucida disappears before completion of the cleavage.
35. In frogs, the fertilisation membrane is formed after fertilisation.
36. The blastula of *Amphioxus* is spherical.
37. *Peripatus* is oviparous.
38. Coelom is formed from mesoderm.
39. Yolk sac placenta is derived from yolk sac only.
40. Telolecithal egg produces discoblastula.
41. The presence of extra embryonic membranes in all vertebrate class is similar.
42. The yolk sac stores vitellus, which is the main source of nutrition in nonplacental embryos.

43. Chorion covers the amnion, yolk sac and allantois.
44. In fishes and amphibians, only the yolk sac is present.
45. In human beings, the yolk sac is vestigial.
46. Amnion, chorion, allantois and yolk sac are parts of the placenta of developing mammals.
47. The placenta does not grow throughout pregnancy.
48. The placenta develops from the same sperm and ovum cells that form the foetus.
49. Human placenta is now being used as a component of antiwrinkle creams.
50. While preparing for gastrulation, the embryo must become asymmetric.
51. Cleavage leads to increase in genetic material.
52. The grey crescent of frog marks the ventral side of the future embryo.
53. During the organogenesis of the brain in chick embryo, the neural folds close simultaneously throughout their length.
54. In chick, extra embryonic membranes are composite structures.
55. The muscles of vertebrates develop from mesoderm.
56. Differentiation of cells occurs during blastula.
57. In frogs, gastrulation is completed by epiboly and emboly.
58. Micromeres of frog contain little yolk.
59. The details of gastrulation vary from species to species.
60. Chalazae are spiral rope-like strands that anchor the yolk in thick egg white.
61. Induction affects pattern of gene expression through physical contact or chemical signal.

Answers to True or False

1. True	2. True	3. True	4. False	5. True	6. True	7. False	8. True
9. True	10. False	11. True	12. False	13. True	14. True	15. False	16. True
17. True	18. True	19. False	20. False	21. True	22. True	23. True	24. False
25. False	26. True	27. True	28. True	29. False	30. True	31. False	32. True
33. False	34. False	35. True	36. False	37. False	38. True	39. False	40. True
41. False	42. True	43. True	44. True	45. True	46. True	47. False	48. True
49. True	50. True	51. True	52. False	53. False	54. False	55. True	56. False
57. True	58. True	59. True	60. True	61. True			

Give Reasons

1. Meroblastic cleavage does not divide the egg completely.
   – Because divisions are restricted to only a part of the egg.
2. The human placenta is also called chorionic placenta.
   – Because chorion takes part in the formation of the placenta.

3. Hens lack the blastula stage.
   - Because of the huge amount of yolk in their eggs.
4. Mammalian zygotes require immediate implantation on the uterine wall.
   - Because mammalian eggs contain a small amount yolk, and therefore require immediate implantation on uterine wall in order to receive nutrients.
5. Every cell descended from zygote contains complete genome of the organism.
   - Because every cell descended from zygote is produced by mitosis division and thus contains complete genome of the organism.
6. The endoderm of the yolk sac secretes digestive enzymes.
   - Because the embryo is unable to utilise yolk directly, so it is converted into a soluble form with the help of digestive enzymes.
7. In placental mammals, the allantois is present but does not perform the excretory function.
   - Because the wastes of embryo are collected by the mother's body through the placenta.
8. Amnion is an adaptation to dry land.
   - Because it prevents the embryo from desiccation.
9. Gastrulation is an important dynamic process in the development of the zygote.
   - Because it involves movement of the cells from the surface of the blastula to their definite positions in the embryo, leading to formation of three germinal layers, viz., ectoderm, mesoderm and endoderm.
10. Allantois is an adaptation to dry land.
    - Because in terrestrial oviparous animals like reptiles and birds, the metabolic waste products cannot be immediately excreted in the surrounding water like fishes and amphibians do, it was essential to develop such structure to store the embryonic wastes till hatching. As such, the allantois took over such work.
11. Mast cells have been found in the chick embryo.
    - Because perhaps these cells have a role in angiogenesis of the extra embryonic vascular system.
12. Reptiles, birds and mammals are called amniotes.
    - Because they have an amnion as well as other extra embryonic membranes during development.
13. During gastrulation, the metabolic activities of the cells are increased.
    - Because of great morphogenetic activities of the blastomeres.
14. Micromeres of the animal pole divide faster than the macromeres.
    - Because of the presence of yolk, the macromeres divide slowly.
15. In *Amphioxus*, the epithelium is thicker at the vegetal pole, while thinner at the animal pole.
    - Because blastomeres of the vegetal pole are larger than the blastomeres of the animal pole.

Made in the USA
Monee, IL
05 November 2026

67807984R00162